HOMELAND SECURITY
OPERATIONAL ANALYSIS CENTER

# Recovery Plan for the Communications and Information Technology Sector After Hurricanes Irma and Maria

Laying the Foundation for the Digital
Transformation of Puerto Rico

AMADO CORDOVA, RYAN CONSAUL, KARLYN D. STANLEY, AJAY K. KOCHHAR, RICARDO SANCHEZ, DAVID METZ

Published in 2020

# Preface

On August 8, 2018, the government of Puerto Rico submitted its economic and disaster recovery plan to Congress, as required by the Bipartisan Budget Act of 2018. Under contract with the Federal Emergency Management Agency (FEMA), the Homeland Security Operational Analysis Center (HSOAC) provided substantial support in developing the plan by soliciting and integrating inputs from a wide variety of stakeholders, contributing analysis where needed, and assisting in drafting the plan. The plan included an overview of damage and needs, courses of action (COAs) to meet those needs, costs of the COAs, and potential funding mechanisms for those costs.

To support federal agencies evaluating and funding recovery actions, HSOAC is releasing this detailed volume for the communications and information technology sector. The analysis in this report was performed during the time period of February 2018 to July 2018. The purpose of this document is to provide decisionmakers greater detail on the conditions in Puerto Rico prior to the 2017 hurricane season, damage from Hurricanes Irma and Maria, COAs that were identified to help the sector (and, more broadly, Puerto Rico) recover in a resilient manner, potential funding mechanisms, and considerations for implementers as they move forward.

This document will likely be of interest to other stakeholders funding or implementing recovery activities in Puerto Rico, including commonwealth and municipal agencies, nongovernmental organizations, and the private sector.[1] Furthermore, this body of material contributes to the larger literature about disaster recovery and resilience and may be of interest to other communities planning for or recovering from similar disasters.

This research was sponsored by FEMA and conducted within the Strategy, Policy, and Operations Program of the HSOAC federally funded research and development center (FFRDC).

HSOAC is releasing similarly detailed research volumes for other sectors. More information about HSOAC's contribution to planning for recovery in Puerto Rico, along with links to other reports being published as part of this series, can be found at www.rand.org/hsoac/puerto-rico-recovery.

## About the Homeland Security Operational Analysis Center

The Homeland Security Act of 2002 (Section 305 of Public Law 107-296, as codified at 6 U.S.C. § 185), authorizes the Secretary of Homeland Security, acting through the Under

---

[1] Puerto Rico, as an unincorporated territory of the United States, has a commonwealth government that deals with territory-wide issues. The roles of the commonwealth government are similar to, but not identical to, the roles of a state government.

Secretary for Science and Technology, to establish one or more FFRDCs to provide independent analysis of homeland security issues. The RAND Corporation operates the HSOAC as an FFRDC for the U.S. Department of Homeland Security (DHS) under contract HSHQDC-16-D-00007.

The HSOAC FFRDC provides the government with independent and objective analyses and advice in core areas important to the department in support of policy development, decisionmaking, alternative approaches, and new ideas on issues of significance. The HSOAC FFRDC also works with and supports other federal, state, local, tribal, and public- and private-sector organizations that make up the homeland security enterprise. The HSOAC FFRDC's research is undertaken by mutual consent with DHS and is organized as a set of discrete tasks. This report presents the results of research and analysis conducted under Task Order 70FBR218F00000032, "Puerto Rico Economic and Disaster Recovery Plan: Integration and Analytic Support."

The results presented in this report do not necessarily reflect official DHS opinion or policy.

For more information on HSOAC, see www.rand.org/hsoac. For more information on this publication, visit www.rand.org/t/RR2599.

# Contents

# Figures

# Tables

# Summary

The two hurricanes that made landfall in Puerto Rico in September 2017 wreaked havoc. The population lost power, water, communication systems, and essential services. In the communications and information technology (IT) sector, a widespread failure of the networks either prevented or severely limited effective communications between residents of Puerto Rico who were in need and the first responders who were responsible for providing emergency services. The devastation of critical infrastructure (CI) in the energy, water, transportation, and communications and IT sectors was so extreme that it took many months to fully restore services in some of these sectors. The hurricanes led to loss of life, extensive housing damage, a complete failure of the energy grid, and months-long interruption of basic services for the residents of Puerto Rico.

On February 9, 2018, Congress passed the Bipartisan Budget Act of 2018.[1] This legislation required the Governor of Puerto Rico, with support and contributions from the Secretary of the Treasury, the Secretary of Energy, and other federal agencies having responsibilities defined under the National Disaster Recovery Framework (NDRF), to produce within 180 days an economic and disaster recovery plan that defined the priorities, goals, and outcomes of the recovery effort.

The Federal Emergency Management Agency (FEMA) requested the Homeland Security Operational Analysis Center (HSOAC) to support the development of the recovery plan by soliciting and integrating the inputs from many stakeholders and by contributing analysis when necessary. The recovery plan includes contributions from a broad range of stakeholders, including civil society representatives, commercial telecommunications providers, emergency communications services experts, government agencies in Puerto Rico and the continental United States (CONUS), and many others. On August 8, 2018, the Governor of Puerto Rico delivered the recovery plan, entitled *Transformation and Innovation in the Wake of Devastation: An Economic and Disaster Recovery Plan for Puerto Rico*, to Congress.

This sector report is one of several documents prepared to describe the research conducted in support of the development of the recovery plan for Puerto Rico. It pertains to the communications and IT sector.[2] The sector report is divided in seven chapters, summarized below.

---

[1] Public Law No. 115-123, Bipartisan Budget Act of 2018, February 9, 2018.

[2] "Information technology" is defined as "the study or use of computers, telecommunication systems, and other devices for storing, retrieving, and transmitting information" (Oxford Reference, undated).

## The Communications/IT Sector in Puerto Rico

This introductory chapter provides context for the communications and IT sector of Puerto Rico, which has a rich history. As early as 1897, public telephone service was offered there. Telecommunications services in Puerto Rico were managed for many years by the government-run Puerto Rico Telephone Company (PRTC), which was created in 1974.[3] PRTC completed a fiber-optic network in 1994, introduced cellular service the same year and began providing internet services two years later. The Telecommunications Act of 1996 permitted new entrants, competitive local exchange carriers (CLECs), to compete with incumbent local exchange carriers (ILECS) such as PRTC. A year later, the Puerto Rico legislature enacted legislation to allow the sale of PRTC. América Móvil fully acquired PRTC in 2007 and later rebranded many of PRTC's services under the name "Claro."

Today, Puerto Rico depends on mobile telecommunications. Whereas only about 20 percent of households have landline telephone service (this penetration is much lower than in CONUS), approximately 99 percent of the residents of Puerto Rico have access to cellular service.[4] The majority of the residents of Puerto Rico have embraced mobile (wireless) technology and use their cellular devices as their only means of communications. The large demand has driven the development of an extensive wireless network infrastructure. Claro provides wireline, wireless, broadband digital subscriber line (DSL), and dial-up services, as well as internet protocol television. Besides Claro, new entrants now provide wireless and internet services.[5] They include AT&T, T-Mobile, and PR Wireless, LLC (a joint venture by Sprint and Open Mobile). Liberty Cablevision is the only cable provider in Puerto Rico.[6] Moreover, undersea cable-related infrastructure is used for providing high-speed connectivity to and from the rest of the world. The landing stations for undersea cable are concentrated on the northern side of Puerto Rico.

---

[3] "Telecommunications services" is defined as "the offering of telecommunications for a fee directly to the public, or to such classes of users as to be effectively available directly to the public, regardless of the facilities used" (Legal Information Institute, Cornell Law School, undated). Telecommunications is defined as "transmission, between or among points specified by the user, of information of the user's choosing, without change in the form or content of the information as sent and received" (Legal Information Institute, Cornell Law School, undated). See also U.S. Code, Title 47, Section 153, Definitions, U.S. Government Printing Office, 2011.

[4] FEMA Hurricane Maria Task Force, *DR 4339-PR, Consolidated Communications Restoration Plan*, October 30, 2017, p. 2. FCC, "Trends in Telephone Service," September 2010, Table 16.2, shows that in 2008, 98.2 percent of U.S. mainland housing units had landline telephone. FCC, "2018 Broadband Deployment Report," GN Docket No. 17-199, FCC 18-10, February 2, 2018a, Appendix E.

[5] Claro is the ILEC and ATT is the CLEC.

[6] Internet service is provided by means of cable (Liberty), DSL (e.g., Claro, WorldNet), fiber (e.g., Optico Fiber), and fixed wireless (e.g., IP Solutions, Caribe.net, Inteco, Neptuno Networks).

The wireless infrastructure in Puerto Rico is supported mainly by aerial fiber-optic cable.[7] which is strung up on utility poles and is more susceptible to severe weather, high winds, and hurricane damage than fiber-optic cable buried in conduit.[8]

In addition to an extensive private telecommunications infrastructure, Puerto Rico has five public safety communications systems: the Puerto Rico Police Department (PRPD) P25 system, which provides statewide radio coverage to police, fire, health, and other entities;[9] the PRPD SmartZone, used by police in the greater San Juan metropolitan area; the Emergency Medical Services (EMS) system, used by the Medical Emergency Corps of Puerto Rico during premedical care and emergency transport to hospitals; the Puerto Rico Emergency Management Agency (PREMA) system for day-to-day mobile radio communications between municipalities and PREMA; the interoperability system for communications throughout Puerto Rico in emergency or urgent situations, such as tsunamis, earthquakes, and others. Puerto Rico's public safety systems use both fiber-optic cable and microwave dishes to transmit signals over their communications networks. These public safety systems and networks are critical for the proper operation of 911 emergency response and government response in the aftermath of disasters.

Governance of the communications and IT sector resides in the Telecommunications Bureau (formerly the Puerto Rico Telecommunications Regulatory Board [PRTRB]), the Puerto Rico Department of Public Safety (PRDPS), the Chief Information Officer (CIO), and the Chief Innovation Officer (CINO).[10]

---

[7] FEMA Hurricane Maria Task Force, "Consolidated Communications Restoration Plan," DR 4339-PR, Oct. 30, 2017, p. 2. See also AT&T, Reply Comments of AT&T, *In the Matter of Response Efforts Undertaken During the 2017 Hurricane Season*, PS Docket No. 17-344, February 21, 2018, p.7 ("the majority of AT&T's fiber was riding on the electrical utility poles"). This FEMA reference starts the sentence with the qualification saying that "with the exception of one wireless carrier, which has buried fiber that supports the core of their network." A former industry official explained to the HSOAC team that this carrier was Centennial de Puerto Rico. Centennial may have had indefeasible rights of use (IRUs) with San Juan Cable when it was first building out its network, and it might also have buried some fiber-optic cable in urban areas of San Juan and in Mayaguez and Ponce. Centennial did not bury fiber in suburban, central, or rural areas of Puerto Rico. Like other carriers in Puerto Rico, Centennial primarily used aerial fiber-optic cable for its network. Centennial was purchased by AT&T in 2008. An IRU is a contractual agreement that confers exclusive right of access to equipment, fibers, or network capacity on a telecommunications system to another telecom operator for an agreed-upon period in return for upfront or recurring payments. The word "indefeasible" means that such contractual agreement cannot be undone, voided, or annulled. See EY, "EY's Spotlight on Telecommunications Accounting," No. 2, 2015.

[8] PPC, *The Complete Guide to Fiber to the Premises Deployment: Options for the Network Operator*, 2017.

[9] "Project 25" or "P25" radio systems are based on a suite of standards for digital, two-way land mobile radio (LMR) devices. U.S. Department of Homeland Security, Office of the Inspector General, "SAFECOM Guidance Frequently Asked Questions: Understanding P25 Standards and Compliance," undated.

[10] The CIO and the CINO offices are both part of Puerto Rico Innovation and Technology Service (PRITS). CIO's focus is on information technology (IT) implementation; CINO's focus is on innovation and related process evolution.

## Damage and Needs Assessment

As Chapter 2 details, Hurricane Maria decimated the telecommunications infrastructure that is the backbone for commercial and public safety communications in Puerto Rico. Residents were unable to access critical emergency services. The government could not effectively coordinate response operations. The private sector could not communicate with CONUS suppliers or corporate headquarters and could not perform certain operations critical to its business.

Public safety communications were severely damaged by the hurricane, including 50 percent of the PRPD's primary network (the P25 radio system) and most equipment of the interoperability system for emergency communications throughout Puerto Rico. Numerous sites of the EMS radio network were not operational.

Commercial telecommunications infrastructure was also severely damaged. Eighty percent of core above-ground fiber-optic cable and 85 to 90 percent of last-mile[11] fiber-optic cable were destroyed,[12] 48,000 utility poles were damaged, and the primary submarine cable landing station was flooded. Moreover, 95 percent of cellular sites were out of service as of September 21, 2017. A report of the Federal Communications Commission's (FCC's) Public Safety and Homeland Security Bureau stated that "after the hurricanes, and in particular after Hurricane Maria, 95.2 percent of cell sites in Puerto Rico were out of service. All *municipios* in Puerto Rico had greater than 75 percent of their cell sites out of service."[13] Forty-eight out of the 78 *municipios* in Puerto Rico had 100 percent of their cell sites out of service.[14]

Immediate steps were taken to restore communications.[15] Thus, for example, FEMA's Emergency Support Function #2 (ESF-2) deployed tactical communications equipment to police, fire, and EMS offices.[16] Commercial wireless carriers used cellular equipment on light trucks (COLTs), cells on wheels (COWs), and portable generators to restore service temporarily.[17] Four wireless carriers opened up cellular service roaming arrangements to ensure service to the maximum number of people with the coverage available. AT&T used a "flying cell on wings" (a

---

[11] In telecommunications jargon, the "last mile" refers to the final leg of the networks that deliver telecommunication services to retail end-users.

[12] This refers to 80 percent of the approximately 350 linear miles of core above-ground fiber and 85 to 90 percent of the approximately 1,500 linear miles of last-mile fiber. See also AT&T, 2018, p.7.

[13] Puerto Rico does not have "counties" but rather 78 *municipios,* or "county-equivalents." Government of Puerto Rico, *pr.gov—Official Portal of the Government of Puerto Rico*, Directory of Municipalities of Puerto Rico, 2019.

[14] Public Safety and Homeland Security Bureau, FCC, *2017 Atlantic Hurricane Season Impact on Communications Report and Recommendations*, Public Safety Docket No. 17-344, August 2018a, pp. 5–6.

[15] More are discussed in Chapter 2.

[16] "ESF-2" refers to FEMA's Emergency Support Function #2, which applies to communications. "ESF-2, Communications Annex," June 2016.

[17] See CTIA, Comments of CTIA, *In the Matter of Response Efforts Undertaken During 2017 Hurricane Season*, PS Docket No. 17-344, January 22, 2018, p. 6; Reply Comments of AT&T, 2018, p. 7.

drone cell site) to temporarily provide data and voice transmission, and text messaging over a 40-mile radius.[18]

By early April 2018, the FEMA Communications and IT (Comms/IT) sector team had finalized a plan to prepare Puerto Rico for the upcoming hurricane season, which would start in June. The plan recommended rapid actions that would allow for continuity of communications for government leadership and other CI stakeholders—including hospitals, public safety answering points, emergency medical services and others—in the event of another disaster. As part of executing this plan, as of July 25, 2018, many steps had been undertaken to ensure that public safety and essential government services would be operational in the aftermath of another disaster.

As of October 19, 2018, the PRPD microwave network had been 70 percent restored to its pre-Maria status, and both the PREMA and the EMS microwave networks had been restored to over 85 percent of their pre-Maria status. A senior official of the PRDPS expressed his satisfaction with the status of the public safety telecommunications systems as of that date. He indicated that response efforts had successfully provided for multiple redundancies in government and public safety communications systems.[19]

The Public Safety and Homeland Security Bureau of the FCC found that "wireless service was restored gradually over a six-month period, considerably longer than for any other storm. After six months, 4 percent of cell sites remained out of service (i.e., completely inoperable) in Puerto Rico."[20] By March 21, 2018, only 4.3 percent of all commercial cell sites in Puerto Rico remained out of service. However, most of the private telecommunication sites had to rely on fuel-based generators for many months due to the failure of the electric grid. As of April 10, 2018, about 30 percent of the cellular sites of one of the main telecommunications carriers were still operating using generators.[21] Finally, as of June 5, 2018, 99.8 percent of cellular sites (2,653 of 2,659) were operational.[22]

In spite of the efforts described above, infrastructure critical to supporting telecommunications remains vulnerable to an upcoming disaster. Aerial fiber-optic cable on poles, as opposed to fiber-optic cable buried in conduit, is particularly vulnerable to harsh weather conditions. Aerial

---

[18] Magdalena Petrova, "Phone Service Can Mean Life or Death After a Disaster and AT&T and Verizon Are Using Drones That Could Help," CNBC, August 24, 2018.

[19] Official of the Puerto Rico Department of Public Safety, telephone communication with the authors, October 19, 2018.

[20] Public Safety and Homeland Security Bureau, 2018a, p. 15.

[21] Representative of a Puerto Rico telecommunications carrier, interview with the authors, April 10, 2018.

[22] The FCC also collects information from wireless carriers about how many cell sites are using generators, but this information is not publicly available.

fiber has been used during repair and restoration.[23] In addition, a constant supply of electricity is essential for operating commercial cell and public safety telecommunication sites. As of early October 2018, it was unclear how many of the 2,653 operational cell sites previously referred to were still operating with fuel-based generators.[24] Use of generators presents multiple problems.[25]

The damage caused by Hurricanes Maria and Irma demonstrates the importance of having survivable technologies that are redundant and resilient to disaster in place. The networks have been restored to close to their pre-Maria status, but they still need to be hardened to achieve adequate resiliency. Moreover, the communications and IT sector will need to bring many systems up to federal telecommunications standards,[26] take advantage of technological advances to upgrade systems and equipment, improve maintenance procedures on CI, and maintain a cadre of qualified communications and IT personnel. While FEMA, the government of Puerto Rico, and the private sector have undertaken significant steps to ensure continuity of government (COG), public safety emergency functions, and commercial telecommunications in preparation for another disaster similar to Hurricane Maria, significant work remains in implementing medium and long-term solutions to address telecommunications infrastructure challenges. We find that in the communications and IT sector two top-level needs have not yet been addressed:

1. The need to implement a state-of-the art, survivable, resilient communications infrastructure for continuity of essential government functions and for the provision of public safety services to the residents of Puerto Rico. This infrastructure should be upgraded and brought up to CONUS standards and should be well maintained and resourced.
2. The need to implement a state-of-the art, survivable, resilient communications infrastructure to provide commercial telecommunication services, including voice and data services, to the residents and the private sector of Puerto Rico. This infrastructure should support affordable access to broadband internet service and emerging technologies throughout Puerto Rico.

The recovery efforts present a unique opportunity to pursue the digital transformation of Puerto Rico. Widely available broadband internet services and a robust IT infrastructure could digitally transform public and commercial services to foster Puerto Rico's economy, prosperity, and well-being. Therefore, we see a third top-level need for the communications and IT sector:

---

[23] Some of the repair and restoration efforts by private telecom carriers were being pursued using aerial fiber instead of fiber in buried conduit due to its lower costs. Representative of a major Puerto Rico telecommunications carrier, interview with the authors, May 15, 2018.

[24] This information is not currently available publicly; however, beginning in 2019, the aggregate number of cell site generators in use after a disaster may be made available as part of the FCC's Wireless Resiliency Cooperative Framework.

[25] First, their intended use is to be stored at cell sites for intermittent backup in case of an emergency, not to run continuously for weeks or months; when they do, they often fail. Second, supplying diesel or gasoline to cell site generators across the island is both logistically challenging and very expensive. Third, many of them were stolen.

[26] Puerto Rico lacked systems that met federal telecommunications standards.

3. The need to develop, deploy, and sustain a modern and resilient information technology infrastructure to further the economic and social vitality of Puerto Rico. This infrastructure should host applications and web services that foster government and private-sector innovation, increase economic opportunity, and improve the quality of life for the residents of Puerto Rico.

## Methodology for Developing the Recovery Plan

The methodology outlined in Chapter 3 consisted of two parallel paths, one top-down and one bottom-up. The top-down path started with the vision and goals provided by the government of Puerto Rico, followed by the development of strategic objectives based on the vision and goals. For the communications and IT sector the bottom-up path started with the three top-level needs cited above. As part of the bottom-up path 33 specific recovery courses of action (COAs) were developed to satisfy these needs.

COAs were developed through a collaborative process and by the joint efforts of a wide range of stakeholders, including federal agencies, government agencies of Puerto Rico, and the Puerto Rico telecommunications private industry. Moreover, the HSOAC Comms/IT team received indirect feedback from Puerto Rico municipalities and from additional stakeholders—such as civil society partners—through the work of the HSOAC teams responsible for developing COAs in the community planning and capacity building sector and in the municipality sector. Details of these processes are provided in Chapter 3. The HSOAC Comms/IT team also participated in the 100 Resilient Cities workshops that were convened in March and April 2018 by the government of Puerto Rico and were attended by a broad range of stakeholders, including civil society and industry representatives.

The FEMA Comms/IT sector lead engaged and coordinated the efforts of the subject matter experts (SMEs) who composed FEMA's Comms/IT solutions-based team (SBT). HSOAC team members met numerous times with the SBT to seek their input and ensure a comprehensive understanding of their findings. The HSOAC team also sought insights from officials of the government of Puerto Rico across multiple agencies, including the Telecommunications Bureau, PRDPS, and offices of CIO and CINO. Puerto Rico communications and IT leadership held meetings twice a week for several months, led by the FEMA Comms/IT sector lead, and hosted at the offices of the Telecommunications Bureau.[27] All important decisions pertaining to the communications and IT sector in Puerto Rico were discussed in this forum. The HSOAC team had a member deployed to Puerto Rico for six months; he attended the majority of these meetings during his deployment.

The team was further informed by discussions with federal regulators at the FCC. In addition to federal and commonwealth partners, the HSOAC team received input from Puerto Rico

---

[27] These weekly meetings also took place after the recovery plan was submitted.

telecommunications providers and engaged with wireless infrastructure and cellular carrier associations. Finally, the team partnered with People-Centered Internet (PCI) to develop COAs that focus on information technology for the digital transformation of Puerto Rico.[28]

## Communications/IT Recovery

Chapters 3, 4, 5, and 6 explain the 33 COAs developed for the communications and IT sector. Chapter 3 describes a precursor COA that addresses effective governance in this sector. Chapter 4 describes 11 COAs that pertain primarily to improvements in the public telecommunications infrastructure to achieve effective emergency services and COG after a disaster. Chapter 5 includes eight COAs that require a strong partnership with the private sector or that take advantage of existing government programs in order to achieve a robust telecommunications infrastructure or to expand the deployment of broadband internet. Chapter 6 describes 13 COAs that focus on information technology infrastructure and on information technology for the digital transformation of Puerto Rico.

A detailed description of each of the 33 communications and IT COAs is provided in Appendix C. While technical interrelationships among COAs are not apparent in that appendix, they are explained in Chapters 4, 5, and 6. Table S.1 provides summary descriptions of all COAs and indicates, in brackets, which chapters discuss them.

To connect the COAs with the strategic objectives, we developed summary portfolios describing alternative approaches to recovery, each with a different level of investment. In Chapter 7 we describe the way forward in the communications and IT sector as represented in the recovery plan to meet the strategic objective of modernizing the telecommunications system. The approach is designated "Smart and Resilient Island IT and Communications" and has a total estimated cost of $3.2 billion. In addition to presenting the portfolios, this last chapter discusses potential sources of funding for the communications and IT COAs and general considerations for successful implementation. It also presents our concluding remarks.

**Table S.1. Short Descriptions of Courses of Action**

| COA ID | Title | Short Description |
|--------|-------|------------------|
| CIT 1 | Land Mobile Radio System [Chapter 4] | This COA consists of preparing a plan for Puerto Rico's future public telecommunications infrastructure, which considers several alternatives and their expected timelines. It also addresses the important need of maintaining a workforce that can ensure the readiness of the public safety telecommunications infrastructure, as well as the need for workforce development in the form of four-year electrical engineering bachelor of science degrees |

[28] PCI was hired by HSOAC as a subcontractor to provide consulting services.

| COA ID | Title | Short Description |
|--------|-------|-------------------|
| CIT 2 | Puerto Rico GIS Resource and Data Platform [Chapter 6] | This COA addresses the need for a comprehensive, real-time, and readily available information system that provides geolocated information to governmental public safety officials, emergency response teams, and community planning agencies |
| CIT 3 | Upgrade and Enhance 911 Service [Chapter 4] | This COA upgrades and enhances 911 services so that 911 centers will have state-of-the art equipment. It also consolidates the functions of 911 centers with first-responder dispatch in the same facility |
| CIT 4 | Rural Area Network Task Force [Chapter 4] | This COA establishes a task force to address the special circumstances of individuals residing in disconnected regions. It further seeks to bridge a gap in information systems to service the specific needs of individuals in these regions should a disaster occur through real-time advisories and guidance |
| CIT 5 | Implement Public Safety/Government Communications Backup Power [Chapter 4] | This COA consists of implementing backup power systems for public tower sites; hospitals; police, fire, and EMS stations; municipal city halls; and government centers identified by the Public Buildings Authority to allow for COG operations and emergency response |
| CIT 6 | Modernize the Emergency Operations Center [Chapter 4] | This COA upgrades and modernizes the Emergency Operations Center (EOC), currently colocated with PREMA headquarters |
| CIT 7 | Establish an Alternate Emergency Operations Center [Chapter 4] | This COA establishes an alternate EOC to serve as a backup location for emergency management activities should the primary EOC in San Juan become inoperable; it will also establish a 911 center in the alternate EOC to serve as a backup to the 911 centers in San Juan |
| CIT 8 | Mobile EOC Vehicle [Chapter 4] | This COA consists of procuring a mobile emergency operations center (MEOC) similar to the National Incident Management System (NIMS) Type-2 vehicles that arrived after the hurricanes |
| CIT 9 | Auxiliary Communications— Volunteer Radio Groups and Organizations [Chapter 4] | This COA concerns leveraging existing volunteer radio groups and organizations, which were able to provide critical communications in support of hospitals and municipalities in the aftermath of Hurricane Maria |
| CIT 10 | Transoceanic Submarine Cable [Chapter 5] | This COA introduces new, very high bandwidth undersea cable(s) to Puerto Rico, situated away from San Juan, including one landing point for the midterm, followed by additional ones in the long-term, to increase capacity and route options for telecommunications |
| CIT 11 | Procure a Mobile Emergency Communications Capability [Chapter 4] | This COA consists of the procurement of deployable assets that can be safely cached and quickly installed throughout Puerto Rico to restore voice and data communications for disaster response, emergency services, and government activities |
| CIT 12 | Perform Site Structural Analysis for All Government Telecom Towers (Both Public and Privately Owned) [Chapter 5] | This COA consists of performing site structural analysis for the telecommunications towers used for emergency services; these towers are part of the CI of Puerto Rico and are susceptible to damage by severe weather, such as hurricanes |
| CIT 13 | Streamline the Permitting and Rights of Way Processes for Towers and the Deployment of Fiber-Optic Cable [Chapter 5] | This COA concerns the government's consolidating and streamlining the permitting and rights of way (ROW) processes for towers and the deployment of fiber-optic cable by telecommunications providers |
| CIT 14 | Consolidated Government Information Systems [Chapter 6] | This COA establishes and implements an open, modular, standards-based platform for information systems and consolidates government systems |
| CIT 15 | Undersea Fiber Ring System | This COA strengthens the ability to communicate within and |

| COA ID | Title | Short Description |
|--------|-------|-------------------|
| | [Chapter 5] | around Puerto Rico over high-speed and high-capacity links through a network infrastructure for a communications ring system |
| CIT 16 | Government Digital Reform Planning and Capacity Building [Chapter 6] | This COA increases the human capacity of PRITS by adding expertise and staff on data science, data architecture, IT architecture, and cybersecurity; it also makes use of that additional expertise and staffing to create a roadmap for the Digital Transformation of Puerto Rico |
| CIT 17 | Puerto Rico Data Center [Chapter 6] | This COA addresses the need for a state-of-the-art, standards-compliant data center in Puerto Rico |
| CIT 18 | Data Storage and Data Exchange Standards for Critical Infrastructure [Chapter 6] | This COA improves Puerto Rico's ability to plan for and comprehensively support CI needs by establishing data storage and data exchange standards for CI |
| CIT 19 | Municipal Hotspots [Chapter 5] | This COA leverages and expands an existing program that is currently funded by the Telecommunications Bureau and currently supports 58 municipal wi-fi hotspots throughout Puerto Rico |
| CIT 20 | Continuity of Business at PRIDCO Sites [Chapter 4] | This COA provides critical communication systems required to maintain key business activities at Puerto Rico Industrial Development Company (PRIDCO) sites when primary communications methods are degraded or unavailable |
| CIT 21 | Government-Owned Fiber-Optic Conduits to Reduce Aerial Fiber-Optic Cable and Incentivize Expansion of Broadband Infrastructure [Chapter 5] | This COA provides a strong incentive to private carriers to reduce the amount of aerial fiber-optic cable and to deploy fiber-optic cable in buried conduit instead, since the latter is more resilient to disasters; for this COA the Puerto Rico government will perform the required trenching and will lay empty conduit, which are the two most expensive steps of burying fiber-optic cable |
| CIT 22 | Use Federal Programs to Spur Deployment of Broadband Internet Island-Wide [Chapter 5] | This COA consists of leveraging an existing program called E-Rate, which supports telecommunications services including broadband internet services for qualified schools and libraries, with funding supplied by the FCC |
| CIT 23 | Data Collection and Standardization for Disaster Preparedness and Emergency Response [Chapter 4] | This COA provides for the continued maintenance and expansion of status.pr, a site launched by the government of Puerto Rico to update the media, public, and first responders about conditions across Puerto Rico |
| CIT 24 | Establish Puerto Rico Communications Steering Committee [Chapter 3] | This COA establishes a new Communications Steering Committee through Executive Order by the Governor. This committee will provide strategic guidance, policy, direction, and standards associated with Puerto Rico communications networks; it will also ensure proper planning and collaboration to effectively and efficiently recover and maintain the communications infrastructure, as well as mitigate interoperability and duplication of effort issues |
| CIT 25 | Evaluate and Implement Alternative Methods to Deploy Broadband Internet Service Throughout Puerto Rico [Chapter 5] | This COA concerns the government's working with a blue-ribbon panel of experts to evaluate and implement alternative methods to deploy broadband internet service throughout Puerto Rico |
| CIT 26 | Wi-fi Hotspots in Public Housing and Digital Stewards Program [Chapter 6] | This COA establishes wi-fi hotspots in public housing and a "Digital Stewards" program to be modeled after similar initiatives in Detroit, Michigan, and Red Hook, New York |

| COA ID | Title | Short Description |
|--------|-------|-------------------|
| CIT 27 | Study Feasibility of Digital Identity [Chapter 6] | This COA undertakes a study of existing models for "digital identities," in order to create secure digital identities that will facilitate digital transactions and reduce transaction costs, while decreasing the potential for fraud and identity theft |
| CIT 28 | Innovation Economy/Human Capital Initiative [Chapter 6] | This COA creates a public-private initiative to provide digital skills training, entrepreneurship programs, and access to new digital technologies for people throughout Puerto Rico through a network of innovation hubs and entrepreneur centers, training partnerships with schools, and outreach via mobile laboratories to rural and underserved areas |
| CIT 29 | Health Care Connectivity to Strengthen Resilience and Disaster Preparedness [Chapter 6] | This COA provides a robust, resilient, multimodal "mesh" communications network to clinics to complement connectivity that is available through the telecom infrastructure or to provide redundancy when such infrastructure is damaged; it also uses the increased connectivity and IT to ensure real-time access to clinical data—including mobile and telehealth—and to support situational awareness, behavioral health, environmental monitoring, and social services |
| CIT 30 | Resiliency Innovation Network Leading to Development of a Resiliency Industry [Chapter 6] | This COA creates a Resiliency Innovation Network (RIN) across Puerto Rico to build on the existing Puerto Rico Science, Technology, and Research Trust university facilities to teach, test, and refine existing resiliency products and services and to develop new ones to enhance capability and stimulate new commercial ventures |
| CIT 31 | Resilience/e-Construction Learning Lab [Chapter 6] | This COA establishes a Resilience/e-Construction Learning Lab for a one-year pilot project to digitize assessment, permitting, and reporting processes in one Puerto Rico municipality and present findings to inform the feasibility and cost-benefit analysis for an e-Permitting and e-Construction ecosystem throughout Puerto Rico |
| CIT 32 | Digital Citizen Services [Chapter 6] | This COA expands the scope of PRITS to include a focus on citizen-centered services and prioritize a "one-stop-shop" experience for accessing government services and information in an easy-to-use fashion; it implements best practices for ensuring digital inclusion and accessibility, such as the ability to access government services from mobile devices in a secure way |
| CIT 33 | Government Digital Process Reform [Chapter 6] | This COA revises governmental methods and policies to drive how systems can or should be utilized with the intent of improving government services provided to citizens |

# Acknowledgments

The authors would like to thank the many experts who contributed to this report. At the Telecommunications Bureau within the Puerto Rico Public Service Regulatory Board, we would like to thank Sandra Torres, Chairwoman; Zaida Cordero Lopez, Associate Commissioner; and Maria Fullana Hernandez.[1] We would also like to thank Alexandra Fernandez Navarro, a former Commissioner of the Puerto Rico Telecommunications Regulatory Board, who is now a Commissioner of the Puerto Rico Public Service Regulatory Board.

We would like to extend our thanks to senior officials of other Puerto Rico agencies who provided us with insights and information, including Luis Carlos Gonzalez, Puerto Rico Department of Public Safety; Waldo Acevedo, Puerto Rico Innovation and Technology Service; Diana Pelegrina, Puerto Rico Deputy Chief Information Officer; Luis Arocho, Puerto Rico Chief Information Officer; Glorimar Ripoll, Puerto Rico Chief Innovation Officer; and Felix Garcia-Bermudez, Puerto Rico Statewide Interoperability Coordinator.

The authors were privileged to work with many subject matter experts from the FEMA and other federal agencies. Although it is not possible to name them all, we would like to thank especially Maggie Holmes, FEMA Comms/IT sector Lead, David Samaniego, previous FEMA Telecom sector Lead, and the members of the FEMA Comms/IT Solutions-Based Team. The members of this team were: Martin Pittinger, Office of the Chief Information Officer of the Department of Homeland Security (DHS); Roberto Mussenden, Federal Communications Commission; Chris Alexander, Office of Emergency Communications of the DHS; Scott Jackson, National Telecommunications and Information Administration; John McClain, National Cybersecurity and Communications Integration Center; and Steve Broniarczyk, FEMA. We also extend our thanks to the Hurricane Maria Communications Task Force, Charles Hoffman, and Patrick Hall of FEMA.

Michael Carowitz, who heads the Puerto Rico Hurricane Task Force at the FCC, and his task-force colleagues were especially helpful to our research. We would also like to acknowledge the important contributions of Gabriela Gross and her team in the Wireline Competition Bureau at the FCC, as well as Catriona Ayer, Deputy Director of the Universal Service Administration Corporation, and her colleagues.

Telecommunications industry representatives met with the Homeland Security Operational Analysis Center (HSOAC) team at the Puerto Rico Telecommunications Regulatory Board in May 2018 to provide their input and ideas about the proposed courses of action (COAs). We would like to thank the companies that participated for their valuable insights, especially AT&T,

---

[1] Previously, the Telecommunications Bureau within the Puerto Rico Public Service Regulatory Board was the Puerto Rico Telecommunications Regulatory Board.

FirstNet, Claro, Liberty Puerto Rico, Neptuno, T-Mobile, and WorldNet Telecom, and Chairwoman Sandra Torres for organizing the meeting.

We wish to acknowledge the contributions of People-Centered Internet, a subcontractor of HSOAC, particularly Marci Harris.

Finally, we would like to thank Dr. Edward Balkovich of RAND Corporation, Professor Jon Peha of Carnegie Mellon University, and Professor David Turetsky of the College of Emergency Preparedness, Homeland Security, and Cybersecurity at the University at Albany, State University of New York, for their thoughtful review of the report and its underlying analysis.

# Abbreviations

| | |
|---|---|
| API | application programming interface |
| ARRA | American Recovery and Reinvestment Act |
| AUXCOM | auxiliary community communication |
| CAD | computer-aided dispatch |
| CDBG | Community Development Block Grants |
| CDBG-DR | Community Development Block Grant Disaster Recovery |
| CI | critical infrastructure |
| CIIO | chief innovation and information officer |
| CINO | chief innovation officer |
| CIO | chief information officer |
| CLEC | competitive local exchange carrier |
| CNC | computer numerical control |
| COA | course of action |
| COG | continuity of government |
| COLT | cellular equipment on light trucks |
| Comms/IT | Communications and Information Technology (sector team) |
| CONUS | continental United States |
| COOP | continuity of operations |
| COW | cells on wheels |
| CPCB | Community Planning and Capacity Building (sector team) |
| DHS | U.S. Department of Homeland Security |
| DoD | U.S. Department of Defense |
| DSL | digital subscriber line |
| DTOP | Department of Transportation and Public Works |
| EAS | Emergency Alert System |
| EDA | Economic Development Administration |

| | |
|---|---|
| EMPG | Emergency Management Performance Grant |
| EMS | emergency medical service |
| EOC | Emergency Operations Center |
| ESF-2 | Emergency Support Function #2 |
| ESINet | Emergency Services IP Network |
| FCC | Federal Communications Commission |
| FEMA | Federal Emergency Management Agency |
| FFRDC | federally funded research and development center |
| FHWA | Federal Highway Administration |
| FirstNet | First Responder Network Authority |
| FOMB | Financial Oversight and Management Board |
| FQHC | federally qualified health center |
| GAO | Government Accountability Office |
| GIS | geographic information system |
| HDPE | high-density polyethylene |
| HHS | U.S. Department of Health and Human Services |
| HIPAA | Health Insurance Portability and Accountability Act |
| HMGP | Hazard Mitigation Grant Program |
| HPMS | Highway Performance Monitoring System |
| HSOAC | Homeland Security Operational Analysis Center |
| HUD | U.S. Department of Housing and Urban Development |
| ILEC | incumbent local exchange carrier |
| IoT | internet of things |
| IPAWS | Integrated Public Alert and Warning System |
| IT | information technology |
| ITT | International Telephone & Telegraph |
| KPI | key progress indicators |
| LMR | land mobile radio |
| Mbps | megabits per second |

| | |
|---|---|
| MEOC | mobile emergency operations center |
| MOU | memorandum of understanding |
| MSAT | Mobile satellite terminal |
| NCR | Natural and Cultural Resources |
| NDRF | National Disaster Recovery Framework |
| NG911 | Next Generation 911 |
| NIMS | National Incident Management System |
| NTIA | National Telecommunications and Information Administration |
| OEC | Office of Emergency Communications |
| PA | public assistance |
| PCI | People-Centered Internet |
| PRDPS | Puerto Rico Department of Public Safety |
| PREMA | Puerto Rico Emergency Management Agency |
| PREPA | Puerto Rico Electric Power Authority |
| PRIDCO | Puerto Rico Industrial Development Company |
| PRINCE | Puerto Rico and the US Virgin Islands Interoperable Communications Network Engagement |
| PRIS | Puerto Rico Institute of Statistics |
| PRITS | Puerto Rico Innovation and Technology Service |
| PRPB | Puerto Rico Planning Board |
| PRPD | Puerto Rico Police Department |
| PRSTRT | Puerto Rico Science, Technology, and Research Trust |
| PRTC | Puerto Rico Telephone Company |
| PRTRB | Puerto Rico Telecommunications Regulatory Board |
| PSAP | public safety answering point |
| PSTN | public switched telephone network |
| R&D | research and development |
| RCOEI | Resiliency Center of Education and Innovation |
| RF | radio frequency |

| | |
|---|---|
| RHI | Red Hook Initiative |
| RIN | Resiliency Innovation Network |
| ROI | return on investment |
| ROW | rights of way |
| SBT | solutions-based team |
| SME | subject matter expert |
| SMFO | single-mode fiber-optic |
| STEM | science, technology, engineering, and mathematics |
| SWIC | Statewide/Territory-wide Interoperability Coordinator |
| TIC | technologias de informacion y comunicacion (technologies of information and communication) |
| UHF | ultra high frequency |
| UPR | University of Puerto Rico (Universidad de Puerto Rico) |
| USAC | Universal Service Administration Corporation |
| USDA | U.S. Department of Agriculture |
| USDS | United States Digital Service |
| VHF | very high frequency |
| VTC | video-teleconference |
| WEA | wireless emergency alert |

# 1. Introduction: The Communications/IT Sector in Puerto Rico

## Overview

Hurricanes Irma and Maria made landfall on Puerto Rico in September 2017, causing unprecedented, widespread and devastating damage. Congress mandated that the Governor of Puerto Rico, with support and contributions from federal agencies, produce an economic and disaster recovery plan within 180 days.[1] This report is one of several documents prepared to describe analysis conducted during the development of the 180-day plan; it pertains to the communications and information technology (IT) sector. The overall economic and recovery plan, *Transformation and Innovation in the Wake of Devastation: An Economic and Disaster Recovery Plan for Puerto Rico*, was delivered to Congress on August 8, 2018.[2]

In this introductory chapter, we provide context regarding the communications and IT sector in Puerto Rico. The chapter begins with a brief history of this sector in Puerto Rico, followed by a description of the key communications and IT assets and functions during both normal and emergency operations. The chapter subsequently addresses the prestorm conditions as well as the functional, operational, and funding challenges. Finally, the chapter provides an overview of the regulatory, governance, public safety, and commercial communications entities in Puerto Rico.

## Brief History of Communications/IT in Puerto Rico

In 1897, the Sociedad Anonima de Telefono offered public telephone service for the first time in San Juan, Ponce, and Mayaguez. In 1914, Sosthenes and Herman Behn founded Ricotelco in San Juan. Six years later, Sosthenes Behn established International Telephone & Telegraph (ITT), which served as a holding company for Ricotelco and several other Caribbean telephone companies.

More than 50 years later, in 1974, the government of Puerto Rico acquired Ricotelco from ITT for $200 million and created the Puerto Rico Telephone Company (PRTC). In 1980, PRTC began deploying fiber-optic capability, and in 1994, it completed its fiber-optic network PRTC introduced cellular services in 1994 and began providing internet services in 1996.[3]

---

[1] Congress passed the Bipartisan Budget Act of 2018 on February 9, 2018 (Public Law No. 115-123). This legislation required the Governor of Puerto Rico, with support and contributions from the Secretary of the Treasury, the Secretary of Energy, and other federal agencies having responsibilities defined under the National Disaster Recovery Framework (NDRF), to produce within 180 days an economic and disaster recovery plan that defines the priorities, goals, and outcomes of the recovery effort. This recovery plan is also referred to as the "180-day plan."

[2] Central Office for Recovery, Reconstruction and Resiliency, *Transformation and Innovation in the Wake of Devastation: An Economic and Disaster Recovery Plan for Puerto Rico*, August 8, 2018.

[3] Gabriel Parra Blessing, "At PR Telephone, the Past Is Prologue," *Caribbean Business*, October 27, 2005.

At this point, PRTC was a government-owned and government-run monopoly provider of telecommunications services in Puerto Rico, but its monopoly soon ended. The U.S. Congress enacted the Telecommunications Act of 1996, which required incumbent local exchange carriers (ILECs) such as PRTC to interconnect with new rivals, competitive local exchange carriers (CLECs), at cost-based rates for unbundled network elements.[4] The Puerto Rico Legislative Assembly enacted the Puerto Rico Telecommunications Act of 1996 and created the Puerto Rico Telecommunications Regulatory Board (PRTRB) to oversee Puerto Rico's telecommunications industry and the introduction of this new competition.[5]

In 1997, the Puerto Rico legislature enacted legislation to allow the sale of PRTC. In 1998, GTE purchased PRTC from the government of Puerto Rico for $2.25 billion. In 2000, Bell Atlantic (now Verizon Communications) purchased GTE. In 2001, PRTC launched Verizon Wireless and introduced digital subscriber line (DSL) service in 2001. PRTC was a subsidiary of Verizon Communications until it was fully acquired by América Móvil in 2007. América Móvil rebranded many of PRTC's services in 2011 and 2013 to the name "Claro." Currently, Claro provides residential wireline, wireless, broadband DSL, television and dial-up services, internet protocol television, as well as services to business, government, and enterprises in Puerto Rico.

## Key Communications/IT Assets

The telecommunications infrastructure of Puerto Rico is made up of different technologies that interconnect to provide voice and data communications. To function at optimal performance, all of these technologies require continuous and sufficient power from the electrical grid.[6] Figure 1.1 provides a simplified illustration of the different network architectures that interconnect in Puerto Rico and are used by both public and commercial providers: wireline service provided by Claro (the public switched telephone network [PSTN]); cellular service provided by AT&T, T-Mobile, Claro, and PR Wireless; broadband internet service provided by Liberty Cable; land mobile radio (LMR) used for public safety by the police; microwave point-to-point communications used for backhaul to the PSTN by the cellular carriers and police; and satellite communications that can be used for voice and data connections during an emergency and for nonemergency operations. Figure 1.1 also illustrates one of the submarine cable landing stations that connect calls to and from Puerto Rico. In addition, wireless infrastructure (cell towers) provided by Crown Castle, American Tower, Innovatel, the SBA Communications Corporation,[7]

---

[4] Telecommunications Act of 1996, Pub. LA. No. 104-104, 110 Stat. 56 (1996).

[5] Act 213 of 1996, Puerto Rico Telecommunications Act, September 12, 1996, as amended, revised May 10, 2012.

[6] Fuel-based generators and batteries can power telecommunications sites (as defined in this chapter) but only for short durations. They should be used only as backup for fast but temporary power restoration while the electrical grid is being repaired.

[7] SBA Communications Corporation is an independent owner and operator of wireless communications infrastructure, including towers, buildings, rooftops, distributed antenna systems and small cells. It was founded in 1989.

**Figure 1.1. Telecommunications Network Examples**

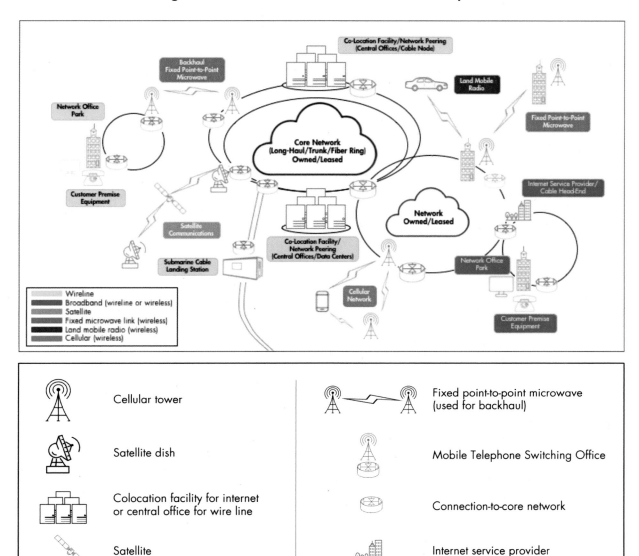

Source: Homeland Security Operational Analysis Center (HSOAC) analysis of U.S. Department of Homeland Security (DHS) and government of Puerto Rico documentation.

and others can be used to mount radios and antennas. Figure 1.1 illustrates the interconnectivity of the system only; it does not reflect the actual configuration of telecommunications networks in Puerto Rico.

In addition to the residential and business wireline, wireless, broadband DSL, and dial-up services and the internet protocol television provided by Claro,[8] wireless and internet services in Puerto Rico are provided by several other sources. Until very recently, there were 5 providers of

---

[8] Claro, which purchased the Puerto Rico Telephone Company, the previously government-owned provider of telecommunications services.

mobile telephony in Puerto Rico: AT&T, T-Mobile, Claro, Sprint, and Open Mobile. In mid-November 2017, Sprint and Open Mobile formed a joint venture, PR Wireless, LLC, which was approved by the Federal Communications Commission (FCC) and PRTRB.[9] The FCC's "2018 Broadband Deployment Report" shows that approximately 99 percent of the residents of Puerto Rico have access to cellular service.[10]

According to the Federal Emergency Management Agency (FEMA) Hurricane Maria Task Force, Puerto Rico before the hurricanes was already

> a mobile society with only approximately 20 percent of households having landline telephone service. Residents instead have embraced mobile (wireless) technology and use their cellular devices as their only means of communications. Due to this fact, Puerto Rico has an extensive wireless network infrastructure, which provides a high-level of capacity to support the customer base . . . most of the wireless infrastructure on the island is supported by aerial fiber backhaul.[11]

The task force further reported that "wireline is primarily fiber based and supports mostly commercial facilities and backhaul for the wireless infrastructure. There is virtually no copper infrastructure in Puerto Rico."[12]

Internet service is provided using technologies such as cable (Liberty), DSL (e.g., Claro, WorldNet), fiber (e.g., Optico Fiber), and fixed wireless (e.g., IP Solutions, Caribe.net, Inteco, Neptuno Networks).[13] Liberty Cablevision of Puerto Rico is the only cable provider. According to its website, "Liberty provides video, broadband internet, and fixed-line telephony services to residential and/or business customers in Puerto Rico."[14] Moreover, PREPA Networks, LLC is a network infrastructure provider that offers services to telecommunication carriers and business sectors. It delivers wholesale and enterprise telecommunication services throughout Puerto Rico, and international connectivity to the U.S.- Miami, New York and the Caribbean Islands.[15]

The map in Figure 1.2 illustrates that broadband service is not available throughout the entirety of Puerto Rico. Although the map is from 2014, it reflects the disparities of internet

---

[9] "Sprint, Open Mobile Complete Joint Venture to Serve PR/USVI," News Is My Business, December 5, 2017.

[10] FCC, 2018a, Appendix E.

[11] FEMA Hurricane Maria Task Force, 2017, p. 2. A former industry official explained to the HSOAC Communications/IT team that Centennial de Puerto Rico may have had indefeasible rights of use (IRUs) with San Juan Cable when it was first building out its network, and it might also have buried some fiber-optic cable in urban areas of San Juan and in Mayaguez and Ponce. Centennial did not bury fiber in suburban, central, or rural areas of Puerto Rico. Like other carriers in Puerto Rico, Centennial primarily used aerial fiber-optic cable for its network. Centennial was purchased by AT&T in 2008.

[12] FEMA Hurricane Maria Task Force, 2017, p. 2.

[13] Broadband Now, Internet Providers in San Juan, Puerto Rico, 2014–2018. Fixed wireless technology provides broadband internet connectivity using, for example, a wireless local area network (LAN).

[14] Liberty Cablevision of Puerto Rico, homepage, undated.

[15] "About PREPA Networks," PREPA Networks, website, December 10, 2019.

**Figure 1.2. Density of Broadband Providers in Puerto Rico**

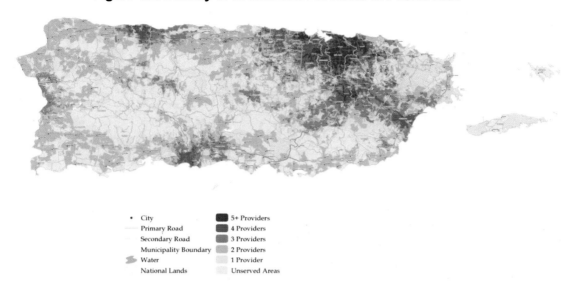

| | |
|---|---|
| • City | ■ 5+ Providers |
| — Primary Road | ■ 4 Providers |
| Secondary Road | ■ 3 Providers |
| Municipality Boundary | 2 Providers |
| Water | 1 Provider |
| National Lands | Unserved Areas |

SOURCE: Connect Puerto Rico, "Galería de Mapas Puerto Rico," Washington, D.C.: Connected Nation, 2014.

service that were referred to a letter dated September 2, 2016, from the head of the PRTRB to Orrin Hatch, Chairman of the U.S. Senate Committee on Finance, stating that

> FCC data shows that home broadband access and adoption in Puerto Rico is lower than any state. Nationwide, approximately 10% of the U.S. population does not have access to fixed broadband service of 25 Mbps download/3 Mbps upload, the broadband threshold the FCC deems necessary for high-quality video, voice, and data applications. In Puerto Rico, 62% of the population (over 2.6 million people) does not have access to 25/3 service. Moreover, in Puerto Rico's rural areas, 98% of the population does not have access to 25/3 service.[16]

Table 1.1 outlines these statistics.

**Table 1.1. Puerto Rico Residents' Access to Fixed Advanced Telecommunications Capability**

| | All Areas | | Urban Areas | | Rural Areas | |
|---|---|---|---|---|---|---|
| | Pop. Without Access | Percentage of Pop. | Pop. Without Access | Percentage of Pop. | Pop. Without Access | Percentage of Pop. |
| **Puerto Rico** | 2,259,097 | 62% | 1,325,638 | 50% | 933,414 | 98% |

SOURCE: Federal Communications Commission, "2016 Broadband Progress Report," FCC 16-6, January 29, 2016, Appendix D.

---

[16] Javier Rua Jovet, Chairman, Puerto Rico Telecommunications Regulatory Board, letter to the Hon. Orrin Hatch, Chairman, U.S. Senate Committee on Finance, September 2, 2016, p. 2. The data rate of the broadband service is given in megabits per second (Mbps).

Figure 1.3 shows an aggregate of all of Puerto Rico's telecommunication sites (as defined below), which are a key communications and IT asset. The map contains both the public safety sites and the private companies' sites;[17] it shows the following:

- Telecommunications infrastructure is densest (more sites per unit area) in and around metropolitan San Juan (north side of Puerto Rico).
- The next-densest level of telecom infrastructure is near the coasts, where the terrain is flat. Not reflected on the map is the fact that the island's fiber-optic backhaul tends to follow the road that parallels the coast and goes around Puerto Rico.
- Certain areas near the coasts (e.g., Mayaguez on the west end) have a higher density of telecom sites, which may correlate to a higher concentration of industry in those areas. Puerto Rico's west coast, in particular, has many pharmaceutical companies, which is one of its most important industry sectors.
- The center of Puerto Rico—mountainous terrain that is far from San Juan and the coasts—has the lowest density of telecom sites. This may be due to the obvious difficulties of installing sites in that terrain and to a smaller population, both of which have low incentives for private investment. The areas that lack internet access or that have insufficient wireless coverage tend to be where telecom site density is low.

For the purposes of this report, the *telecommunication site* refers to one of three installations:

1. A composite of the following: a tower; one or more antennas (often many of them) affixed to that tower; a small building structure near the tower that houses electronic equipment; cables running from the antennas to the equipment inside the small building; means to power all the equipment (i.e., connections to the power grid, a generator, and often both); and means to provide air conditioning to the building when the equipment requires it.[18] Currently, most generators run on gasoline or diesel in Puerto Rico. This is the type of installation referred to when we discuss public safety telecom sites and many of the private telecom sites (which typically have smaller towers than public safety sites).
2. A rooftop installation, which is a composite of antennas; associated electronic equipment; connecting cables; and means to provide power (i.e., connection to the electric grid or a generator). These components are placed on top of a public or private building. This type of site is more widely used where tall buildings are available, as in metropolitan areas.
3. A wireline central office, which is a building equipped with switches, routers, and many other devices to support the outside plant and to connect to other central offices. The central office depends on electricity to power its equipment. "Outside plant" refers to buried and aerial cables, poles, manholes, and other structures and equipment in customer premises. Interoffice facilities include equipment and cables used to interconnect central offices.[19]

---

[17] We are allowed to release this information to the public domain because it is provided at the aggregate level, and there is no possibility of identifying the sites that refer to any particular company.

[18] As discussed in this report, the electronic equipment on a telecommunication site may come from multiple providers, which make use of the different elements of the infrastructure (tower, building structure, land, and access road). Moreover, these infrastructure elements may have multiple owners.

[19] In addition to cellular sites and central offices, other important components of Puerto Rico's wireline infrastructure are cable head-end and neighborhood nodes. These components require power, and, in general, a cable system

**Figure 1.3. Private and Public Telecommunications Sites in Puerto Rico**

SOURCE: Puerto Rico Telecommunications Regulatory Board.

Puerto Rico's public safety communications consists of the following main state-level radio systems:

- *Puerto Rico Police Department (PRPD)P25 Trunk System:* This system operates from 19 radio frequency (RF) sites distributed around Puerto Rico. It provides statewide radio coverage to police, fire, health, and other entities.

- *PRPD SmartZone:* This system operates from two RF sites. For police use only, it provides communications capabilities in the greater San Juan metropolitan area.

- *Emergency Medical Service (EMS) System:* This is a conventional ultra high frequency (UHF) analog system operating from five RF sites. It provides communications capabilities to the Medical Emergency Corps of Puerto Rico, which is responsible for premedical care and emergency transport to hospitals

- *Puerto Rico Emergency Management Agency (PREMA) system*: This is a conventional very high frequency (VHF) system that provides day-to-day mobile radio communications between municipalities and PREMA.

- *Interoperability System*: This is a cross-band interoperability system that operates from eight RF sites and can support VHF, UHF, and 700 and 800 MHz radio traffic. The repeaters operate in P25 digital mode. This system provides statewide communications in emergency or urgent situations, such as tsunamis, earthquakes, and many others. The system is designed strictly as an interoperability platform and not for day-to-day usage.

---

needs power all the way to the home to provide internet connectivity. However, Figure 1.3 does not include these components as part of its depiction.

To support these systems, Puerto Rico uses two primary microwave networks as backbones: the PRPD network and the EMS network.[20] These systems and networks are critical for the proper operation of 911 emergency response.

Presently, submarine cable–related infrastructure constitutes the primary, high-capacity communications link to and from Puerto Rico. Both submarine cabling and the related terrestrial infrastructure are vulnerable to storm damage.[21] Figure 1.4 shows that landing stations are centered on the San Juan area. Therefore, there are no alternative route options available from points outside the northern side of Puerto Rico.[22]

**Figure 1.4. Concentration of Submarine Cable Landing Stations in the San Juan Region of Puerto Rico**

SOURCE: Submarine Cable Map, undated. Cable route information downloaded November 16, 2018.
NOTE: The map shows that all existing routes (to or from other parts of the world) converge onto a single region of Puerto Rico, in and around San Juan. The landing sites are concentrated in one region.

---

[20] DHS Office of Emergency Communications, Interoperable Communications Technical Assistance Program, "Puerto Rico Public Safety Communications Survey Report," February 2018.

[21] Doug Madory, "Puerto Rico's Slow Internet Recovery," Oracle Internet Intelligence, December 7, 2017.

[22] See Submarine Cable Map, undated. Some technical details on the submarine cable infrastructure, including technical details on the landing stations are sensitive and are not included here.

## Functions of Communications/IT

### Normal Operations

Prior to Hurricane Maria's impact, communications in Puerto Rico were primarily achieved wirelessly: Over 80 percent of Puerto Rico customers exclusively used cellular phone or other wireless connections; only 20 percent used landline connections.[23]

As mentioned, five companies and a new joint venture provide mobile telephony to Puerto Rico, and there are at least four companies that provide cell towers to cellular carriers and the wireless industry.[24] Mobile telephony providers lease space on poles owned by the Puerto Rico Electric Power Authority (PREPA) or by Claro to deploy their aerial fiber network. Aerial fiber was used for a very significant amount of backhaul between cell sites and the communications network.[25]

As mentioned in the discussion preceding Table 1.1, broadband internet use in Puerto Rico is much less than it is in the continental United States (CONUS). There have been government-sponsored efforts through Connect Puerto Rico and the Broadband Initiative to increase the availability of internet service and penetration in Puerto Rico since 2010. The National Telecommunications and Information Administration (NTIA) of the U.S. Department of Commerce has awarded over $41 million in broadband-related federal grants to Puerto Rico.[26]

PRTRB, the regulatory agency, has also been active in Connect Puerto Rico and other initiatives to increase broadband internet's wide-spread deployment.

### Emergency Operations

Communications for public safety in Puerto Rico are diverse. As explained by the FCC, "The Wireless Communications and Public Safety Act of (911 Act) took effect on October 26, 1999. The purpose of the 911 Act is to improve public safety by encouraging and facilitating the prompt deployment of a nationwide, seamless communications infrastructure for emergency services."[27] A public safety answering point (PSAP) is a key element of 911 emergency services. A primary PSAP receives calls from a 911 selective router or 911 tandem.[28] A secondary, or

---

[23] FEMA Hurricane Maria Task Force, 2017, p. 2; FCC, 2018a, shows that approximately 99 percent of the residents of Puerto Rico have access to cellular service.

[24] We refer to companies that provide cell towers to cellular carriers and the wireless industry, not to cellular carriers that own their own towers.

[25] FEMA Hurricane Maria Task Force, 2017.

[26] NTIA, Broadband USA, undated.

[27] FCC, "911 and E911 Services," undated.

[28] "A primary PSAP is defined as a PSAP to which 911 calls are routed directly from the 911 Control Office, such as a selective router or 911 tandem." FCC, "911 and E911 Services," undated.

backup, PSAP receives calls that cannot be handled by the primary PSAP.[29] Puerto Rico has both a primary and a backup PSAP, as shown in Figure 1.5.

However, these PSAPs do not have a dispatch capability.[30] Instead, they use PSTN to relay emergency information to police, fire, and emergency medical services in each of Puerto Rico's 78 municipalities. Emergency services are then dispatched by the appropriate municipal organization: one of the 78 municipal police stations, 93 municipal fire stations, or 56 local EMS stations.[31] PREMA, the agency that oversees all emergency activities, has a radio network that permits communications between its headquarters and 12 regional offices.[32] It also manages an Interoperability Radio Network.[33] Puerto Rico's police and firefighters use a separate radio network managed by PRPD.[34]

Figure 1.5 depicts the emergency response process. Because Puerto Rico's PSAPs are not capable of dispatch, they rely on a two-call process. In the first part of the process, the 911 caller makes a phone call to the main PSAP (shown by the arrow at top left). The main PSAP requires either commercial power or a backup generator. If neither is available or if the main PSAP is overwhelmed with calls, the 911 call is routed to an operator in a backup PSAP (shown by the arrow at top right). The PSAP operator answers the call, gathers relevant information from the caller, and enters the caller's name, phone number, nature of call, and location into the computer-aided dispatch (CAD) system.[35] Then the operator decides whether the caller needs police, fire, or medical help, and finds out what district or municipal office is in charge of dispatch in the caller's area.

The second call is from the PSAP operator to the district or municipal office in charge of the appropriate first responders (shown by the arrows in the center of Figure 1.5). Once the district or municipal first responder's office gets the call, that office dispatches a police patrol, ambulance, or fire truck via LMR or one of the microwave networks illustrated at the bottom of Figure 1.5 if the first responders are not reachable by LMR.

---

[29] "A secondary PSAP is defined as a PSAP to which 911 calls are transferred from a primary PSAP." FCC, "911 and E911 Services," undated.

[30] We mean that, once the operator of the 911 center (or PSAP) has received an emergency call from the public, a second phone call has to be made by the operator to contact the proper first-responder district or municipal office (police, fire, or EMS) so that this office performs the dispatch function. This is explained in more detail in the text.

[31] FEMA Hurricane Maria Communications Task Force, 2017, p. 2.

[32] FEMA Hurricane Maria Communications Task Force, 2017, p. 3.

[33] FEMA Hurricane Maria Communications Task Force, 2017, p. 3.

[34] FEMA Hurricane Maria Communications Task Force, 2017, p. 3.

[35] Puerto Rico Fire does not currently use direct CAD transfer from the PSAP.

**Figure 1.5. Puerto Rico's Process for Responding to 911 Emergency Calls**

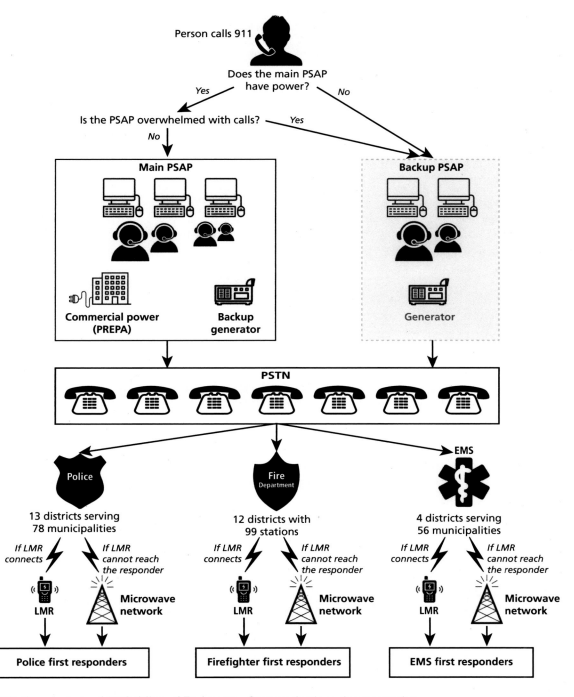

SOURCE: Our analysis of DHS Office of Emergency Communications documentation.
NOTE: LMR = land mobile radio.

## Prestorm Conditions and Challenges

### Context

Puerto Rico had a variety of prestorm conditions that made communications networks vulnerable in the event of a natural disaster. As Table 1.2 illustrates, these conditions—and the ensuing vulnerabilities—led to serious challenges in the provision of public, private, and emergency communication services.

**Table 1.2. Summary of Key Pre-Storm Conditions and Challenges**

| Condition | Vulnerabilities |
|---|---|
| Lack of buried fiber | Aerial fiber was exposed to the elements, which resulted in a greater loss of communications networks than if buried fiber had been utilized |
| PSTN reliance | Without a functioning PSTN, emergency calls could not be properly relayed |
| Maintenance issues | Lack of adequate maintenance to utility poles that carried aerial fiber (including, for example, excess weight of attachments) increased susceptibility to damage/destruction |
| Aging power grid | Communications, including wireless, wireline, and cable, depended on an aged electrical grid; backup power was not available at cell sites for extended periods of time |
| Single points of failure | There was no alternate Emergency Operations Center (EOC) that could serve as a backup location for emergency management activities should the primary EOC in San Juan become inoperable |
| Concentration of submarine cable landing stations | Locating the landing stations in a single area increases the risk to losing communications to and from Puerto Rico if a major disaster occurs in that area |

In brief, the state of telecommunications in Puerto Rico prior to the hurricanes was as follows. Residents of Puerto Rico had very limited access to broadband internet service (the FCC standard is referred to as "25/3" service).[36] Puerto Rico had a patchwork of aging communication systems and networks to support the provision of emergency services. PSAPs did not have a dispatch capability, and instead used PSTN to relay emergency information to police, fire, and EMS in each of Puerto Rico's 78 municipalities. Aerial fiber-optic cable was used for most of the wireline infrastructure, as well as for cellular service backhaul. The aerial fiber-optic cable was deployed predominately on utility poles owned by PREPA, many of which were old and not well maintained. Locating the submarine cable landing stations in a single area provided a single point of failure for communications to and from Puerto Rico. The full telecommunications infrastructure was strongly dependent on a reliable electrical grid.

---

[36] The Chair of the Puerto Rico Telecommunications Board stated in a letter to Senator Orrin Hatch, "In Puerto Rico, 62% of the population (over 2.6 million people) does not have access to 25/3 service. Moreover, in Puerto Rico's rural areas, 98% of the population does not have access to 25/3 service" (Rua Jovet, 2016).

*Functional, Operational, and Funding Challenges for Communications/IT*

Both the FEMA officials and the commercial providers with whom we spoke affirmed that the communications and IT sector faces some fundamental functional and operational challenges. For one, Puerto Rico's geography is a key challenge to communications. According to FEMA officials, the power and communications infrastructure was in a "fragile" state before the storms.[37] They noted that towers were overloaded and not properly maintained and that proper maintenance was limited by the remote location of certain telecommunication sites. This lack of robust maintenance made the sites more susceptible to damage and destruction from debris and fallen vegetation as a result of Hurricane Maria. One wireless carrier commented that the geography of Puerto Rico proved extremely challenging to disaster response, particularly the main mountain range, La Cordillera Central, which divides Puerto Rico.[38]

In addition to geography, Puerto Rico's aging infrastructure, dependence on functioning electrical systems, and ownership issues create a challenging environment in which to sustain communications. One wireless carrier noted that power plants are on average 44 years old in Puerto Rico, compared with an industry average of 18 years.[39] In comments provided to the FCC, CTIA noted that as of early November 2018, 74 percent of the residents of Puerto Rico had wireless coverage despite the fact that less than 43 percent had access to commercial power at that time.[40] This progress was largely due to the use of generator power. Despite these statistics, communications in Puerto Rico are clearly dependent on the electrical grid.[41] An additional complication is that numerous users, including different public safety agencies (police, EMS, PREMA) and private carriers, may lease both a space in a cell tower to install their antennas and a part of the associated building to house their electronic equipment. For those sites, it may be difficult for owners to determine who is responsible for which repair and for the government of Puerto Rico to track this information. Moreover, federal recovery funds (e.g., those provided by FEMA) can be provided only to repair, restore, and upgrade public, as opposed to private infrastructure.[42] As a result, FEMA officials indicated that determining whether infrastructure was public or private added complexity during recovery efforts.

---

[37] Emergency Communications Coordinator for Region 2, U.S. Department of Homeland Security, telephone interview with the authors, March 2, 2018.

[38] T-Mobile, Comments of T-Mobile USA, INC., In the Matter of Response Efforts Undertaken During the 2017 Hurricane Season, PS Docket No. 17-344, January 22, 2018.

[39] T-Mobile, 2018.

[40] CTIA, 2018.

[41] Telecommunications Bureau officials noted that, in their opinion, the electrical power grid was not in a weak state prior to Hurricane Maria; blackouts only occurred sporadically. However, because of the hurricanes, the Puerto Rico power grid failed catastrophically. Officials of the Telecommunications Bureau, interview with the authors, 2018.

[42] Emergency Communications Coordinator for Region 2, U.S. Department of Homeland Security, telephone interview with the authors, March 2, 2018.

The primary funding challenge not just for the communications and IT sector but for many others prior to the hurricanes was that economic conditions in Puerto Rico were already dire. The Executive Director of the Financial Oversight Management Board, Natalie Jaresko, wrote:

> Before the hurricanes, Puerto Rico had over $74 billion in debt, over $53 billion of unfunded pension liabilities, an economy that had contracted nearly 15% over the last decade, and a nearly 50% poverty rate. In addition, Puerto Rico's structural budget deficits were projected to average 50% of recurring revenues. Severe liquidity challenges and persistent budget deficits have contributed to a perilous lack of investment in infrastructure. . . . Puerto Rico is faced with a unique set of circumstances—the largest public entity restructuring in the history of the U.S. combined with the greatest hurricane devastation to strike in 100 years.[43]

One wireless carrier noted that the power grid was ill-prepared to sustain a major hurricane.[44] According to this carrier, funding conditions made the power grid "extremely susceptible to hurricane damage." Specifically, the carrier stated that Puerto Rico's sole power supplier, PREPA, was "bankrupt" and noted that PREPA itself "called its system 'degraded and unsafe' after 'years of under-investment.'" The funding challenges experienced by the energy sector increased the risks of susceptibility to communications given the dependence of communications and IT on a functioning electric grid.

### Prestorm Emergency Preparedness

Prior to Hurricane Maria, PRTRB and commercial carriers took some steps to improve emergency preparedness. PRTRB entered into a cooperation agreement with the Telecommunications Industry Committee for Emergency Management.[45] PRTRB also entered into an interagency agreement with PREPA to facilitate prompt power restoration to critical communications facilities that support response and recovery efforts.[46] In 2016, commercial carriers established the Wireless Network Resiliency Cooperative Framework, a voluntary initiative developed by the private sector to facilitate improved service restoration when

---

[43] Natalie Jaresko, Executive Director, Financial Oversight and Management Board for Puerto Rico, "Examining Challenges in Puerto Rico's Recovery and the Role of the Financial Oversight and Management Board," written testimony before the House Committee on Natural Resources, November 7, 2017.

[44] T-Mobile, 2018.

[45] According to the Telecommunications Bureau, the Telecommunications Industry Committee for Emergency Management is comprised of representatives from telecommunications and cable television companies operating in Puerto Rico to restore and protect telecommunications and cable television infrastructure during disasters or emergencies. See Sandra E. Torres Lopez, Letter from the Puerto Rico Telecommunications Regulatory Board to Honorable Ajit Pai, Chairman of the Federal Communications Commission, January 22, 2018.

[46] Torres Lopez, 2018.

networks go down. [47] According to CTIA, the framework's functions included providing reasonable roaming when feasible, fostering aid among carriers, convening with public safety representatives, developing a consumer readiness checklist, and reporting outages to the FCC.[48]

Prior to Hurricane Maria, the government of Puerto Rico also signed memorandums of understanding (MOUs) with the main telecommunication carriers so that in case of emergency each carrier could provide a certain number of cellular phones for emergency use by the government and the municipalities.

Each agency of the government of Puerto Rico has an emergency plan that explains the procedures to follow. There is also an emergency master plan. These emergency plans list the phone numbers of officials and personnel to be contacted in case of an emergency. However, having emergency telephone numbers was not useful in Puerto Rico after Hurricane Maria. The reason was that attempting to call an emergency phone number in the aftermath of the hurricane was unlikely to be successful due to the fact that 95 percent of the cellular sites were not operating. As an additional preparedness step, PRTRB provided a certain number of satellite phones to the Governor and heads of various commonwealth agencies. After the disaster, FEMA brought additional satellite phones to Puerto Rico for use by the government.

## Structure and Governance of Communications/IT

Tables 1.3, 1.4, and 1.5 provide an overview of several important stakeholders for communications and IT in Puerto Rico. Table 1.3 describes the agencies responsible for communications and IT governance and their

---

[47] On April 27, 2016, CTIA announced a Wireless Network Resilience Cooperative Framework for disasters and emergencies that applies throughout the United States, including Puerto Rico. See House Committee on Energy and Commerce, "CTIA & Pallone Announce 'Wireless Network Resilience Cooperative Framework' for Disasters and Emergencies," press release, April 27, 2016.

[48] A December 2017 Government Accountability Office (GAO) report stated that in April 2016, "an industry coalition consisting of CTIA, a wireless industry association, and five wireless carriers announced the Wireless Network Resiliency Framework (framework) in response to FCC's 2013 notice of proposed rulemaking on wireless network resiliency." The GAO report noted that "in December 2016, FCC said it would continue to engage with industry on the implementation and use of the framework, and FCC has taken some steps to monitor the framework's implementation. Specifically, FCC developed a plan to track certain tasks related to the framework in August 2017. In August 2017, FCC also issued a public notice inviting carriers beyond the five signatory wireless carriers to sign on to the framework." GAO concluded that "there are no specific measures for what the framework hopes to achieve. As a result, FCC lacks specific and measurable terms to monitor the effect of the framework." GAO, "Telecommunications: FCC Should Improve Monitoring of Industry Efforts to Strengthen Wireless Network Resiliency," Report to the Ranking Member, Committee on Energy and Commerce, House of Representatives, GAO-18-198, Washington, D.C., December 2017a, pp. 23, 27–28.

On December 10, 2018, the FCC issued a public notice soliciting input on the efficacy of the 2016 Wireless Resiliency Cooperative Framework. The FCC stated, "This Public Notice continues the Commission's line of inquiry into the Framework's effectiveness and builds upon the record we have received following the 2017 and 2018 Atlantic hurricane seasons" (Public Safety and Homeland Security Bureau, FCC, *Public Safety and Homeland Security Bureau Seeks Comment on Improving Wireless Network Resiliency to Promote Coordination Through Backhaul Providers*, PS Docket No. 11-60, Public Notice, December 10, 2018b, pp. 1–2).

**Table 1.3. Commonwealth Agencies Responsible for Communications/IT Governance**

| Agency | Responsibilities | Authorities |
|---|---|---|
| PRTRB; Junta Reglamentadora de Telecomunicaciones de Puerto Rico | • Regulates telecommunications services in Puerto Rico<br>• Ensures compliance with the Puerto Rico Telecommunications Act of 1996 to include imposing reasonable administrative fines for violations of the act and ordering the ceasing of violations<br>• Conducts inspections, investigations and audits, if necessary, to attain the purposes of the act<br>• Has primary jurisdiction over all telecommunications services and persons who render these services in Puerto Rico<br>• Performs Emergency Support Function #2 (ESF-2) responsibilities[a]<br>• Board's president and four members are appointed by the Governor with advice and consent of the Senate | Established by the Puerto Rico Telecommunications Act of 1996 (this statute has been amended 21 times between 1999 and 2017) |
| Puerto Rico Department of Public Safety (PRDPS; Departamento de Seguridad Publica de Puerto Rico) | • Includes the PRPD; Policía de Puerto Rico, the Puerto Rico Fire Department (Cuerpo de Bomberos), EMS (Cuerpo de Emergencias Médicas), PREMA (Agencia Estatal Para el Manejo de Emergencias y Administración de Desastres), the 911 System (Sistema de Emergencia 911), the Forensic Sciences Institute (Instituto de Ciencias Forenses), and the Office of Special Investigations (Negociado de Investigaciones Especiales)<br>• Oversees and has direct authority over all the organizations listed above<br>• Is led by the secretary of public safety who reports to the Governor | Law of the Department of Public Safety of Puerto Rico (law number 20, dated April 10, 2017, and amended by law number 78, dated February 4, 2018) |
| Office of the Chief Information Officer of Puerto Rico (PRCIO; Principal Ejecutivo de Informacion de Puerto Rico) | • Advises the Governor on creating and implementing public policy concerning the use of technologies of information and communication (TIC; technologias de informacion y comunicacion)<br>• Advises the Governor on the strategic management and best use of TIC for the government<br>• Responsible for an inventory of TIC of the government of Puerto Rico<br>• Establishes and implements strategic plans, policies, standards, and integrated architecture of the TIC of the government of Puerto Rico<br>• Establishes and implements government policies and security measures for the use of the internet and the interagency network<br>• Is led by the chief information officer who reports to the Governor | Reestablished by Governor's Executive Order, dated January 24, 2017 (published in Bulletin OE-2017-014), which was first established in 2009 under Executive Order (published in Bulletin OE-2009-09) but eliminated in 2015 by Executive Order (published in Bulletin OE-2015-19) |

[a] Puerto Rico Executive Order OE-2001-26, June 25, 2001. This executive order from the governor of Puerto Rico assigned the authority to conduct ESF-2 functions in Puerto Rico to the PRTRB. The ESF-2 function coordinates state and federal actions to assist industry in restoring public safety communications systems and infrastructure and first responder networks. In addition to the agencies shown in Table 1.3, the Governor of Puerto Rico appointed a chief innovation officer (CINO) in June 2017. According to the Governor's announcement, CINO is responsible for leading innovation processes in the government of Puerto Rico, making the processes more agile, and facilitating the accessibility of government services to citizens. Due to the priority of effecting a digital transformation of Puerto Rico, the communications and IT sector will play an important role in supporting the CINO's innovation efforts.

responsibilities. Table 1.4 summarizes the key commercial provider stakeholders. Table 1.5 summarizes the key commercial infrastructure stakeholders. Governance challenges for emergency communications were identified by the DHS Office of Emergency Communications (OEC) as critical prestorm issues to be addressed.[49] Specifically, while the Governor created a Puerto Rico Interoperability Committee for Emergency Communications in 2011 through executive order, the committee was ineffective and rarely met.[50] Given the complexities of Puerto Rico's communications systems, coordination across government is important to planning for and implementing an effective emergency communications system.[51] Because of the importance of effective governance to communications and IT, a course of action (COA) included later in this report is the establishment of a Communications Steering Committee to facilitate improved communication and governance among key public safety communications and telecom stakeholders.

In March 2019, the Governor issued an executive order to establish the position of chief innovation and information officer (CIIO).[52] This official also concurrently serves as the Executive Director of the Puerto Rico Innovation and Technology Service (PRITS). The mission of PRITS is "to set in motion and enable the transformation of Puerto Rico that will result in new

**Table 1.4. Key Commercial Provider Stakeholders Identified for Communications/IT**

| Organization | Role |
| --- | --- |
| AT&T | Wireless service provider |
| T-Mobile | Wireless service provider |
| Claro (formerly Puerto Rico Telephone Company) | Provides residential wireline, wireless, broadband DSL, television, and dial-up services, as well as services to business, government, and enterprises in Puerto Rico |
| Sprint | Wireless service provider (joint venture with OpenMobile to create PR Wireless) |
| Open Mobile | Wireless service provider (joint venture with Sprint to create PR Wireless) |
| Liberty Cablevision of Puerto Rico | Only cable provider in Puerto Rico; provides video, broadband internet, and fixed-line telephony services to residential and/or business customers |

**Table 1.5. Key Commercial Infrastructure Stakeholders Identified for Communications/IT**

| Organization | Role |
| --- | --- |
| Crown Castle | Provides wireless infrastructure (cell towers) |
| SBA Communications | Provides wireless infrastructure (cell towers) |
| American Tower | Provides wireless infrastructure (cell towers) |
| Innovattel | Provides wireless infrastructure (cell towers) |

---

[49] DHS OEC, 2018.

[50] DHS OEC, 2018.

[51] DHS OEC, 2018.

[52] Governor of Puerto Rico, Executive Order OE-2019-012, Responsibilities of the Chief Innovation and Information Officer and Chief Technology Officer, March 13, 2019.

knowledge and real impact through innovation, technology and collaborative hands-on approach to our challenges.[53] The chief information officer (CIO) now reports to the CIIO.

In August 2018, the Governor signed House Bill 1408, establishing the Puerto Rico Public Service Regulatory Board reorganization.[54] At that time the PRTRB was renamed the Telecommunications Bureau and organizationally placed under the Public Service Regulatory Board. The prior functions of the PRTRB were transferred to the Telecommunications Bureau, and one of the PRTRB's former commissioners was made a Commissioner of the new Public Service Regulatory Board.[55] In the remainder of this report we will be using the later name, Telecommunications Bureau, instead of PRTRB.

## Summary

We end this chapter by briefly summarizing the state of telecommunications in Puerto Rico prior to the hurricanes. At that time, residents had limited access to broadband internet service, especially in rural areas. The communication systems and networks relied upon for the provisioning of emergency services were a patchwork of aging systems. The 911 centers did not have a dispatch capability, as is typical in CONUS. These centers depended on the commercial PSTN to relay emergency information to police, fire, and EMS in all municipalities within Puerto Rico. The single location area for submarine cable landing stations presented a vulnerability as a single point of failure for communications to and from Puerto Rico. Most of the wireline infrastructure in Puerto Rico, as well as backhaul for cellular service, depended on aerial fiber-optic cable, rather than buried fiber-optic cables. These aerial fiber-optic cables were deployed predominately on electric utility poles owned by PREPA, many of which were old and not well maintained at the time of the hurricanes, partly as a result of the fiscal challenges at PREPA. Additionally, the entire telecommunications infrastructure was strongly dependent on a reliable electrical grid to maintain power to the cell sites and networks during and after an emergency, a typical vulnerability that was exacerbated by conditions at PREPA.

In the next chapter we will present our damage and needs assessment, and in subsequent chapters we will describe the COAs that address the vulnerabilities in the emergency and commercial telecommunications networks and improvements to broadband internet access throughout Puerto Rico. These COAs are designed to assist the government of Puerto Rico in achieving its vision of "building back better."[56]

---

[53] PR Innovation and Technology Service, homepage, undated.

[54] "Governor Signs Law to Establish Puerto Rico Public Service Regulatory Board Reorganization," *Caribbean Business*, August 13, 2018.

[55] "Governor Announces Appointments to Public Service Regulatory Board," *Caribbean Business*, August 20, 2018.

[56] Governor of Puerto Rico, *Build Back Better Puerto Rico: Request for Federal Assistance for Disaster Recovery*, November 13, 2017.

# 2. Damage and Needs Assessment

## Overview

This chapter addresses our assessment of damage to the communications and IT sector in Puerto Rico and the needs for its recovery. The chapter consists of four sections. The first section explains the data assessment methodology, including the sources used, as well as the limitations and uncertainties of the data analysis.

The second section describes the damage caused by the hurricanes and their impact. This section starts with a summary of key impacts and metrics and example statistics. It continues with more detailed descriptions of damage in public safety and commercial communications then presents preliminary dollar damage estimates, and ends with the impact that damage to the telecommunications infrastructure had on the residents of Puerto Rico.

The third section discusses the posthurricane response, including initial emergency response operations, barriers to response, and efforts to prepare for the 2018 hurricane season and status of repairs. In the fourth section, we offer a summary damage and needs assessment, which sets up the communications and IT sector's three top-level needs as determined by our analysis.

## Damage Assessment Methodology

### Data Sources

To report on prestorm conditions and challenges, damage caused by Hurricanes Irma and Maria, and posthurricane response, this chapter relies on documentation and testimonial evidence. Two primary sources of information were provided by FEMA officials: an October 2017 report completed by ESF-2 and a February 2018 report completed by the DHS OEC Interoperable Communications Technical Assistance Program. The chapter cites information from publicly available FCC sources, including comments filed with the FCC by telecommunications carriers in response to a formal notice requesting public comment regarding the 2017 hurricane response. It also leverages information provided by PRPD and the Telecommunications Bureau. In addition, it includes information received through discussions with various officials involved with the hurricane response efforts. Table 2.1 summarizes the key data sources used. As discussed later in this chapter, it does not include sensitive information, such as that marked "for official use only," or proprietary information from commercial carriers.

**Table 2.1. Summary of Key Damage-Assessment Data Sources**

| Dataset | Source | Date Range | Pre storm | Post storm | Recovery | Evaluation Notes |
|---|---|---|---|---|---|---|
| Puerto Rico Public Safety Communications Summary and Recommendations Report | FEMA | Feb. 2018 | N | Y | Y | Good descriptive detail on public safety communications systems |
| *DR-4339-PR Consolidated Communications Restoration Plan* | FEMA | Oct. 2017 | Y | Y | Y | Good descriptive detail on public safety communications systems |
| Letter from the PRTRB to the Chairman of the FCC | Telecommunications Bureau | Jan. 2018 | N | Y | Y | Good descriptive detail on telecommunications |
| CTIA comments on 2017 hurricane season | FCC publicly available | Jan. 2018 | Y | Y | Y | Perspective from industry association with some aggregate data |
| AT&T comments on 2017 hurricane season | FCC publicly available | Feb. 2018 | Y | Y | Y | Commercial perspective with some aggregate data |
| T-Mobile comments on 2017 hurricane season | FCC publicly available | Jan. 2018 | Y | Y | Y | Commercial perspective with some aggregate data |
| Verizon comments on 2017 hurricane season | FCC publicly available | Jan. 2018 | Y | Y | Y | Commercial perspective with some aggregate data |
| Puerto Rico Police Department Damage Report | PRPD | Sept. 2017 | N | Y | N | Photographic evidence of damage |
| Communication status reports | FCC publicly available | Sept. 2017 and Mar. 2018 | N | Y | Y | Data on cellular site service outages |
| Puerto Rico EMS Inspection Report for Communication Sites | Government of Puerto Rico | Post-Maria | N | Y | N | Photographic evidence of damage |
| PREMA Inspection Report for Communication Sites | Government of Puerto Rico | Post-Maria | N | Y | N | Photographic evidence of damage |

For the two DHS reports cited—the *Consolidated Communications Restoration Plan* and the *Puerto Rico Public Safety Communications Summary and Recommendations Report*—we also interviewed the lead official responsible for overseeing these reports.

The following section explains the limitations of and uncertainties in our analysis.

*Data Limitations and Uncertainties*

This chapter relies on available information to describe prestorm conditions, damage inflicted, and initial posthurricane recovery efforts. While much information has been made available through documentary and testimonial evidence from DHS, the government of Puerto Rico and its departments, other experts, and public sources, this chapter does not include sensitive or proprietary information that could otherwise add more clarity.[1] As a result, information in this assessment is largely presented at an aggregated level and not in more granular detail.

The lack of a more granular view may miss opportunities to describe the impact of the damage on specific municipalities. Information on the private telecommunications infrastructure—such a location of sites and level of damage of sites that could be traced to a particular company—is considered proprietary and not releasable in the public domain. Further, although additional quantitative data on key metrics, such as the percentage and number of ILEC central offices that are operational at a certain point in time, might be helpful in assessing damage, these data are also considered company-proprietary information.

This chapter seeks to mitigate these information gaps by using publicly available information from commercial carriers and the FCC. Cellular site outage data reported by the FCC provide an example of how aggregate data might not be representative of certain municipalities. Because certain municipalities continued to experience higher percentages of cellular site outages than the average across Puerto Rico months after the hurricanes, analysis based on aggregated data may miss opportunities to describe the impact of the damage to specific municipalities and status of response activities. Analysis based on aggregated data may also miss insights into factors that made some communities more vulnerable than others.

Furthermore, with regard to damage to the public telecommunications networks, this chapter cannot provide certain specific information on sites, such as their location, names, ownership, or level of damage. This is due to sensitivities associated with critical infrastructure (CI) information.

## Damage Caused by Hurricanes Irma and Maria

*Impact of Damage*

Public safety and commercial communications networks in Puerto Rico were significantly damaged by the hurricanes. Hurricane Maria decimated the infrastructure that is the backbone for commercial and public safety communications. Table 2.2 summarizes how the residents, government, and private sector of Puerto Rico were affected by Hurricane Maria; Table 2.3 presents metrics to describe this impact.[2]

---

[1] For example, some technical details on the submarine cable infrastructure are "for official use only."

[2] These selected metrics are not a comprehensive list.

**Table 2.2. Summary of Key Damage and Impact**

|  | Public Safety Communications | Commercial Communications |
|---|---|---|
| Resident | Inability to request and receive assistance from emergency services | Inability to communicate within and outside Puerto Rico |
| Government | Inability to coordinate response operations and perform emergency services | Inability to communicate across agencies and with CONUS and to perform certain government functions |
| Private sector | Inability to receive assistance from emergency services | Inability to perform certain business operations |

Puerto Rico's economic and disaster recovery plan described the devastation wrought by Hurricane Maria:

> Damage to critical infrastructure resulted in cascading failures of the lifeline systems of energy, telecommunications, water, and transportation. Because the disaster occurred at the end of a very active hurricane season, federal resources for disaster response were stretched. . . . Given the scale of the disaster, the limited response resources, and the failure of lifeline systems, emergency services were compromised and residents lacked electricity, food, and water for a prolonged period. And with roads impassable, residents had limited access to medical care. After the hurricane, people lost their jobs, schools were closed, government services and private enterprise could no longer operate effectively, landslides caused flooding hazards, and wastewater polluted marine environments. While the hurricanes touched virtually every segment of the population, older adults, children, individuals with disabilities or chronic illnesses, and women were disproportionately affected by this disaster.[3]

Under the circumstances, Hurricane Maria's impact on Puerto Rico's communications networks was unprecedented, widespread, and devastating. According to the early damage estimates and recovery plan released by the government of Puerto Rico in *Build Back Better*, the "widescale communications 'blackout' significantly increased the risk to public safety, hampered emergency response and recovery efforts, and shutdown Puerto Rico's economy."[4] Failures in the public safety communications systems left citizens unable to access critical emergency services at a time when such services were a necessity for many. Residents able to make emergency calls to 911 dispatchers suffered frustration when emergency services failed to

---

[3] Central Office for Recovery, Reconstruction and Resiliency, 2018, p. vii.

[4] Governor of Puerto Rico, *Build Back Better Puerto Rico: Request for Federal Assistance for Disaster Recovery*, November 13, 2017.

**Table 2.3. Summary of Key Metrics and Example Statistics**

| Metric | Description | Statistic | Source |
|---|---|---|---|
| 1 | Operational cell sites | 5% as of September 21, 2017 | FCC, Communications Status Report for Areas Impacted by Hurricane Maria, September 21, 2017 |
| 2 | Operational central offices | 95.8% as of March 19, 2018 / Unknown by the authors | |
| 3 | Use of generators | Unknown, but government officials indicated that pre-hurricane the electrical grid serviced Puerto Rico (and communications infrastructure) with only occasional blackouts / 60% of communications infrastructure using generators as of January 22, 2018 (115 days after storm) | Comments by the PR Telecom Regulatory Board to the FCC on January 22, 2018, re: Public Safety and Homeland Security Bureau questions in the proceeding under PS Docket No. 17-344 |
| 4 | 911 calls | 2.5 million in 2015 / Unknown by the authors | DHS OEC/Interoperable Communications Technical Assistance Program; Puerto Rico Public Safety Communications Summary and Recommendations Report |
| 5 | Access to broadband service | 62% of the population of Puerto Rico without access to broadband service[a] / Unknown by the authors | FCC 2016 Broadband Progress Report, FCC 16-6, App. D, rel. January 29, 2016 |

a "In Puerto Rico, 62% of the population (over 2.6 million people) does not have access to 25/3 (broadband) service. Moreover, in Puerto Rico's rural areas, 98% of the population does not have access to 25/3 service." This is an access gap of national importance—1 in 15 of all Americans without access to 25/3 fixed broadband service live in Puerto Rico" (Rua Jovet, 2016, p. 2).

## Metrics Definitions

- Metric 1. Percentage and number of cellular telecommunication sites that are operational at a certain point in time.
- Metric 2. Percentage and number of central offices of the incumbent local exchange carrier that are operational at a certain point in time.
- Metric 3. Percentage of telecommunications sites (incl. central offices) that are operational at a certain point in time but that require power generators.
- Metric 4. Number of calls received per day, month, or year by the Puerto Rico 911 emergency response system at a certain point in time.
- Metric 5. Percentage and number of residents of Puerto Rico who have access to broadband service faster than 25 Mbps (download speed).

respond due to the lack of emergency communications linkages across municipalities and agencies. The lack of commercial communications clearly had a profound impact on daily life. Even as commercial communications were restored, enduring difficulties remained for many residents and businesses. For example, in some cases, voice- and text-messaging services were restored before data services. Without data services and access to the internet, FEMA officials told us, residents' ability to apply for relief through FEMA was limited. Additionally, according to comments from private-sector officials, lack of communications and internet connectivity impeded businesses' abilities to contact employees, fulfill orders, and run web-based enterprise software. Given the dependence of the communications system on electrical power, individuals without power could not fully benefit from restored communications networks if their mobile devices were not charged. Finally, the government of Puerto Rico had limited means by which to communicate essential information to its citizens.

Communications in Puerto Rico have been highly dependent on cellular service since before the hurricanes, and it appears that CONUS is trending in the same direction.[5] This suggests that the priority for restoration and rebuilding of the telecommunications network in Puerto Rico should be making the cellular network more resilient, rather than focusing on landline service. Similarly, investing resources to make the cellular network more resilient may be especially important, given that smartphones can provide access to the internet and Puerto Rico currently has a very low penetration of fixed broadband internet service.[6]

### Communications/IT Damage: Public Safety Communications

Although some elements of the public safety communications network were minimally impacted and remained operational, other elements sustained major damage, which prevented residents of Puerto Rico from accessing public safety services. Specifically, while both PSAP sites were operational once fuel-based generators were utilized, they had limited ability to relay calls to public safety agencies because of the failure of the PSTN. This situation was compounded by the additional call volume following Maria. Most agency stations across municipalities were unable to relay information. The following list summarizes the number of stations able to directly relay call information.

---

[5] Puerto Rico reflects recent trends observed by the Pew Research Center in a 2018 report, which found that "the vast majority of Americans—95%—now own a cellphone of some kind. The share of Americans that own smartphones is now 77%, up from just 35% in Pew Research Center's first survey of smartphone ownership conducted in 2011." The same study also found that "as the adoption of traditional broadband service has slowed in recent years, a growing share of Americans now use smartphones as their primary means of online access at home. Today one-in-five American adults are 'smartphone-only' internet users—meaning they own a smartphone, but do not have traditional home broadband service" (Mobile Fact Sheet, Pew Research Center, Washington, D.C., February 5, 2018).

[6] The Chair of the PRTRB stated: "Adoption of broadband by Puerto Rico households, at any speed and technology, also trails the rest of the United States. According to the U.S. Census, only 45% of Puerto Rico households have purchased broadband Internet, far lower than the national average of 75%" (Rua Jovet, 2016, p. 2).

- Police: 12 of 78 stations
- Fire: 0 of 93 stations
- EMS: 14 of 56 stations.[7]

Public safety LMR systems were severely crippled due to their dependence on infrastructure that had been destroyed by Hurricane Maria. The primary PRPD radio system provided less than half of fire stations with limited voice communications. Figure 2.1 maps the locations of the PRPD's primary network—the P25 radio system.[8] According to a damage report on the PRPD provided to the authors (September 2017), 50 percent of these sites—mainly in the southern half of Puerto Rico—were damaged. Although Figure 2.1 includes police headquarters (Departamento Estado) and the airport (Aeropuerto) as sites, PRPD did not provide information on the extent of damage caused at these locations. The airport site is red in the figure below because it was not in operation prior to Hurricane Maria.

**Figure 2.1. Puerto Rico Police Department Radio Network Sites**

Source: DHS OEC, 2018.

The EMS (referred to in Figure 2.2 as CEM for its Spanish acronym) radio network also sustained critical damage. As Figure 2.2 shows, several elements of the system were not operational, and numerous links were either intermittent or not functional. Regions where repeater systems or linkages were not working were unable to effectively communicate medical emergencies in situations where the EMS dispatcher's LMR could not directly reach the first

---

[7] FEMA Hurricane Maria Communications Task Force, 2017.

[8] DHS OEC, 2018.

**Figure 2.2. Emergency Medical Service Radio Network Status Post–Hurricane Maria**

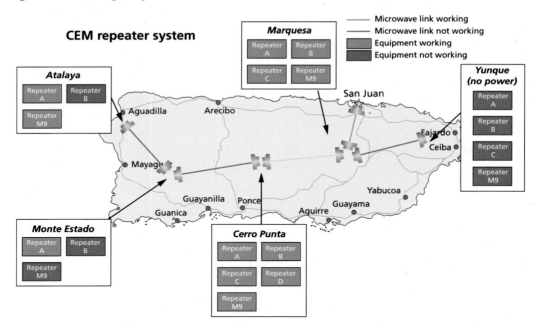

SOURCE: FEMA Hurricane Maria Communications Task Force, *DR-4339-PR Consolidated Communications Restoration Plan*, October 30, 2017.

responder. In Figure 2.2 (and Figure 2.3), the boxes in green reflect equipment that was operating after the hurricane; boxes in red reflect equipment that was not. The red lines represent microwave links that were not functioning.

Additionally, much of Puerto Rico lost the use of another key public safety communications system, the State Interoperability System. Multiple agencies and the municipalities use this system during urgent situations such as hurricanes, tsunamis, and earthquakes. Figure 2.3 shows the significant damage to this system from Hurricane Maria.

Additional examples of the damage sustained by public safety telecommunications sites (Figures 2.4, 2.5, and 2.6) come from photographs taken by the government of Puerto Rico during visits to various sites on September 20, 2017.

The physical damage caused by Hurricane Maria was compounded by the prehurricane condition of the electric grid. As an AT&T representative explained at the FCC docket dedicated to the 2017 hurricane season,

> The electric grid was outdated and extremely susceptible to hurricane damage, resulting in widespread and lengthy outages. The majority of the electric grid was composed of above-ground electric utility poles, which were brought down by sustained high winds or fallen trees. As a result, the communications infrastructure was also severely affected by the storm because the majority of AT&T's fiber was riding on electrical utility poles. In many areas, the fiber backhaul was simply gone.[9]

---

[9] AT&T, 2018, pp. 8–9. See also T-Mobile, 2018, p. 9.

**Figure 2.3. Puerto Rico State Interoperability System Status Post–Hurricane Maria**

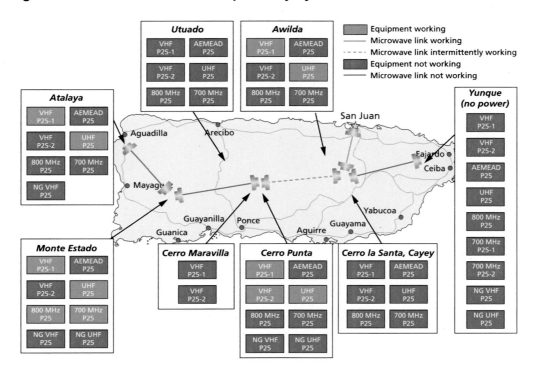

SOURCE: FEMA Hurricane Maria Communications Task Force, 2017.

**Figure 2.4. Example of Damage Caused by Hurricane Maria to Public Telecommunication Site as of September 20, 2017**

SOURCE: Government of Puerto Rico.

**Figure 2.5. Example of Damage Caused by Hurricane Maria to Public Telecommunication Site as of September 20, 2017**

SOURCE: Government of Puerto Rico.

**Figure 2.6. Example of Damage Caused by Hurricane Maria to Public Telecommunication Site as of September 20, 2017**

SOURCE: Government of Puerto Rico.

Figure 2.7 shows an example of downed poles.

**Figure 2.7. Typical Downed Distribution Pole**

SOURCE: Burns & McDonnell.
NOTE: Note the pole in the center of the wooded area that is broken off at the base and conductors (wires) laying in the grass at its base. This pole was one of 14 in a row that were downed in this wooded area.

### Communications/IT Damage: Commercial Communications

Commercial communications were almost nonexistent on Puerto Rico immediately after Hurricane Maria. Since the vast majority of residents uses mobile phones and that infrastructure was damaged or inoperable, most people lacked the ability to communicate to anyone outside their immediate location. According to the Telecommunications Bureau, as of September 24, 2017, Hurricane Maria had damaged 91 percent of the private telecommunications infrastructure (primarily antennas and fiber), affecting the government, retail, banks, pharmaceutical, developers, food, transportation, and other businesses.[10] Figure 2.8 reflects the fact that 95 percent of cellular sites were out of service as of September 21, 2017. All municipalities

---

[10] Torres Lopez, 2018.

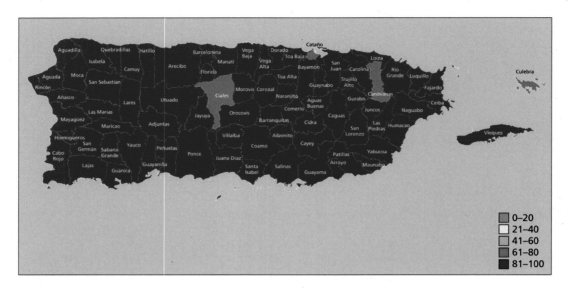

SOURCE: FCC, "Communications Status Report for Areas Impacted by Hurricane Maria," March 21, 2018c.
NOTE: The map shows service outages as of 11:01:12 a.m. Atlantic Standard Time. The number and associated color represent percentage ranges for cell sites out of service.

saw more than 75 percent of their cell sites go dark, and 48 of the 78 municipalities lost 100 percent of their cell sites.[11]

FEMA's ESF-2 reported that 80 percent of the approximately 350 linear miles of core aboveground fiber was destroyed, as was 85 to 90 percent of the approximately 1,500 linear miles of last-mile fiber.[12] According to a Puerto Rico homeland security official, it was reported that 48,000 utility poles had been damaged by the storm.[13] The reliance on aerial fiber, as opposed to buried fiber, put communications at risk to disruption.

As indicated in Chapter 1, all of the submarine cable landing stations in Puerto Rico are in the San Juan metro area on the northern coast. According to news reports, Hurricane Maria damaged the primary submarine cable landing station that provides services to the major U.S. telecom carriers, thereby significantly affecting communications to and from Puerto Rico.[14]

---

[11] FCC, "Communications Status Report for Areas Impacted by Hurricane Maria," March 21, 2018c.

[12] FEMA Hurricane Maria Communications Task Force, 2017. The core network serves to interconnect traffic from cellular, wireline, and broadband internet; last mile fiber provides transmission from the core network to the retail end-user.

[13] Emergency Communications Coordinator for Region 2, U.S. Department of Homeland Security, telephone interview with the authors, March 2, 2018.

[14] Rachel Becker, "Trying to Communicate After the Hurricane: 'It's as if Puerto Rico Doesn't Exist,'" *The Verge*, September 29, 2017.

A participant in a HSOAC focus group explained the harrowing situation following Hurricane Maria: "As the days went by it descended into chaos. You needed money . . . because it was all cash. The ATMs weren't working, the phones were cut off."[15]

### Preliminary Damage Cost Estimates

In October 2017, FEMA officials estimated (rough order of magnitude) that basic restoration of only public safety communications systems to prestorm status would require less than $8 million. This estimate included repair and/or replacement of radio system components, tower-mounted components, and network transport components. The estimate did not include rebuilding numerous public safety towers since these towers are privately owned.[16]

Separate from the restoration of public safety communications systems, in January 2018, the Telecommunications Bureau estimated the private telecommunications damage at more than $1.5 billion.[17] This estimate was included in *Build Back Better*, which provided aggregated damage information and initial cost estimates to rebuild, to reconstruct the fiber network and replace damaged cell sites.[18] Additionally, the FCC chairman announced in March 2018 a plan to invest $750 million through the creation of the Bringing Puerto Rico Together Fund. The funding would address short-term restoration of service and long-term improvement and expansion of broadband.[19] While the announced investment does not represent direct estimates related to the costs of the damage, it does provide some corroboration of the scale of the investment needed to restore and recover communications after Hurricane Maria.

## Posthurricane Response

### Barriers to Response

Physical infrastructure, geographic, bureaucratic, and other barriers increased the complexity of the response environment in Puerto Rico. One wireless carrier cited three main causes of communications outages due to the hurricanes: loss of commercial power, loss of backhaul transport communications, and total tower loss.[20] In addition, backhaul facilities of many carriers are leased, not owned, and the loss of the core network, which was not under the control of the cellular carriers, was another important issue. Communications dependence on electrical power was and remains a barrier to the response efforts. Private ownership of towers further

---

[15] Central Office for Recovery, Reconstruction and Resiliency, 2018, p. 47.

[16] FEMA Hurricane Maria Communications Task Force, 2017.

[17] Torres Lopez, 2018.

[18] Governor of Puerto Rico, 2017.

[19] FCC, "Chairman Pai Unveils $954 Million Plan to Restore and Expand Networks in Puerto Rico and U.S. Virgin Islands," March 6, 2018b.

[20] AT&T, 2018.

complicated response efforts, while Puerto Rico's geography posed additional challenges to physically accessing sites for response activities. According to the Telecommunications Bureau, multiple incidents of copper theft, fiber-optic cuts, and theft of generators and fuel also delayed recovery efforts and increased costs.[21] Effectively addressing these numerous challenges requires functioning governance to coordinate the multiple key stakeholders whose efforts are critical to a sustained recovery. In this case, however, while the Puerto Rico Interoperability Committee for Emergency Communications had been created through Executive Order by the Governor in 2011, it was not effectively utilized to make the communications and IT sector more resilient before the hurricanes of 2017.[22]

### Initial Emergency Response Operations

According to ESF-2's October 2017 report, immediate steps were taken to restore public safety communications. Due to a generator failure at the primary PSAP, the U.S. Army Corps of Engineers provided a generator to the location and fueled it weekly immediately after the storm. As a result, the primary PSAP was operational, although its ability to relay calls was limited. ESF-2 also deployed tactical communications equipment to 13 police district headquarters, 12 fire district offices, and EMS offices. The purpose of this effort was to restore basic voice communications to facilitate transmission of 911 calls to municipalities. PREMA requested that EMS stations have internet connectivity, which required the offices to relocate places where ESF-2 solutions were already deployed. To restore LMR communications, ESF-2 used additional communications systems such as very small aperture terminals (VSATs), radios, satellite phones, and mobile satellite terminals (MSATs). Civil Authority Information Support units from the Department of Defense (DoD) were assigned to provide alert and warning capabilities as part of restoring the emergency alert system.[23] According to a lead FEMA official, ESF-2 utilized a significant amount of equipment from CONUS. Much of this equipment remained deployed in the event of challenges during the 2018 hurricane season.[24]

Commercial carriers reported taking a number of steps to restore wireless service to customers, such as using cellular equipment on light trucks (COLTs), cells on wheels (COWs), and portable generators. In addition to leveraging this equipment, four wireless carriers opened up roaming to ensure service to the maximum number of people with the coverage available.[25] Carriers also used nontraditional means to restore service. For example, AT&T used a "flying cell on

---

[21] Torres Lopez, 2018.

[22] We are referring to the Puerto Rico Interoperability Committee for Emergency Communications created through Executive Order by the Governor in 2011. See Chapter 1 and DHS OEC, 2018.

[23] FEMA Hurricane Maria Communications Task Force, 2017.

[24] Emergency Communications Coordinator for Region 2, U.S. Department of Homeland Security, telephone interview with the authors, March 2, 2018.

[25] These four Puerto Rico wireless carriers allowed roaming at no extra cost after Hurricane Maria. Roaming is the ability of a cellular customer to automatically make/receive voice calls and send/receive data when traveling outside the geographical coverage area of his cellular provider's network. Roaming typically incurs additional charges for a cellular customer. CTIA, 2018.

wings"—a drone cell site—to temporarily provide data and voice- and text-messaging services over a 40-mile radius. In another effort, AT&T and T-Mobile partnered with Alphabet's Project Loon to provide LTE-based[26] services to tens of thousands of citizens through the use of network balloons. Such efforts helped to restore wireless service despite catastrophic infrastructure damage and sustained electrical outages. However, according to a lead FEMA official, a lack of coordination between PREPA and commercial carriers resulted in some restoration setbacks. Specifically, in order to restore electricity, PREPA cut lines or removed fiber-optic cable from poles for which telecommunication lines had already been repaired by commercial carriers. Such activities were stopped, and coordination improved as restoration activities progressed.[27]

### Preparation for Hurricane Season and Status of Repairs

By April 9, 2018, the FEMA Comms/IT sector team had finalized a plan to prepare Puerto Rico for the upcoming hurricane season, which would start in June. The plan consisted of two documents addressing critical communication preparedness for the 2018 hurricane season. The first was a report written by the FEMA Comms/IT Segment 1 solutions-based team. This report recommended rapid actions that would allow for continuity of communications for government leadership and other CI stakeholders—including hospitals, PSAPs, EMS, and others—in the event of another disaster. The second was a memo from the FEMA Comms/IT Lead to the Federal Coordinating Officer outlining the seven critical actions that would improve the readiness of critical communications for the 2018 hurricane season.[28] These actions were

- install an alert and early warning system for the Guajataca Dam[29]
- transfer UHF high power mobile radios/antenna kits to EMS, Department of Health, PREMA, and municipalities
- obtain and install high-speed satellite data and telephone systems for public safety dispatch facilities, PRPD, PREMA, Department of Health, and the Telecommunications Bureau

---

[26] LTE is an abbreviation for Long Term Evolution. It is a fourth generation wireless communications standard designed to provide up to ten times the speed of third generation networks for mobile devices such as smartphones, tablets, netbooks, notebooks, and wireless hotspots.

[27] Emergency Communications Coordinator for Region 2, U.S. Department of Homeland Security, telephone interview with the authors, March 2, 2018.

[28] The FEMA Communications/IT Lead was also referred to as the FEMA Communications/IT Division Director. David G. Samaniego, Division Director, Federal Emergency Management Agency, Communications/IT Sector, FEMA-4339-DR-PR, Memorandum for Mike Byrne, Federal Coordinating Officer/Disaster Recovery Manager, FEMA-4336/4339-DR-PR, April 8, 2018.

[29] The Guajataca Dam is a large dam in Puerto Rico. It had been previously identified by the Army Corps of Engineers as a failure risk. Hurricane Maria's heavy rains caused the dam to overflow its banks on September 22, 2018, damaging its spillway. Additional damage to the spillway occurred during the 20 days that it took the Army Corps to stabilize the dam. Ongoing work is attempting to stabilize both the dam and the spillway in preparation for another Maria-like event. It is estimated that 70,000 people could be in danger if the dam fails. See Luke Abaffy, "Feds Fight to Firm Up Puerto Rico Dam," ENR Southeast, February 26, 2018.

- maintain and support existing iridium satellite phones already deployed with state and municipal agencies and provide additional phones as well as care packages [30]
- maintain and support existing MSAT terminals already deployed with public safety agencies and provide additional terminals[31] (this action would also include establishing a procedure to open a conference bridge for satellite users upon activation of emergency procedures)
- validate backup power generators in coordination with PREMA, Telecommunications Bureau, Universal Service Administration Corporation (USAC), and PRPD for critical communication sites and deploy assets as necessary if shortfalls are identified
- deploy a commercial HF network in about 50 critical facilities in the event that all other systems failed.

The memo indicated that "the above seven [sets of actions] were determined to be highest priority in providing alternate, resilient, critical communications capabilities to be in place by the start of the 2018 hurricane season."[32] In that same memo, the FEMA Comms/IT lead recommended that the FEMA emergency personnel and equipment deployed to Puerto Rico should remain in place through the 2018 hurricane season. The memo presented a total cost estimate of $2,456,000 for six of the actions; the cost of backup power generators for critical communications sites was unknown and not included.

As July 25, 2018, more than 90 mobile satellite terminals had been installed throughout Puerto Rico. These installations included the two 911 centers (primary and alternate), five EMS offices located throughout Puerto Rico, each one of the PREMA offices located in the 12 Puerto Rico emergency zones, 12 Puerto Rico Fire Department offices, and nine PRPD offices. Furthermore, over 70 LMR base station radios had been installed in local emergency management offices and in municipal police and fire stations and other municipal offices; and over 70 G2-satellite phones had been provided to hospitals and water treatment, police, fire, and other public safety offices. By the end of July more than 150 Iridium satellite phones had also been delivered to multiple Puerto Rico officials in San Juan and in the municipalities, and more than 470 hand-held radios had been provided to public safety and emergency personnel.[33] The provision of this equipment was intended to ensure public safety and operation of essential government functions in the aftermath of another major disaster. Finally, also as of the end of July 2018, restoration of the PRPD, EMS, and PREMA microwave networks was underway. More specifics about public safety networks' restoration is provided later in this section.

---

[30] The satellite phone "care package" includes batteries, AC and DC chargers, antennas and adapter, calling instruction sheet, and solar charger.

[31] Hughes terminals were one of the options suggested by the FEMA Communications/IT sector. These terminals allow satellite-based access to the internet.

[32] Samaniego, 2018.

[33] FEMA's Excel spreadsheet, *Communications/IT Master Equipment*, Puerto Rico, July 25, 2018. This reference also shows the addresses and geolocations (latitude and longitude) of many of the offices where the equipment was located. That information is sensitive.

An important aspect of preparation for the hurricane season was training of public service and emergency personnel by FEMA subject matter experts (SMEs). For example, training of PREMA and other emergency personnel on the Integrated Public Alert and Warning System (IPAWS) was conducted in Puerto Rico at the end of June 2018. IPAWS is a nation-wide system that provides public safety officials with an effective way to alert and warn the public about serious emergencies.[34] The plan to restore emergency alert systems to the Guajataca dam included the installation of nine to 15 alarms compatible with IPAWS. As of October 2018 at least one of those alarms had been tested.

Concerning FEMA expenditures for restoration of the public service communications networks and for preparation for the hurricane season, categories A and B obligations amounted to $1,770,772 as of the end of June.[35]

FEMA approved funding for the restoration of the public microwave networks. As of October 19, 2018, the PRPD microwave network had been 70 percent restored to its pre-Maria status, and both the PREMA and the EMS microwave networks had been restored to over 85 percent of their pre-Maria status.[36] A senior officer of PRDPS expressed his satisfaction with the status of the public safety telecommunications infrastructure in Puerto Rico; he indicated that response efforts had successfully provided for multiple redundancy in government and public safety communications systems.[37] The same officer noted that only one of the sites of the PRPD microwave network—an important one at Vieques—was still powered by a fuel-based generator, and that the energy company planned to have electricity restored to that site in November.

---

[34] IPAWS integrates the Emergency Alert System (EAS), the Wireless Emergency Alerts (WEA), the National Oceanic and Atmospheric Administration Weather Radio, and other public alerting systems. See, for example, FEMA, Integrated Alert and Warning System, website, undated. FEMA was designated to implement the policy for a public alert and warning system. FEMA, "Integrated Alert and Warning System (IPAWS)," Executive Order 13407, undated-c. This source states that by the end of 2011, over 70 primary entry point stations will be operational throughout the United States, including Puerto Rico. PEP stations are radio broadcast stations that serve as primary sources for a national EAS message.

[35] FEMA Comms/IT sector PowerPoint presentation, *Incident Complex – Puerto Rico: Rico: Recovery Sectors Solutions Weekly Meeting. Puerto Rico*, July 3, 2018. This source states that this is the total dollar amount for Category A and B Project Worksheets of the Communications/IT sector as of June 29, 2018. Categories A and B in FEMA terminology refer to "emergency work" or "work that must be performed to reduce or eliminate an immediate threat to life, protect public health and safety, and to protect improved property that is significantly threatened due to disasters or emergencies declared by the President" (FEMA, Disaster Page Definitions, undated-a). For a detailed explanation of FEMA public assistance program, which includes Categories A and B work and expenditures, see FEMA, *Public Assistance Applicant Handbook FEMA P-323*, March 2010.

[36] Senior official of the Puerto Rico Department of Public Safety, telephone communication with authors, October 19, 2018. The official indicated that the last ten percent of the restoration effort will focus on antennae alignment and software programming and it would be a little slower. These actions were part of the ongoing response efforts but took place after our analysis for the recovery plan was finished (July 2018) and after the recovery plan was delivered to Congress (August 2018).

[37] Senior official of the Puerto Rico Department of Public Safety, telephone communication with authors, October 19, 2018.

Concerning repairs conducted by the private sector, Figure 2.9 illustrates the progress made in restoring cell sites to operation. According to the FCC communications status report from March 21, 2018, just 4.3 percent of cell sites in Puerto Rico remained out of service on that date. However, even months after the storm, some municipalities were still experiencing a higher percentage of cell-site outages than others; for example, Maunabo only had three of 12 cell sites, or 25 percent, restored, according to the FCC's March 2018 report.[38] Figure 2.9 shows the progress made in restoring operation to cell sites (as compared with Figure 2.8).

The FCC's reports do not provide details such as whether the sites have been fully restored to prestorm status or whether they continue to use fuel-based generators for power. However, according to a Telecommunications Bureau filing to the FCC made on January 22, 2018, 60 percent of the communications infrastructure in Puerto Rico was using generators more than 115 days after the hurricane.[39] Additional information provided by a representative of one of the main telecommunications carriers indicated that as of early April 2018, about 30 percent of this carrier's sites were still using fuel-based generators.[40]

**Figure 2.9. The Percentage of Cellular Sites Out-of-Service by Municipality as of March 21, 2018**

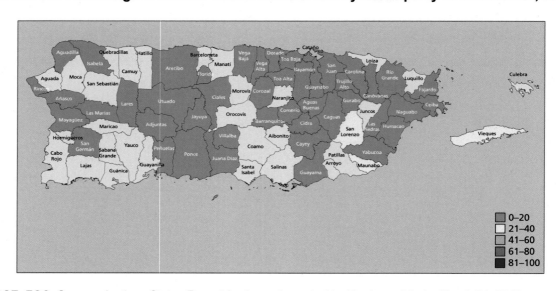

SOURCE: FCC, Communications Status Report for Areas Impacted by Hurricane Maria, March 21, 2018.
NOTE: The numbers and associated color represent percentage ranges for cell sites out of service. Grey areas indicate either an absence of data or lack of cell sites out of service. FCC does not clarify in its reports.

---

[38] FCC, 2018c.

[39] Puerto Rico Telecommunications Regulatory Board, "Comments of the Puerto Rico Telecommunications Regulatory Board," *In the Matter of Response Efforts Undertaken During the 2017 Hurricane Season*, PS Docket No. 17-344, January 22, 2018.

[40] Representative of a Puerto Rico telecommunications carrier, interview with authors, April 10, 2018.

As of June 5, 2018, 99.8 percent of cellular sites in Puerto Rico (2,653 of 2,659) were operational according to an official government website.[41]

After Hurricane Maria, the FCC established a docket to obtain public comment concerning the impact of the 2017 Atlantic hurricane season on communications and to explore ways to prepare for future disasters.[42] In an ex parte filing in that proceeding, Free Press, a public interest advocacy organization, suggested that the FCC consider strengthening regulations requiring carriers to pre-position equipment before hurricanes make landfall. The filing explained that two Puerto Rico carriers, Claro and AT&T, took different approaches that yielded different response capabilities and results. They compared the activities of Claro Puerto Rico, which pre-positioned generators, diesel fuel, batteries, and vehicles several days before the hurricane struck Puerto Rico, with those of AT&T, which, according to the Free Press filing, did not pre-stage recovery assets. The Free Press filing noted that an independent panel reviewing the impact of Hurricane Katrina on communications infrastructure found that the lack of pre-positioned backup equipment was one of the "significant impediments" in the recovery of communications networks. Free Press urged the FCC to consider establishing additional requirements that carriers have equipment and pre-positioned supplies on site, and other measures to ensure rapid response to natural disasters.[43]

Similarly, in a filing submitted to the FCC's hurricane season docket, Public Knowledge, a nonprofit public policy organization, proposed that each carrier should have an emergency plan and should be required to communicate that plan to the FCC before natural disasters occur. It recommended that the Commission require additional backup power for cell towers, establish suitable metrics to measure whether natural disaster recovery efforts are adequate, and make carrier resiliency information available to the public.[44] At the conclusion of the proceeding, the FCC urged the adoption of recommended best practices for governments, service providers, and related stakeholders identified in Public Safety Docket No. 17-344.[45]

---

[41] The website status.pr, undated, indicates that the last update on the status of the island's cellular sites was on June 5, 2018.

[42] See Public Safety and Homeland Security Bureau, 2018c.

[43] Joseph Torres, Senior Director of Strategy and Engagement, Free Press, "Ex Parte Notice in WC Docket Nos. 18-143, 10-90, 14-58 and 17-287; and PS Docket No. 17-344," letter to Marlene H. Dortch, Secretary, FCC, July 20, 2018, pp. 1, 2, and 4. See also Nancy Victory, Chair, Independent Panel Reviewing the Impact of Katrina on Communications Networks, "Report and Recommendations to the Federal Communications of the Independent Panel Reviewing the Impact of Katrina on Communications Networks," letter to Kevin J. Martin, Chair, FCC, June 12, 2006.

[44] Comments of Public Knowledge, *In the Matter of Public Safety and Homeland Security Bureau Seeks comment on Response Efforts Undertaken During the 2017 Hurricane Season*, PS Docket No. 17-344, January 22, 2018, at pp. 1, 2, and 9.

[45] Public Safety and Homeland Security Bureau, 2018a, p. 26. See Section V of that report, "Lessons Learned and Next Steps," p. 28, for a summary of the FCC's recommended best practices.

As part of its response to the 2017 hurricane season, the FCC established an internal Hurricane Recovery Task Force on October 6, 2017, which was comprised of SMEs from bureaus and offices throughout the Commission. According to an August 2018 report of the Public Safety and Homeland Security Bureau at the FCC, the task force "provided outside agencies, such as the Puerto Rico Telecommunications Regulatory Board (PRTRB), commercial providers, and entities seeking experimental authorizations or offering solutions, with a central point to engage the Commission on hurricane recovery issues."[46] It also made several recommendations for addressing longer-term recovery challenges, which the FCC is moving forward to implement. On July 27, 2018, FCC Chairman Pai announced the formation of a working group within the Broadband Deployment Advisory Committee to "address infrastructure and disaster recovery issues."[47] On December 10, 2018, the Public Safety and Homeland Security Bureau released a public notice seeking public comment on improving wireless network resiliency, specifically by promoting coordination between wireless carriers and backhaul providers. It stated that "the Bureau now seeks comment on how to ensure that wireless carriers and backhaul providers better coordinate with each other, as well as with other stakeholders, both before and during an emergency event, and as part of post-event restoration efforts."[48]

## Remaining Needs for the Communications/IT Sector

The status of repairs and restoration efforts in Puerto Rico were thus as follows: First, as of June 5, 2018, 99.8 percent of cellular sites (2,653 of 2,659) were operational; second, as of July 25, 2018, the FEMA Comms/IT sector had prepared and executed a plan of rapid actions to allow for continuity of communications for commonwealth and municipal leadership and CI stakeholders such as hospitals, PSAPs, EMS, and others for the 2018 hurricane season; third, as of October 19, 2018, repairs and restoration of the Puerto Rico public safety microwave networks had achieved over 70 percent (PRPD network) and over 85 percent (PREMA and EMS networks) of their pre-Irma/Maria status.

However, infrastructure critical to supporting communications such as aerial fiber remains vulnerable to an upcoming disaster. Aerial fiber has been and continues to be used for a significant amount of backhaul between cell sites and the communications network, including

---

[46] Public Safety and Homeland Security Bureau, 2018a, p. 26.

[47] Public Safety and Homeland Security Bureau, 2018a, p. 26. See also FCC, *FCC Solicits Nominations For New Disaster Response and Recovery Working Group of the Broadband Advisory Committee*, Public Notice, GN Docket No. 17-83, DA 18-837, August 9, 2018e.

[48] Public Safety and Homeland Security Bureau, Federal Communications Commission, *Public Safety and Homeland Security Bureau Seeks Comments on Improving Wireless Network Resiliency to Promote Coordination Through Backhaul Providers*, PS Docket No. 11-60, December 10, 2018b.

during the restoration efforts.[49] Aerial fiber on poles, as opposed to fiber buried in conduit, is particularly vulnerable to harsh weather conditions. In addition, a constant supply of electricity is essential for operating cell and public safety telecommunication sites. As of early October 2018, it was not known how many of the 2,653 operational cell sites were still operating with fuel-based generators.[50] Use of these generators presents numerous problems. One is that they are intended for intermittent backup in case of an emergency, not to run continuously for weeks or months; when they do, they tend to fail. Another is that supplying diesel or gasoline to cell site generators across Puerto Rico is both logistically challenging and very expensive.[51] Moreover, fuel-based generators may also impact air quality or cause fuel spills affecting groundwater, flora, and fauna. In the short term, an additional problem was that generators were stolen.

The disaster caused by Irma and Maria demonstrates the importance of having survivable technologies in place that are redundant and resilient. In addition to achieving adequate resiliency, recovery efforts in the communications and IT sector will require bringing many systems up to federal telecommunications standards, taking advantage of technological advances to upgrade systems and equipment, improving maintenance procedures on CI, and maintaining a cadre of qualified communications and IT personnel. In particular, repair and restoration of the power grid and other infrastructure will take a sustained and significant investment of resources—both public and private. For example, commercial telecommunications providers may require incentives to rebuild their fiber-optic systems underground instead of mounting cable on poles exposed to extreme weather.[52] While FEMA, the government of Puerto Rico, and the private sector have undertaken significant steps to ensure continuity of government (COG),

---

[49] FEMA Hurricane Maria Communications Task Force, 2017. The FEMA Task Force report states on page 2: "Puerto Rico is a mobile society with only approximately 20% of households having landline telephone service. Residents instead have embraced mobile (wireless) technology and use their cellular devices as their only means of communications. Due to this fact, Puerto Rico has an extensive wireless network infrastructure. . . . most of the wireless infrastructure on the island is supported by aerial fiber backhaul." This indicates that, even if all else in the telecommunications network were hardened except for aerial fiber, an event like Hurricane Maria probably would still devastate the telecommunications network. Some of the repair and restoration efforts by private telecom carriers were being pursued using aerial fiber instead of buried fiber due to its lower costs. Representative of a Puerto Rico telecommunications carrier, interview with the authors, April 10, 2018.

[50] We learned from the FCC that although information on the use of generators by cell sites is collected by the FCC, as of the writing of this report, it is not publicly available.

[51] See, e.g., Verizon, Comments of Verizon, *In the Matter of Response Efforts Undertaken During the 2017 Hurricane Season*, PS Docket No. 17-344, January 22, 2018, p. 18.

[52] For example, in a 2013 notice of proposed rulemaking, the FCC addressed measures to promote transparency to consumers about how mobile wireless service providers compare in keeping their networks operational in emergencies. The NPRM stated that "practices regarding the provision of backup power supplies at otherwise similar cell sites appear to vary among mobile wireless service providers, which may contribute to the ability of some mobile wireless service providers to provide more continuous and reliable service during the storm than others." Among other issues, the Commission wished to evaluate "whether the proposed reporting and disclosures would provide consumers with useful information for making comparisons about mobile wireless products and services." Specifically, the FCC intended to evaluate "whether such disclosures, by holding providers publicly accountable, could incentivize improvements to network resiliency while allowing providers flexibility in implementing such improvements." See FCC, *In the Matter of Improving Resilience of Mobile Wireless Communications Networks*, PS Docket No. 13-239, and *In the Matter of Reliability and Continuity of Communications Networks, Including Broadband Technologies*, PS Docket No. 11-60, September 27, 2013, pp. 1–3.

public safety emergency functions, and commercial communications in the aftermath of a disaster, significant work remains in implementing medium and long-term solutions to address public safety communications and telecommunications infrastructure challenges. We conclude that in the communications and IT sector two top-level needs have not yet been addressed:

1. The need to implement a state-of-the art, survivable, resilient communications infrastructure for continuity of essential government functions and for the provision of public safety services to the residents of Puerto Rico. This infrastructure should be upgraded and brought up to CONUS standards and should be well maintained and resourced.
2. The need to implement a state-of-the art, survivable, resilient communications infrastructure to provide commercial telecommunication services, including voice and data services, to the residents and the private sector of Puerto Rico. This infrastructure should support affordable access to broadband internet service and emerging technologies throughout Puerto Rico.

The plan to address these two needs should take into consideration the particularities of Puerto Rico, which is mountainous, hot, and humid, experiences yearly storms and ocean surges, and is located in a zone that is prone to hurricanes. Furthermore, the rural population faced far greater challenges than more populated areas near San Juan.[53] Participation from key commonwealth and municipal public safety communications and telecom stakeholders will increase Puerto Rico's resiliency by fostering coordination to better prepare for, respond to, and recover from future extreme weather events.

The challenges faced by Puerto Rico come with an opportunity to reimagine its communications networks and implement a more resilient system that leverages cutting-edge technology to better serve Puerto Rico's citizens. With a reliable power grid and communications infrastructure in place and with widespread broadband access, information technology (IT) can lead to the digital transformation of Puerto Rico and spur economic growth—as Puerto Rico officials recognized even before the storm. Broader access to high-speed internet will provide opportunities for citizens to develop new skills and to better connect with each other and the rest of the world. Widely available broadband deployment and a robust IT infrastructure could also provide an opportunity to digitally transform public and commercial services to foster Puerto Rico's economy, prosperity, and well-being.

We thus conclude with a third top-level need for the communications and IT sector:

---

[53] For example, Puerto Rico residents in central and rural areas faced greater challenges in receiving aid from police, fire, and EMS following Hurricane Maria. First, as Figure 1.5 illustrates, the process to request and then dispatch fire, police, or EMS services requires several steps, two of which depend on proper operation of the PSTN. Second, as Figure 1.3 illustrates, the central and rural areas of the island had significantly less telecommunications infrastructure, so less redundancy and resilience of the PSTN.

3. The need to develop, deploy, and sustain a modern and resilient information technology infrastructure to further the economic and social vitality of Puerto Rico. This infrastructure should host applications and web services that foster government and private-sector innovation, increase economic opportunity, and improve the quality of life for the residents of Puerto Rico.

# 3. Methodology

## Overview

The efforts to recover from the devastation of Maria and Irma present a unique opportunity for transforming Puerto Rico, implementing cost-effective solutions, making use of best practices and novel ideas, spurring economic growth, and fostering the well-being of the residents of Puerto Rico, Therefore, we emphasize that the economic and disaster recovery plan for Puerto Rico not only focuses on restoring the infrastructure to its prehurricane status, but also aims toward these other goals. In the communications and IT sector, providing wide access to broadband internet services and leveraging IT for the digital transformation of the island have the potential of becoming strong catalysts for economic growth and well-being. A strong and reliable communications and IT sector is fundamental to successful recovery.

The previous chapter provided our assessment of the damage caused by Hurricanes Irma and Maria to the communications and IT sector of Puerto Rico and summarized the top-level needs we assessed for this sector. This chapter presents the methodology for developing the economic and disaster recovery plan with a focus on the communications and IT sector. The methodology followed two parallel paths: top-down and bottom-up.

The first half of the chapter describes the top-down path. This path includes the vision and goals provided by the government of Puerto Rico and the development of strategic objectives based on the vision and goals. The top-down path was not sector-specific; it thus applied to all twelve sectors involved in the plan—namely, energy, communications and IT, water, transportation, housing, public buildings, natural and cultural resources, health and social services, education, municipalities, community planning and capacity building, and economic.

The second half of the chapter describes the bottom-up path for the communications and IT sector. This path started with the top-level communications and IT needs identified in the previous chapter. As the second step in this path, the team constructed COAs, which are detailed recovery actions or policies to address those needs. Subsequent chapters of this report will explain each one of these COAs in detail.

The top-down and the bottom-up paths meet at the level of the "portfolios" that were designed to meet each strategic objective. In general, each portfolio consists of COAs from one or more sectors.

## Top-Down Path

### *Visions, Goals, and Strategic Objectives*

Figure 3.1 summarizes the top-down path pursued during the development of the recovery plan. The top of the figure shows Puerto Rico's vision and goals. The HSOAC teams, FEMA,

**Figure 3.1. Top-Down Path for the Economic and Disaster Recovery Plan for Puerto Rico**

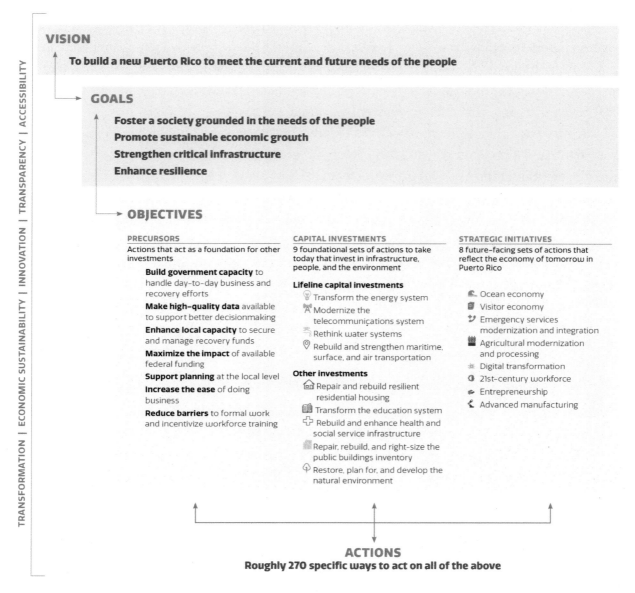

SOURCE: Central Office for Recovery, Reconstruction and Resiliency, 2018.

and representatives from the government of Puerto Rico then worked together during coordination meetings and workshops and generated a series of strategic objectives, which are also shown in the figure. As part of the top-down path, a comprehensive review of many completed and forthcoming plans for the island was performed, including the plan in *Build Back Better*.[1]

---

[1] Governor of Puerto Rico, *Build Back Better Puerto Rico: Request for Federal Assistance for Disaster Recovery*, November 13, 2017.

As shown in Figure 3.1, the strategic objectives of the recovery plan were divided into three groups: precursors, capital investments, and strategic initiatives.

The seven precursors shown in the figure were identified as critical for the success of the government of Puerto Rico's vision for recovery and for the priorities discussed in the plan. They consist of building government and local capacity, making high-quality data available, increasing the ease of doing business in the island, maximizing the impact of federal funding, supporting local-level planning, and reducing barriers to formal work.

The nine capital investment objectives are intended to build a stronger Puerto Rico and lay down the foundation upon which Puerto Rico will grow. They consist of strategic objectives aimed at rebuilding the physical infrastructure and improving it beyond its prehurricane conditions to provide the services that Puerto Rico communities and businesses need. Capital investments include energy, telecommunications, water, transportation, housing, public buildings, education, health and social services, and the natural environment.

The eight strategic initiatives aim at catalyzing specific areas of growth, including economic development, digital transformation, entrepreneurship, and workforce development.

Further details on precursors, capital investments, and strategic initiatives can be found in *Transformation and Innovation in the Wake of Devastation: An Economic and Disaster Recovery Plan for Puerto Rico.*

## Communications/IT in Support of Strategic Objectives

There are many interdependencies among the strategic objectives. For this report, we note that the recovery actions for the communications and IT sector (which are the subject of most of the remainder of this report) are critical to many of the capital investments as well as strategic initiatives depicted in Figure 3.1. Clearly, *Modernize the Telecommunications System* is a strategic objective that requires recovery actions undertaken under the communications and IT sector,[2] but *Transform the Energy System, Rethink Water Systems, Rebuild and Strengthen Maritime, Surface, and Air Transportation, Transform the Education System, and Rebuild and Enhance Social Service Infrastructure* all require communications and IT recovery actions as well.

Moreover, recovery actions in the communications and IT sector are, together with those in other infrastructure sectors, the foundation of all the strategic initiatives. Two strategic initatives in which communications and IT play a major role are *Digital Transformation* and *Emergency Services Modernization and Integration. Digital Transformation* emphasizes building digital capabilities and the skilled workforce needed to fundamentally tranform key industry and government processes. Expanding broadband internet access to Puerto Rico's citizens can help

---

[2] *Modernize the Telecommunications System* was the strategic objective that addressed satisfying all the needs of the communications and IT sector, which are identified in this report. Moreover, portfolio 3 of this strategic objective included all 33 communications and IT COAs. See Chapter 7 for a description of the three portfolios of the strategic objective *Modernize the Telecommunications System.*

foster these digital skills, while using a "digital stewards" program to deploy increased wi-fi access in public housing (as explained in Chapter 6) can train residents in essential skills needed for a digital world. Other avenues, such as entrepreneurship programs, innovation hubs, and mobile labs, may provide opportunities to nurture the digital literacy needed to proliferate the "human cloud"—a skilled digital workforce that can work with companies from around the world in Puerto Rico. Efforts to grow Puerto Rico's digital human capital could be an important foundation to expand the use of new technologies and innovative processes in Puerto Rico and reinforce the message that Puerto Rico is "open for business."

The strategic initiative *Emergency Services Modernization and Integration* aims at enhancing public safety and first responders' ability to deliver reliable, modern, and integrated emergency services. It addresses improving the capacity to respond to major emergencies at the state and municipal levels. Communications and IT contribute to this strategic initiative by upgrading the communications systems used by emergency response personnel, reconstituting and enhancing basic commercial communications, and establishing data centers to support decisionmaking, particularly during the response phase directly after an event.[3]

The linking step between the top-down and the bottom-up paths is the one pertaining to portfolios. The HSOAC Comms/IT team focused on the strategic objective *Modernize the Telecommunications System* for this report because it aligned best with the communications and IT sector and addressed all the needs of the sector. For this strategic objective the team produced three portfolios, which are described in Chapter 7 of this report. The way forward to meet the needs of this sector, incorporated into the recovery plan by the government of Puerto Rico, is entitled "Smart and Resilient Island IT and Communications" and has a total cost of $3.2 billion.

## Bottom-Up Path

### Courses of Action

In parallel to the top-down path, HSOAC organized researchers into the following sector teams: Energy, Communications and IT (Comms/IT), Water, Transportation, Housing, Public Buildings, Natural and Cultural Resources (NCR), Health and Social Services, Education, Municipalities, Community Planning and Capacity Building (CPCB), and Economic. Each one of these HSOAC teams focused on the needs of its specific sector, conducted background research and analysis, engaged with many stakeholders and SMEs, and developed COAs to address the needs of their sector for recovery and economic development.

In this second part of this chapter we describe the specific path pursued by the HSOAC Comms/IT sector team, which developed COAs to address the three top-level needs identified in Chapter 2. The mapping of these COAs to needs is shown in Table 3.1. For the purposes of this recovery plan, a COA represents a collection of potential activities, which may include suggested

---

[3] Central Office for Recovery, Reconstruction and Resiliency, 2018.

changes to a set of policies, a project, and actions and related methods for addressing a specific need. The following questions were used as guidance to compose COAs.

- What is the need that the COA is addressing?
- What are the potential benefits that the COA intends to achieve?
- What are the potential pitfalls that could arise from implementing this COA?
- Are there sectors other than communications and IT that may be affected by this COA?
- What is the likely time scale to achieve benefits from the COA?
- Who is responsible for implementing the COA?
- What are the COA's estimated costs?
- What are potential funding mechanisms for this COA?
- Are there other COAs that are necessary precursors to this one?

Thirty-three COAs were developed for this sector. Table 3.1 shows the names and labels for each of them. It also shows the matching between the three top-level needs we identified in Chapter 2 and the COAs. An X in the column that corresponds to one of the needs means that such COA is important to satisfy that need. Notice that several COAs are important to satisfy more than one need. COAs are described in later chapters, and appear in full detail in Appendix C. Interrelationships between COAs are explained in Chapters 4, 5 and 6.

**Table 3.1. Communications/IT Courses of Action of the Recovery Plan for Puerto Rico and Matching of Courses of Action to Communications/IT Needs**

| COA ID | Title | Need 1 | Need 2 | Need 3 |
|--------|-------|--------|--------|--------|
| CIT 1 | Land Mobile Radio System [Chapter 4] | X | | |
| CIT 2 | Puerto Rico GIS Resource and Data Platform [Chapter 6] | X | X | X |
| CIT 3 | Upgrade and Enhance 911 Service [Chapter 4] | X | | |
| CIT 4 | Rural Area Network Task Force [Chapter 4] | X | | |
| CIT 5 | Implement Public Safety/Government Communications Backup Power [Chapter 4] | X | | |
| CIT 6 | Modernize the Emergency Operations Center [Chapter 4] | X | | |
| CIT 7 | Establish an Alternate Emergency Operations Center [Chapter 4] | X | | |
| CIT 8 | Mobile EOC Vehicle [Chapter 4] | X | | |
| CIT 9 | Auxiliary Communications—Volunteer Radio Groups and Organizations [Chapter 4] | X | | |
| CIT 10 | Transoceanic Submarine Cable [Chapter 5] | X | X | X |
| CIT 11 | Procure a Mobile Emergency Communications Capability [Chapter 4] | X | | |
| CIT 12 | Perform Site Structural Analysis for All Government Telecom Towers (Both Public and Privately Owned) [Chapter 5] | X | | |
| CIT 13 | Streamline the Permitting and Rights of Way Processes for Towers and the Deployment of Fiber-Optic Cable [Chapter 5] | | X | X |
| CIT 14 | Consolidated Government Information Systems [Chapter 6] | | | X |
| CIT 15 | Undersea Fiber Ring System [Chapter 5] | X | X | X |

| COA ID | Title | Need 1 | Need 2 | Need 3 |
|--------|-------|--------|--------|--------|
| CIT 16 | Government Digital Reform Planning and Capacity Building [Chapter 6] | | | X |
| CIT 17 | Puerto Rico Data Center [Chapter 6] | X | | X |
| CIT 18 | Data Storage and Data Exchange Standards for Critical Infrastructure [Chapter 6] | X | | X |
| CIT 19 | Municipal Hotspots [Chapter 5] | X | X | X |
| CIT 20 | Continuity of Business at PRIDCO Sites [Chapter 4] | X | | |
| CIT 21 | Government-Owned Fiber-Optic Conduits to Reduce Aerial Fiber-Optic Cable and Incentivize Expansion of Broadband Infrastructure [Chapter 5] | X | X | X |
| CIT 22 | Use Federal Programs to Spur Deployment of Broadband Internet Island-Wide [Chapter 5] | | X | X |
| CIT 23 | Data Collection and Standardization for Disaster Preparedness and Emergency Response [Chapter 4] | X | | X |
| CIT 24 | Establish Puerto Rico Communications Steering Committee [Chapter 3] | X | X | X |
| CIT 25 | Evaluate and Implement Alternative Methods to Deploy Broadband Internet Service Throughout Puerto Rico [Chapter 5] | | X | X |
| CIT 26 | Wi-fi Hotspots in Public Housing and Digital Stewards Program [Chapter 6] | | | X |
| CIT 27 | Study Feasibility of Digital Identity [Chapter 6] | | | X |
| CIT 28 | Innovation Economy/Human Capital Initiative [Chapter 6] | | | X |
| CIT 29 | Health Care Connectivity to Strengthen Resilience and Disaster Preparedness [Chapter 6] | X | | X |
| CIT 30 | Resiliency Innovation Network Leading to Development of a Resiliency Industry [Chapter 6] | | | X |
| CIT 31 | Resilience/e-Construction Learning Lab [Chapter 6] | | | X |
| CIT 32 | Digital Citizen Services [Chapter 6] | | | X |
| CIT 33 | Government Digital Process Reform [Chapter 6] | | | X |

The communications and IT Recovery Plan includes one cross-cutting governance COA, which merits its own discussion here, rather than in later chapters. This COA, *CIT 24 Establish Puerto Rico Communications Steering Committee*, proposes the establishment of a new Communications Steering Committee through Executive Order by the Governor to include but not be limited to officials from the Telecommunications Bureau, PREMA, CIO Office, CINO Office, Department of Public Safety (including PRPD, EMS, and the Fire Department), and municipal officials as relevant. Input from the private sector would also be sought. The discrete responsibilities and tasks of the committee would be outlined in a charter signed by the Governor. The purpose of the committee would be to better organize planning efforts and coordinate among key public safety and commercial communications stakeholders. The committee would provide strategic guidance, policy direction, and standards associated with Puerto Rico communications networks. It would also better ensure proper planning and collaboration to effectively and efficiently recover and maintain the communications infrastructure, as well as mitigate interoperability and duplication of effort issues.[4] A Congressional Research Service Report on *The Hurricane Sandy Rebuilding Strategy* noted that " one of many lessons learned in responding to the Gulf Coast Hurricanes of 2005 (Katrina, Rita, and Wilma), was that the long-term disaster recovery process from a catastrophic disaster could be just as difficult and daunting, if not more so, than the immediate response process."[5] One result was development of the NDRF to "serve as a guide to recovery efforts after major disasters and emergencies."[6] The Hurricane Sandy Rebuilding Task Force was created as a complement to the NDRF.[7] The Communications Steering Committee would play a similar role in guiding Puerto Rico to achieving rebuilding success.

### Primary Stakeholders

As specified by U.S. legislation, the recovery plan was to be prepared by the government of Puerto Rico with help from federal agencies. Therefore, the primary stakeholders and senior partners in the development of the plan in general and of the COAs in particular were officials from government of Puerto Rico offices representing the leadership of the communications and IT sector in Puerto Rico. The four main offices were

- the Telecommunications Bureau
- the Department of Public Safety (PRDPS)
- the Chief Information Officer (CIO)
- the Chief Innovation Officer (CINO).

Puerto Rico communications and IT leadership held meetings twice a week for several months. These meetings were led by the FEMA Comms/IT sector lead and were hosted at the offices of the Telecommunications Bureau. All important decisions pertaining to the communications and IT sector in Puerto Rico were discussed in this forum.

The FEMA Comms/IT sector lead oversaw and implemented immediate recovery activities across the island to bring communications and IT back to prestorm status. The FEMA Comms/IT sector lead also coordinated the efforts of the SMEs in FEMA's Comms/IT solutions-based team (SBT). The solutions-based team (SBT) assessed long-term solutions to position communications and IT for success.

In the next section we summarize the process to develop the COAs.

---

[4] The committee chair would rotate among the committee members twice a year. The rotating chairperson would need authority to set direction and facilitate decisions. The committee would not be involved in tactical decisions such as reforms that deal with centralizing rights of way (ROW) and permitting approval authority and are detailed in COA CIT 13.

[5] Jared T. Brown, "The Hurricane Sandy Rebuilding Strategy: In Brief," Washington, D.C.: Congressional Research Service, February 10, 2014, p. 8.

[6] Brown, 2104, p. 8. See also National Disaster Recovery Strategy, 6 U.S.C. Section 771.

[7] Brown, 2104, p. 8. See also National Disaster Recovery Strategy, 6 U.S.C. Section 771.

## Course of Action Development Process

COAs for recovery and economic growth were developed through a collaborative process and the joint efforts of a wide range of stakeholders, including federal agencies, agencies of the government of Puerto Rico, and the Puerto Rico telecommunications private industry. The HSOAC Comms/IT sector team leveraged the wide range of expertise of all these stakeholders. Moreover, we received indirect feedback from Puerto Rico municipalities and from additional stakeholders through the work of other HSOAC teams, as explained in the next section of this chapter.

An important foundation for many of the COAs was the work of the SBT, which was assembled by the FEMA Comms/IT sector lead and was comprised of experts with years of operational experience in public safety/emergency communications and infrastructure. Its findings were invaluable in the development of these COAs. The HSOAC team met numerous times with the SBT to seek its input and ensure a comprehensive understanding of its findings. An HSOAC team member was "embedded" with the SBT during the several weeks of daily brainstorming sessions undertaken by the SBT. We also thoroughly reviewed the findings of the SBT's 2018 report, which was one of the most important references for this work.[8]

In addition, the HSOAC team sought insights from government officials across multiple agencies, including the four Puerto Rico communications and IT agencies referred to in the previous section. These meetings provided critical context to ensure the feasibility and operational utility of COAs and furthered our understanding of key documents such as *Build Back Better*. Discussions with federal regulators at the FCC informed us on how current federal programs could be leveraged by Puerto Rico in its recovery efforts.

In addition to input from federal and commonwealth partners, input from Puerto Rico telecommunications providers helped in the development of the COAs that focused on partnering with the private sector. To ensure a comprehensive perspective, HSOAC also engaged with wireless infrastructure and telecommunications carrier associations and obtained written comments submitted by them to the FCC. Such perspectives were critical not only to the development of COAs that addressed necessary investments in commercial communications, but also helped us understand the impact that COAs related to emergency communications may have on the private sector.

Additionally, we partnered with People-Centered Internet (PCI) to develop innovative COAs to address communications challenges experienced by Puerto Rico's citizens and communities.[9] This work led to the development of many COAs concerning information technology for the digital transformation of Puerto Rico. We utilized findings from domestic and international pilot programs that could further innovation and transform the economy through technology.

---

[8] FEMA Communications/IT Solutions-based Team, "Puerto Rico Communications/IT Solutions-based Team Report," San Juan, PR, June 30, 2018.

[9] PCI was hired by HSOAC as a subcontractor to provide consulting services.

## Engaging Municipalities and Civil Society Stakeholders

We participated in workshops organized by 100 Resilient Cities.[10] In these workshops, which took place in March and April 2018 in Puerto Rico, the team received feedback from several civil society stakeholders including ReImagina Puerto Rico,[11] the Rockefeller Foundation, the Ford Foundation, the Open Society Foundation, El Centro para una Nueva Economía, and faculty members from the University of Puerto Rico (UPR).[12]

Moreover, we received indirect feedback from Puerto Rico municipalities and from additional stakeholders through the work of the HSOAC teams responsible for developing COAs in the CPCB and municipality sectors.[13]

The HSOAC CPCB sector team worked closely with a social services organization in Puerto Rico and recruited residents of Puerto Rico from 20 municipalities to participate in focus groups. During April and May 2018, a total of 20 focus groups were conducted. Their purpose was to collect information about community experiences before, during, and after the hurricanes and gather feedback that could be useful for developing COAs. All focus groups were conducted in Spanish by a representative of the social services organization and observed by Spanish-speaking HSOAC members.[14] The HSOAC CPCB team also partnered with a social services organization to recruit local leaders, academics, and key stakeholders who could describe the effects of Hurricane Maria in their particular communities. These individuals were located in multiple municipalities across Puerto Rico. They were interviewed in either English or Spanish (depending on interviewee's preference) by a representative of the social services organization or a member of the HSOAC team.[15] Finally, the HSOAC CPCB sector team also interviewed individuals from urban and rural areas across Puerto Rico who had migrated to Florida as a direct result of Hurricane Maria and its aftermath.[16] Among the findings from the HSOAC CPCB team we report two, namely, disaster preparedness plans were in place in Puerto Rico prior to Maria but they were not adequate to address a storm of this magnitude, particularly in the communications and IT sector, and concerning disaster preparedness components, communications was the component most commonly cited as being inadequately

---

[10] See for example, 100 Resilient Cities "San Juan's Resilience Challenge," undated.

[11] See for example, ReImagina Puerto Rico, homepage, undated.

[12] Reports on Puerto Rico recovery by these civil society stakeholders can be found at ReImagina Puerto Rico, "Final Reports," undated; and 100 Resilient Cities, "San Juan's Resilience Challenge," undated.

[13] See RAND Corporation, "Supporting Puerto Rico's Disaster Recovery Planning," webpage, undated.

[14] The size of focus groups ranged from eight to 12 participants. The duration of each focus group meeting was approximately one hour. Verbal informed consent was obtained from each participant.

[15] A total of 44 interviews was completed during April and May 2018. The duration of each interview was approximately 30–45 minutes.

[16] A master protocol and questions were developed for the interviews and the focus groups. This protocol provided a framework for gathering data in a consistent manner in order to could facilitate comparison across focus groups and interviews. RAND Corporation, undated.

implemented. The CPCB team's findings were instrumental in designing a broad set of communications and IT COAs to address achieving a resilient telecommunications infrastructure for emergency services (see Chapters 4 and 5 of this report).

The HSOAC Municipalities sector team conducted site visits to each of Puerto Rico's 78 municipalities to conduct a Municipal Survey.[17] This survey included questions about the municipality's capacity to provide services, how they are structured and operate, and the damage they incurred during the hurricanes, as well as specific questions provided by the HSOAC Comms/IT team.[18] The HSOAC Municipalities sector team also worked with the government of Puerto Rico to invite each of Puerto Rico's 78 municipalities to participate in one of 12 regional roundtables.[19] Furthermore, this team met weekly with UPR faculty and UPR SMEs, who provided extensive advice and data, and historical context and contributed a number of white papers. They also met with key civil society stakeholders in Puerto Rico, including Abre Puerto Rico and Estudios Tecnicos.[20] This team's analysis of municipal web presence showed that many municipalities do not have a working official website. Another of their findings was that many important online services were available in only a quarter or fewer municipalities; these services included a centralized 311-type customer service system,[21] electronic payment systems for taxes, utility bills, and fees, and electronic portals to facilitate common interactions with government, such as applying for business licenses. The Municipalities team's findings were instrumental in designing communications and IT COAs that leverage IT for the benefit of Puerto Rico communities and municipalities, including improving the capacity and ability to deliver key services (see Chapter 6 of this report).

Chapters 4, 5, and 6 are devoted to explaining the communications and IT COAs, grouped as follows:

- Resilient Public Telecommunications for Emergency Services and Continuity of Government (Chapter 4)

---

[17] The survey consisted primarily of a standardized data collection instrument developed by the International City County Management Association in consultation with HSOAC, FEMA, and the UPR.

[18] Each site visit was conducted by an interdisciplinary team led by professional city and county management staff identified by ICMA or by a UPR professor.

[19] Municipalities were grouped geographically, and each roundtable included five to seven municipalities. Roundtable discussions were moderated by a senior HSOAC researcher and by senior faculty from UPR using a semistructured discussion guide; HSOAC staff and UPR students took notes on the proceedings.

[20] Abre Puerto Rico is a nonpartisan organization that focuses on cultivating informed and active citizens in Puerto Rico. The organization compiles and shares government data using interactive technology for analysis and dissemination. Abre Puerto Rico, "About Us," website, 2019.
Estudios Tecnicos is a Puerto Rican economic, market strategies and planning consulting firm. Estudios Tecnicos Inc., "About Us," website, 2019.

[21] 311 is a nonemergency telephone service used by many cities. Through 311 people can find information about services, make complaints, or report problems such graffiti and road damage. Please see, for example, Colin Wood, "What is 311?" Digital Communities, *Government Technology*, August 4, 2016.

- Partnering with the Private Sector for a Robust Telecommunications Infrastructure and for Broadband Internet Deployment (Chapter 5)
- Information Technology for Critical Infrastructure and for the Digital Transformation of Puerto Rico (Chapter 6).

# 4. Resilient Public Telecommunications for Emergency Services and Continuity of Government

## Overview

As we discussed in Chapter 2, a plan to allow for continuity of communications for government leadership and CI stakeholders such as hospitals, PSAPs, and EMS was in place by the end of July 2018. However, Puerto Rico's public safety telecommunications infrastructure remains vulnerable to another catastrophic disaster of the magnitude of Hurricane Maria. Emergency communications systems need to be strengthened to facilitate quick and effective response in the event of a natural disaster, such as a hurricane, tsunami, earthquake, flooding, or manmade disaster, including a terrorist attack. Capabilities need to be developed to achieve robust and resilient emergency communications so that emergency services and government functions are effective and responsive in the aftermath of a disaster. Moreover, a strong and reliable communications and IT sector is fundamental to the successful recovery of other sectors and of Puerto Rico's economy. The communications and IT COAs presented in this and subsequent chapters supported the development of the congressionally mandated economic and disaster recovery plan, *Transformation and Innovation in the Wake of Devastation: An Economic and Disaster Recovery Plan for Puerto Rico*, which was delivered to Congress on August 8, 2018.

In this chapter we start looking toward the development of a communications and IT sector that is modern, robust, resilient, and efficient. Achieving these qualities will require undertaking prolonged investment. COAs funded by FEMA and other federal agencies can establish a foundation for the future development of the sector. However, sustaining the development of the public safety telecommunications infrastructure will require bringing systems up to federal telecommunications standards, taking advantage of technological advances to upgrade systems and equipment, improving maintenance procedures on CI, maintaining a cadre of qualified communications and IT personnel, and undertaking an integrated governance approach.

The COAs presented in this chapter address the need for a resilient, modern, and robust public telecommunications infrastructure that will support the provision of emergency services and COG in the aftermath of a disaster. They are also intended to bring about improvements in efficiency and effectiveness in the provision of emergency services during normal operations.

## Emergency Centers

This section discusses COAs that aim at improving the command and control of emergency services during and after a disaster, as well as improving the dispatch of emergency personnel during both normal operations and emergencies.

We start this section by addressing upgrades and improvements to one of the most important types of emergency center in the island, namely, the "911 center," or PSAP. As explained in Chapter 1, Puerto Rico has a primary and a backup PSAP. The primary PSAP receives calls from a 911 selective router or 911 tandem. The secondary, or backup, PSAP receives calls that cannot be handled by the primary PSAP. Once the PSAP operator has received an emergency call from the public, a second phone call has to be made by the operator to contact the proper first responder district or municipal office (police, fire, or EMS) to perform the dispatch function. Stated otherwise, calls from the public have to be relayed over the PSTN to the appropriate public safety agency before dispatch.[1]

As described in Chapter 2, after the hurricanes Puerto Rico PSAPs remained largely operational, functioning on fuel-based generators when commercial power was lost.[2] However, in spite of being operational, the PSAPs had limited ability to relay calls to public safety agencies because PSTN operation was limited due to hurricane damage. To reduce this vulnerability of the emergency response system due to potential damage of the PSTN in a future disaster, the FEMA Hurricane Maria Task Force recommended that the functions of PSAP and first-responder dispatch be consolidated at the same center with comparable or improved resilience.[3]

The task force also recommended upgrades to the current PSAPs, to include upgrading to Emergency Services IP Network (ESINet) connectivity and implementing Next Generation 911 (NG911).[4] Moving to NG911 will bring about features such as automatic location information/automatic numbering information and the ability to share photo, video, and GPS location with first responders. These features should improve the general effectiveness of the 911 service.

Based on the above considerations, we designed four COAs that are described in this section. The first COA, *CIT 3 Upgrade and Enhance 911 Service* consists of consolidating

---

[1] There is no radio communication in the PSAPs, and all dispatch is done from the first responder agency once it receives the incident information from the PSAP.

[2] FEMA Hurricane Maria Communications Task Force, 2017.

[3] FEMA Hurricane Maria Communications Task Force, 2017.

[4] Puerto Rico already had text-to-911 capability prior to Maria. Concerning ESINet, twenty-first-century public safety agencies will be required to connect to and/or build a private ESInet as defined by NENA-STA-010 to continue to provide services to their constituents. An ESINet is a "managed IP network that is used for emergency services communications, and which can be shared by all public safety agencies. It provides the IP transport infrastructure upon which independent application platforms and core functional processes can be deployed, including, but not restricted to, those necessary for providing NG911 services." See National Emergency Number Association (NENA), Interconnection & Security Committee, NG9-1-1 Architecture Subcommittee, Emergency Services IP Network Design Working Group, *Emergency Services IP Network Design (ESIND) Information Document*, April 5, 2018. NG911 is an internet Protocol-based 911 system. It allows digital information (e.g., voice, photos, videos, text messages) to flow from the public through the 911 network and to first responders. See NTIA, "Next Generation 911," undated.

PSAP and dispatch and upgrading the PSAPs to include ESINet connectivity as well as NG911 capabilities.[5]

The next three COAs relating to emergency centers pertain to the physical locations of and technical upgrades on disaster management command and control centers. Command and control in the aftermath of a disaster is typically conducted from an EOC, a central location in which executive decisions can be made and interagency coordination can be conducted with the purpose of overseeing the emergency response and the provision of public safety. In an EOC the local government and representatives from support agencies coordinate planning, preparedness, and response activities.

FEMA offers some specific guidance on EOCs:

> EOCs are locations where staff from multiple agencies typically come together to address imminent threats and hazards and to provide coordinated support to incident command, on-scene personnel, and/or other EOCs. EOCs may be fixed locations, temporary facilities, or virtual structures with staff participating remotely. The purpose, authorities, and composition of the teams that staff EOCs vary widely, but generally, the teams consolidate and exchange information, support decision making, coordinate resources, and communicate with personnel on scene and at other EOCs.[6]

Further, FEMA described primary EOC staff functions as "collecting, analyzing, and sharing information; supporting resource needs and requests, including allocation and tracking; coordinating plans and determining current and future needs; and in some cases, providing coordination and policy direction."[7]

Puerto Rico currently has an EOC that is colocated with the headquarters of the PREMA (or, in Spanish, Agencia Estatal para el Manejo de Emergencias y Administración de Desastres). The FEMA Comms/IT SBT concluded that the current EOC does not meet industry best practices and standards[8] and thus recommended modernizing its systems and equipment.[9]

An upgraded EOC, with improved data and communications systems, would better support response operations, incident management, decisionmaking processes, and enhance the ability of emergency authorities to manage disaster response and recovery. Therefore, the second COA in

---

[5] Per request by a reviewer of the Puerto Rico Recovery Plan, we added to this COA that implementation of the COA requires coordination with government agencies in the housing sector for the adoption of Enhanced 911 (E911) address conversion of rural route addresses. E911 is support for wireless phone users who dial 911. There is also a COA in the housing sector dealing with E911 conversion, *HOU 11 Develop a Common Address System*. Thus, implementation of CIT 3 will require coordination with implementation of HOU 11.

[6] FEMA, National Incident Management System, October 2017.

[7] FEMA, National Incident Management System, October 2017.

[8] FEMA Communications/IT Solutions-based Team, "Puerto Rico Communications/IT Solutions-based Team Report," June 30, 2018.

[9] FEMA also provides a checklist to assist state and local governments in performing an assessment of the hazards, vulnerabilities, and risk to their existing EOCs, and this checklist should be used as a guide to conduct upgrades. FEMA, Emergency Operations Center Assessment Checklist, undated-b.

this section, *CIT 6 Upgrade and Modernize the Island's Emergency Operations Center*, addresses these needs. It is based on FEMA guidance and on recommendations by the FEMA Comms/IT SBT.[10] According to the SBT report, many systems in the EOC require modernization as the equipment is at, or past, end-of-life. This report indicates that EOC modernization will require new equipment such as automation hardware, WebEOC[11] software, and video-teleconference (VTC) facilities.[12] In addition to modernizing the existing EOC, the FEMA SBT recommended the establishment of an alternate EOC. Its primary benefit is to serve as a backup location for emergency management activities, should the primary EOC in San Juan become inoperable. Therefore, the third COA in this section is *CIT 7 Establish an Alternate Emergency Operations Center.*

In designing a backup location for emergency management, other benefits can be achieved (those are also goals of CIT 7). First, the COA envisions establishing a PSAP colocated with the alternate EOC to serve as a backup to the primary PSAPs in San Juan. Second, the alternate EOC will be designed to serve as a COG and continuity of operations (COOP) site for non-EOC activities. [13]

The fact that the only two Puerto Rico PSAPs are located in San Juan creates a vulnerability of the island's 911 system to a disaster seriously impacting San Juan. Such a disaster may also disrupt emergency management to be conducted from the existing EOC. Thus, a consideration for this COA is to establish the alternate EOC outside of the San Juan metropolitan area. This ensures an extra layer of redundancy by providing a backup location where emergency and other critical functions (such as PSAP) can be carried out when the sites in San Juan are degraded or disabled.[14] The alternate EOC will improve the availability of the government of Puerto Rico's operational infrastructure by designating a site that has redundant, fault-tolerant telecommunications systems, including satellite backup. It is also important that the alternate EOC not be located in a vulnerable area such as a floodplain or tsunami zone.

Finally, support to COG and COOP activities must be provided in coordination with the municipal governments. Therefore, the municipalities should be part of the requirements gathering process and be involved in any training and exercises conducted at the EOC.

---

[10] FEMA Communications/IT Solutions-based Team, 2018.

[11] WebEOC is a web-based information management system that allows the sharing of information in real time during and after an incident or emergency. It allows for document sharing, photo uploading, and map displays. Users can access the information from a computer connected to the internet. See Juvare, "Emergency Management Technology—Powered by Juvare Exchange," WebEOCx, website, 2019.

[12] Appendix C of this report provides a preliminary cost estimate and potential funding sources.

[13] COG is the principle of establishing defined procedures that allow a government to continue its essential operations in case of nuclear war, major terrorist attack, or any other major manmade or natural catastrophic event. COOP is a U.S. federal government initiative to ensure that agencies are able to continue performance of essential functions under a broad range of circumstances.

[14] FEMA Communications/IT Solutions-based Team, 2018.

The fourth COA pertaining to emergency systems, *CIT 8 Mobile EOC Vehicle* comes from the FEMA SBT's recommendation that Puerto Rico procure a mobile emergency operations center (MEOC) similar to the National Incident Management System (NIMS) Type-2 Mobile Communications Command Center vehicles that arrived on the island after the hurricanes.[15] A mobile EOC is a useful tool when coordinating an on-site response to a local emergency or disaster. It provides independent communications over civilian and military frequencies, cellular and/or satellite, and generate its own power. It may also contain computers to run incident management software and VTC equipment. A mobile EOC is designed to function when other infrastructure is disabled or is in remote locations. In addition to being used for emergency purposes, a mobile EOC can function as a command center to monitor special events.

Upgrading and modernizing the primary EOC (CIT 6), establishing an alternate EOC (CIT 7), and procuring a mobile EOC (MEOC) (CIT 8) need to be addressed together during the implementation phase, taking into consideration their estimated costs, the availability of an adequate site for the alternate EOC/PSAP (CIT 7), and the fact that the fastest to implement is the MEOC.[16] An option to address the dependencies among these COAs was discussed with the FEMA Comms/IT sector and with officials of the government of Puerto Rico. This option was not introduced in the recovery plan delivered to Congress because these stakeholders thought it would be better addressed during the implementation phase of the recovery plan. This option is briefly discussed in Appendix B of this report.

## Ensuring Resilient Power for Public Safety Communications and Essential Government Operations

Lack of sustained electrical power had a significant impact on the telecommunications network during Hurricane Maria. As efforts continue to recover the energy infrastructure, the government of Puerto Rico might choose to invest in backup power sources, using standardized equipment where possible or appropriate, in order to provide the public safety and government communications networks with alternate power sources in the event of damage to or destruction of the energy infrastructure.[17]

The need for backup power is a result of an unreliable energy grid. As Puerto Rico makes improvements to the energy grid and reliability increases, there may still be a need for backup

---

[15] FEMA Communications/IT Solutions-based Team, 2018.

[16] Another consideration is that for certain disasters having a mobile EOC near the location of the damage may make management and execution of the response more effective. This assumes that roads leading to the disaster area in which the mobile EOC can transit will remain accessible.

[17] Power backup of the private telecommunications sites is the responsibility of the private carriers; we do not address it in this COA.

power to increase resilience in communications and IT. Using fuel-based generators as backup power has several shortcomings, which we discussed in Chapter 2.[18]

For the various reasons indicated above, the recovery plan proposes *CIT 5 Implement Public Safety/Government Communications Backup Power*. This COA would implement backup power systems for public tower sites, hospitals, police, fire, and EMS stations, municipal city halls, and government centers identified by the Public Buildings Authority to allow for the COG operations and emergency response. Implementing backup power at these locations will also help ensure that essential functions could continue under emergency conditions. This COA will bring about several benefits. First, having a diversified portfolio of power sources including but not limited to renewable energy sources such as solar, wind, and water could aid in ensuring resilience for public safety and government communications networks in the event of a catastrophic loss of power across Puerto Rico.[19] Second, renewable sources including solar, wind, and water systems have the potential to be self-sustaining and thereby freeing resources for use of systems that must rely on nonrenewable sources. Third, use of standardized equipment where possible or appropriate could aid maintenance and repairs, provided that Puerto Rico has access to relevant spare parts and components, according to FEMA officials. Careful consideration would be needed to determine the most appropriate mix of alternate sources to meet a given region's needs.[20] Standardizing this equipment could ease the logistical burden on Puerto Rico and make maintenance more efficient. In addition, first responders could use small-suitcase, solar-based cell towers to provide redundancy and immediate restoration during natural disasters.

Investments in alternate or renewable energy sources, such as solar, wind, and water, may be inappropriate in some regions, too expensive, or less efficient than the current fuel-based generators and require a more structured maintenance program to ensure effective operation. Additional assessments regarding the survivability of such technologies and their ability to withstand future extreme weather events are important. Due to these considerations, implementation of this COA will require close coordination with the efforts being made in the energy sector.

Two energy sector COAs are relevant to CIT 5: namely, *ENR 5 Harden Grid Assets to Support Critical Infrastructure* and *ENR 16 Provide Backup Generation to Priority Loads*. The former has the goal of hardening grid assets to support CI and public services in order to ensure

---

[18] Generators are not designed to run for extended periods of time. Supplying diesel or gasoline to generators across the island is logistically challenging and very expensive. Many generators have also been reported stolen. Additionally, the use of generators has a spillover effect on the transportation system given that Puerto Rico needs to leverage its transportation networks to ensure fuel is appropriately distributed. Finally, fuel-based generators may also impact air quality or cause spills that can affect groundwater, flora, and fauna.

[19] The transmission, distribution, and management of the grid are as important as the generation of energy. See RAND Corporation, undated, for additional information.

[20] The failure modes of different energy sources would need to be uncorrelated to achieve increased resilience. Moreover, if a common energy distribution system is implemented to power multiple sites, we need to ensure that it does not lead to a common failure mode.

that hospitals can care for the injured, first responders can dispatch 911 calls, and water pumps can continue to operate. The latter recommends providing backup generation to priority loads to ensure the sustained delivery of public services in the absence of the bulk power system. This COA includes targeted energy solutions for households with electricity-dependent medical needs and prioritization of backup generation for facilities that provide the greatest public benefit.

To make preliminary cost estimates for CIT 5, we assumed that each site was powered by a fuel-based generator instead of a system based on renewable sources.[21] Details on the cost estimates are provided in Appendix C.

## Communications Systems to Service Puerto Rico Communities

### Auxiliary Communications—Volunteer Radio Groups and Organizations

In the aftermath of Hurricane Maria, amateur radio operators and volunteer radio groups provided critical communications in support of hospitals and municipalities.[22] The COA *CIT 9 Auxiliary Communications—Volunteer Radio Groups and Organizations* will leverage these existing resources in Puerto Rico. This COA will leverage a workforce comprised of volunteers that is not yet formally engaged in disaster services. By formally establishing auxiliary communications through volunteer radio groups and related organizations, CIT 9 seeks to enhance the capacity of disaster response services through a coordinated, structured engagement among uniformly trained, highly skilled, and certified communications volunteers or volunteer groups.

This COA will cultivate such a volunteer workforce to leverage auxiliary community communications (AUXCOM) during disasters by engaging with volunteer radio groups and organizations. CIT 9 will support, incentivize, uniformly train, and encourage volunteer radio operations (e.g., the government of Puerto Rico will set up backup stations, distribute radios, or have representatives attend AUXCOM working group meetings), following on, or improving upon, models implemented in disaster-prone regions. Activities for emergency operations will be effectively extended at minimal cost by utilizing an experienced, intrinsically motivated group of volunteers.

Volunteer radio organizations represent an underutilized capability for communications in Puerto Rico. Although the COA focuses on auxiliary, volunteer communications, experience that results from its implementation can be applied to formally develop other auxiliary programs

---

[21] PlugPower, *Comparing Backup Power Options for Communications*, undated.

[22] National Association for Amateur Radio, "Comments of ARRL, The National Association for Amateur Radio," *In the Matter of Response Efforts Undertaken During the 2017 Hurricane Season*, PS Docket No. 17-344, January 22, 2018. There are more than 4,000 licensed amateur radio operators in the island, though only about 2,000 are active, according to Ismael Miranda, "Sabes qué es un KP4?" October 18, 2013.

across Puerto Rico. From a governmental standpoint, this is important to consider when weighing the COA's benefit relative to cost.

An implementation approach for enabling an AUXCOM force might use an MOU to preestablish necessary KP-4 call-signal assets at emergency management offices across municipalities and in all PREMA zones, PREMA HQ EOC, PRPD districts, fire districts, the 911 call center, and Puerto Rico EMS dispatch.[23] A provision for local or state government personnel to be trained, licensed, and equipped with KP-4 radio systems can also be considered.

Through potentially highly dispersed locations, AUXCOM could augment information available to multiple sectors, including housing, transportation, health and social services, water, and energy, after a disaster-related event.[24] Guidance might also be relayed from informed, authoritative sources via recognized volunteers to individuals residing in heavily affected regions. Coordination with the CPCB sector may be beneficial for implementation of this COA.

If volunteer groups become too large or experience high turnover, volunteers might not receive adequate training and preparation for disaster scenarios. The postdisaster availability and condition of an emergency communications infrastructure poses an operational uncertainty for AUXCOMM activities that potentially undermines the program's utility. However, this uncertainty is shared across emergency services and not unique to AUXCOMM or availability of a prepared workforce.

The EOC referred to in prior COAs should have point(s) of contact for auxiliary or volunteer communications.

### Rural Area Network Task Force

Effective means of communicating during and after a disaster are critical to address the public safety and health care needs of people situated in rural or disconnected areas, especially those who are isolated, have limited mobility, or are elderly, as well as the needs of their caregivers. The COA *CIT 4 Rural Area Network Task Force* concerns the establishment of a rural area task force that will assess what are the best information systems and communications networks (or systems and networks) to service these communities. The systems and networks to be considered need to include devices within homes or at designated local stations, rely on new or recent infrastructure, be practical and efficient, operate in the immediate aftermath of a disaster, and provide key information (e.g., guidance, health issues) prior to and after a disaster, for residents, emergency services, and medical providers alike.

The task force's goal is to advance public safety and health care delivery to loosely connected communities (e.g., "rural areas") by providing comprehensive, real-time situational awareness on weather conditions, emergency guidance, and medical needs. In its assessment,

---

[23] KP-4 is a call sign assigned by the FCC for amateur radio in Puerto Rico.

[24] National Association for Amateur Radio, 2018; National Association for Amateur Radio, *2017 Annual Report*, October 4, 2017.

the task force can leverage several other COAs in this report, including *CIT 9 Auxiliary Communications—Volunteer Radio Groups and Organizations, CIT 1 Land Mobile Radio System* (in particular, to leverage the First Responder Network Authority [FirstNet] when it becomes available), *CIT 23 Data Collection and Standardization for Disaster Preparedness*, and *CIT 29 Health Care Connectivity to Strengthen Resilience and Disaster Preparedness*. One of the important outcomes of the task force's assessment may be that additional communication network(s) or information systems may need to be developed.

This COA seeks to improve the survivability (avoid loss of life) and improve the health of people for whom there is presently a limited communications infrastructure and thus constraints on the quality or timeliness of required health or medical services. Extending the communications infrastructure to poorly connected areas can improve the readiness, planning, and coordinated allocation of resources for services sourced from medical facilities and emergency responders. The government of Puerto Rico will have an improved, detailed awareness of disaster related issues, which will better inform decisions and coordination activities at a governmental level.

The task force will need to formally consider the provisioning and delivery of emergency services to individuals under disaster-scale conditions where emergencies are a pervasive norm. Thus, its focus necessarily needs to go beyond the technical availability of a municipality- or facility-level telecommunications network infrastructure to develop an acute view of public safety and health care information and communication needs for all people situated remotely from quality medical and emergency services in the wake of a disaster. While the availability of a telecommunications network infrastructure is a major and critical component of communications, it can only partly address the challenge of delivering public safety and health care related services to rural regions in a way that maximizes their benefit to the population residing in those areas, especially in the wake of a disaster.

Reliable availability of power and broadband access will both be important for implementing recommendations resulting from the task force's work. It may also be possible to directly leverage COAs such as *CIT 29 Health Care Connectivity to Strengthen Resilience and Disaster Preparedness* and *CIT 9 Auxiliary Communications—Volunteer Radio Groups and Organizations* as part of a resulting strategy.

The task force itself should be comprised of a mixture of highly informed stakeholders, including but not limited to government officials, telecommunications experts, and emergency professionals, who have the ability to assess needs and options across all municipalities. Furthermore, the task force needs to be broadly inclusive, with civil society partners as well as representatives from community and other groups. The expectation is that this work can be reasonably completed in approximately one year and that it will provide a comprehensive set of achievable recommendations that are complementary to existing initiatives, but are centered on ensuring the public safety and health care needs of populations situated in remote areas or areas difficult to access during a disaster.

## Additional Methods to Enable Emergency Communications and Functions in the Aftermath of a Disaster

### Procure a Mobile Emergency Communications Capability

COAs discussed in the chapter thus far address the need to strengthen the public communications infrastructure to facilitate quick and effective emergency response and COG in the event of a disaster. In addition to the capabilities provided by these COAs, the government of Puerto Rico was interested in a mobile emergency communications capability that was flexible and straightforward to implement and integrate into operations in the near term.[25]

The COA *CIT 11 Procure a Mobile Emergency Communications Capability* will develop the capability to quickly reestablish communications for emergency and government operations in the aftermath of a manmade or natural disaster that causes widespread, catastrophic damage to the telecommunications infrastructure. This system, consisting of deployable telecommunications equipment, is envisioned as a quick and temporary but important measure that addresses the loss of primary communications capability. Reliable and interoperable communications are essential to providing effective and responsive disaster recovery, emergency services, and government operations.

CIT 11 envisions the procurement of deployable assets that can be safely cached and quickly installed throughout Puerto Rico to restore voice and data communications for disaster response, emergency services, and government activities. This equipment will provide secure, interoperable, cross-agency communications. The solution may include but is not limited to utilizing civilian and military frequencies and a network of deployable nodes. The system will include portable power generation to ensure independent operations and/or remote deployment.

One potential implementation of this COA consists of a deployable network of mobile communications nodes similar to the network already implemented in several New Jersey counties, called "JerseyNet." JerseyNet was funded by grant from the NTIA and consists of 42 mobile nodes.[26] We extrapolated to what would be needed in Puerto Rico by considering the island's geography and assuming a nominal range for each node.[27] As we explain in the appendix, we estimated two cost bounds for this COA. For the lower bound, we assumed that

---

[25] The government of Puerto Rico stated that the mobile communications capability outlined in this COA is important to ensure resilient communications. The FEMA Communications/IT SBT stated that such capability would duplicate private, commonwealth, and federal capabilities. However, the government of Puerto Rico believes that such capabilities will not be available for some time and that a solution that can be implemented in the near term is needed.

[26] Most of the funds provided by the grant went to the mobile nodes, according to a private communication between HSOAC researchers and NTIA. See NTIA, Broadband USA Grants Awarded, New Jersey, website, undated, for the award. In this website, the grant appears with the name "New Jersey Broadband Network project." State of New Jersey, Office of Homeland Security and Preparedness, website, undated.

[27] Puerto Rico's coastal urban areas, rural mountainous areas, and suburban areas with landscape of pastures and secondary forest.

after a future disaster about 50 percent of the island's telecommunications infrastructure remains operational and thus only 50 percent of the deployable network is needed.[28] This deployable network is assumed to be interoperable with the telecom infrastructure that remains operational after the disaster.

In addition to the procurement of systems, CIT 11 addresses the need for a team of technicians to deploy, maintain, and store the mobile nodes. The team will coordinate with relevant agencies to exercise their ability to deploy and operate the system. This includes engaging in coordinated training exercises to support deployment of the system in the event of a disaster. The team will also develop maintenance and deployment plans, taking into consideration the climate and terrain challenges of Puerto Rico.

Because this COA envisions providing a capability to support government operations, there must also be coordination with the municipalities to ensure both that their needs are understood and that the municipalities understand the limitations of the capability being provided.

### Ensure Continuity of Business at Puerto Rico Industrial Development Company Sites

The Puerto Rico Industrial Development Company (PRIDCO) is a government-owned corporation dedicated to promoting Puerto Rico as an investment destination for companies and industries worldwide.[29] Business enterprises at PRIDCO sites are major contributors to the Puerto Rico and CONUS economies and span multiple domains and industries; a diverse range of industries that provide key services to the U.S. and Puerto Rico economies. Examples are biomedical services and equipment and aerospace services. The 2017 storm season halted operations in several areas and had a cascading impact that reached well beyond Puerto Rico. Critical operational data were no longer accessible due to loss of communications and power at PRIDCO sites.

The COA *CIT 20 Continuity of Business at PRIDCO Sites* will establish multiple alternative business processes to leverage several platforms for telecommunications and information systems including, but not limited to, the use of fiber, satellite, and microwave systems and cloud-based or hosted services and information systems.[30] Enabling a redundant communications network and power systems at PRIDCO sites will help mitigate the risks arising from outages in upstream services, to which they are currently prone. By doing so, it will improve the resiliency of essential business activities delivered by PRIDCO in the aftermath of a disaster.

---

[28] After Maria, more than 90 percent of the telecommunications infrastructure was not operational. Only 5 percent of the cell towers on the island were operational. Similarly, most of the towers that provide backhaul communications for police, fire, and EMS were damaged, destroyed, or without power. See FEMA Hurricane Maria Communications Task Force, 2017.

[29] See PRIDCO, Company Overview, undated.

[30] A business process defines a set of corresponding activities that achieve a business goal or intended outcome, whether through automation or human activity. Correspondingly, there may be several options available for application at a particular facility or site of a business entity.

The COA will also provide the critical communication systems required to maintain key business activities at PRIDCO sites when primary communications methods are degraded or unavailable to provide continuity of services related to disaster recovery.

The improved availability of communications at PRIDCO sites can potentially be extended to local and rural communities during disasters. PRIDCO sites are distributed geographically to improve the economy as a whole, especially in areas that are not yet population centers. Through improved, on-site telecommunications networks, PRIDCO may be able to further support these communities in the event of a disaster.[31] Recovery efforts across sectors—including health and social services, housing, transportation, energy, and communications and IT—could be accelerated by leveraging PRIDCO services if they are available during a disaster or outage.

This COA will need reliable power systems and may require infrastructural upgrades to accommodate modernized communication systems.

### Data Collection and Standardization for Disaster Preparedness

Presently, Puerto Rico lacks quality data to inform the public and the policymaking process. Updated data on day-to-day information are helpful under normal circumstances, but critically essential after a disaster. A vital resource for public data, data.pr.gov, has historically been maintained by the Puerto Rico Institute of Statistics (PRIS), an independent entity. However, PRIS has been the subject of reorganization and challenges to its independent role in data collection efforts since 2018.

Moreover, historically, it is also not uncommon for Puerto Rico to fall through the cracks of studies that look at the condition of U.S. states (some of which do not address Puerto Rico at all) and other studies that examine data for specific countries worldwide (which subsume Puerto Rico into data for the United States as a whole).

A crucial action for Puerto Rico is to prioritize the coherent, standardized collection and publication of a broad range of data for use in disaster response and in general within Puerto Rico.[32] This will enable objective and independent analysis of key disaster-related data for status information; in the long-term, it will also directly inform assessments of conditions in Puerto Rico, within the United States and globally.

In the wake of Hurricane Maria, the government of Puerto Rico recognized a need for publicly available up-to-date information about basic services in Puerto Rico. The Governor's office launched the website www.status.pr to update the media, public, and first responders about

---

[31] PRIDCO representatives, interview with authors, discussion regarding the impact of hurricanes on operational activities for businesses and on local communities, June 1, 2018.

[32] U.S. Federal studies that do not include Puerto Rico on par with states: Census Bureau, National Center for Education Statistics, Bureau of Labor Statistics, Bureau of Justice Statistics, National Center for Health Statistics, Substance Abuse and Mental Health Services Administration, Bureau of Economic Analysis – source: Financial Oversight and Management Board for Puerto Rico, *Annual Report*, 2017. See also Congressional Task Force on Economic Growth in Puerto Rico, *Congressional Task Force on Economic Growth in Puerto Rico Report to the House and Senate*, 114th Congress, December 20, 2016.

conditions across the island. Through personal outreach, crowdsourcing, and tremendous manual effort, status.pr kept track of how many ATMs, gas stations, supermarkets, and pharmacies were open; how many people were without water, power, or communications; how many people were in shelters; and how many pets were displaced. Status.pr became *the* public dashboard for the recovery effort and a lifeline for residents as they went about strategically planning their days to access basic needs.

Media outlets used status.pr frequently to inform their stories and compare with other sources of information. The *Washington Post* created a Twitter bot to provide an hourly update on service delivery in Puerto Rico. That bot pulled information directly from status.pr. [33] A member of the Puerto Rico's diaspora living in Connecticut created an automated tool to check status.pr for hourly updates. FEMA itself referred journalists to status.pr for up-to-date information on the restoration of basic needs in Puerto Rico. [34]

The COA *CIT 23 Data Collection and Standardization for Disaster Preparedness* provides for the continued maintenance and expansion of status.pr, including the formation of data partnerships and opportunities for crowdsourcing or self-reporting information for situational awareness throughout Puerto Rico.

The COA will be carried out by PRITS and will build upon the agency's mission to digitize government data, [35] form data-sharing partnerships, and utilize "smart" devices such as internet of things (IoT) sensors to continue to provide status updates to policymakers, the media, and the public. These activities may include

- partnerships with critical commercial entities such as banks, gas stations, pharmacies, and hospitals, to develop a mechanism for maintaining updated information on outages or service disruptions
- data-sharing partnerships with utilities and government service providers to enable application programming interface (API) delivery of status information to maintain a public-facing dashboard of service delivery and performance measures
- pilot programs to install or access sensors that could provide real-time updates on hours of operation, outages or service disruptions
- work with the private sector and "civic hacking" community to design ways for citizens to participate with status reports or updates on important issues, potentially drawing on existing solutions for citizen reporting.

In addition to further developing the public-facing status.pr, this COA will help PRITS to work closely with the medical community, especially the 86 federally qualified health centers (FQHCs) across Puerto Rico, on potential solutions for mapping and maintaining contact with critical-care patients in an emergency, while maintaining the security of data protected under the

---

[33] Phillip Bump, "We've Created a Twitter Bot that Provides Hourly Updates on the Situation in Puerto Rico," *Washington Post*, October 11, 2017.

[34] Jenna Johnson, "FEMA Removes—Then Restores—Statistics About Drinking Water Access and Electricity in Puerto Rico from Website," *Washington Post*, October 6, 2017.

[35] PRITS's mission to "digitize government data" is explained and expanded in the COAs described in Chapter 6.

Health Insurance Portability and Accountability Act. This effort should specifically explore ways to address patients in need of insulin, oxygen, or dialysis or have other critical medical issues that should receive priority attention during emergency response.

Creating a platform for publicly sharing data in a standardized, user-friendly format provides valuable information for policymakers, the media, and emergency responders, while also providing machine-readable information that can digitally inform and serve citizens. Such information can be applied for capacity-planning and community-building, economic initiatives that may improve the resilience of affected regions, health and social services functions, and assessing energy, water, and transportation needs. The effort also supports transparency and begins to address the disparity of data resources that are available for Puerto Rico.

## The Future Public Telecommunications Network of Puerto Rico

The COAs presented thus far in this chapter together with the response efforts described at the end of Chapter 2 are intended to provide Puerto Rico with a resilient, robust, and standards-compliant public safety telecommunications infrastructure. This infrastructure will support the provision of public safety and the continuation of critical government functions in the event of a disaster in the short to medium terms. In this section, we address the long-term goal of achieving efficient and effective public safety telecommunications for supporting emergency operations both after a disaster and during normal operations. In the next section we will address the important topic of maintaining the workforce needed to ensure the readiness of the public telecommunications infrastructure as well as the development of such workforce.

We pointed out in Chapter 2 that as of October 19, 2018, the three public telecommunication networks in Puerto Rico were well on their way to achieving pre-Irma/Maria status—that is, the PRPD P25 network had reached over 70 percent of that status, whereas the PREMA and EMS networks have achieved over 85 percent of that status.[36] However, even after the public safety networks and associated systems have reached prehurricane status and are fully operational, important technical issues will remain. As described in Chapter 1, public safety radio communications in Puerto Rico consist of five separate radio systems, connected by three separate microwave networks, and operating on three different frequency bands that make use of different modulation protocols. These differences lead to more expensive logistics and maintenance.[37] Moreover, the differences in radio systems, microwave networks, frequency bands, and protocols go beyond logistics and maintenance since they make interoperability difficult to achieve. Lack of interoperability can lead to serious operational inefficiencies in the provision of public safety services both after a disaster and during normal operations.

A good example of such inefficiencies is the need to use two separate microwave networks and systems to allow for communications between many hospitals and their EMS personnel.

---

[36] Per the PRDPS, the last phase of the work on these networks will focus on antenna alignment and software programming.

[37] It is generally easier to maintain one type of system than many.

Hospitals in Puerto Rico use the PRPD P25 microwave network whereas ambulances (EMS personnel) use the EMS network. Therefore, relaying messages between such hospitals and ambulances requires going through EMS and P25 network dispatchers. That is, communications between ambulance and hospital follow a complex path: from ambulance over radio to EMS dispatcher, from EMS dispatcher to health dispatcher, from health dispatcher to hospital; finally from hospital to ambulance following the same path but in reverse.

To avoid the above inefficiencies, the Comms/IT SBT considered upgrading and potentially consolidating the disparate microwave networks into one microwave backbone.[38] The upgrade and consolidation would entail use of P25 radios by the full first responder force. The Comms/IT SBT also recommended following closely the development of and leveraging two ongoing initiatives: the First Responder Network Authority (FirstNet) initiative and the Puerto Rico and the US Virgin Islands Interoperable Communications Network Engagement (PRINCE). These ideas were combined into a COA entitled *CIT 1 Land Mobile Radio System.* This COA consists of preparing a plan for Puerto Rico's public telecom infrastructure that considers all the above alternatives and their expected timelines. This plan will align the buildout and future upgrades of Puerto Rico's public safety networks with the development of PRINCE. As part of this plan, FirstNet will be regularly assessed either as a backhaul service provider or as a complementary service and, eventually, as a potential replacement for the public safety network's system. Finally, the plan will include different options that could be different for the short and long terms and for voice and nonvoice applications (such as using an LMR system, either the commonwealth's or PRINCE for critical voice communications, and FirstNet for data transfer). Once this plan is produced, the following step of the COA is to implement that plan.

In what follows we describe the different alternatives, all of which need to be considered for a successful implementation of this COA.

The upgraded and consolidated system will be based on the P25 standard. Use of this standard presents several advantages:

- A large part of the Puerto Rico first responder force, including the majority of police and fire personnel, already use P25 radios.
- The current police microwave network that services police, fire, and hospitals—which is also the largest microwave network in the island—is also based on the P25 standard.
- P25 radios are very flexible, namely, they can communicate in analog mode with legacy radios and in either digital or analog mode with other P25 radios.
- The P25 standard includes a requirement for protecting digital communications (voice and data) with encryption capability. This capability is optional to the user, and the radio systems manager is able to remotely change encryption keys.
- The P25 standard exists in the public domain, allowing any manufacturer to produce a P25 compatible radio product.

---

[38] FEMA Communications/IT Solutions-based Team, 2018.

- P25 is a recognized standards-based system produced through the joint efforts of the Association of Public Safety Communications Officials International, the National Association of State Telecommunications Directors, selected federal agencies, and DHS OEC and is standardized under the Telecommunications Industry Association.[39]

The main disadvantage of selecting the P25 standard for upgrading or consolidating is the high cost of P25 radios when compared with the standard UHF/VHF radios currently used by some of Puerto Rico first responders.[40]

The Comms/IT SBT recommended that the FirstNet initiative be monitored and regularly assessed as a backhaul service provider and as a complementary service to and potential replacement for Puerto Rico's first responders' system.[41] An initiative of the U.S. Department of Commerce, FirstNet was authorized by Congress in 2012 with the mission to develop, build, and operate the nationwide, broadband network for first responders—to include police, fire and emergency medical services. The network was to be used in disasters, emergencies, and daily public safety service. FirstNet entered into a public-private partnership with AT&T to develop this network throughout the United States. FirstNet will be a 25-year partnership in which the government provides initial funding. Each U.S. state or territory governor has two options vis-à-vis FirstNet: to opt-in or opt-out. In the former case, AT&T would deploy, maintain, and operate the FirstNet network within that state or territory at no cost to the state or territory for 25 years. The Governor of Puerto Rico decided to opt-in, and AT&T is currently making progress in deploying the FirstNet network in Puerto Rico. An important consideration for the design and built of the new Puerto Rico public telecommunications network should be interoperability with FirstNet.

PRINCE, a program by DHS , has the goal of developing a joint system for the different LMR systems used by DHS components in Puerto Rico and the Virgin Islands.[42]

> The Department of Homeland Security is proceeding with reconstituting its Puerto Rico and Virgin Islands tactical communications system jointly amongst all its components. The system will be scalable and will allow the potential for other partners to join the system. As DHS rebuilds its system, it intends to engage with its federal, state and local partners to pursue opportunities to expand the DHS system to include other partnering users and/or to integrate systems to maximize and simplify communications with others outside DHS, and to promote efficient interoperable communications.[43]

---

[39] See, for example, CODAN Communications, website, undated.

[40] Another caveat is that not all P25 radios are compatible with each other. Therefore, the COA would need to ensure compatibility between the new P25 radios and the old P25 radios, which are already in use by police and fire.

[41] FEMA Communications/IT Solutions-based Team, 2018.

[42] FEMA Communications/IT Solutions-based Team, 2018.

[43] U.S. Department of Homeland Security, "DHS Puerto Rico and Virgin Islands (PR/VI) Reconstitution Effort Key Highlights and Talking Points," April 4, 2018.

The DHS's Joint Wireless Programs Management Office briefed its leadership on developing such joint system and received approval to proceed; a preliminary design was developed.[44] Subsequently, at a PRINCE meeting in Puerto Rico, DHS started to socialize Puerto Rico's participation. The recommendation vis-à-vis PRINCE was to follow its development and make an arrangement with DHS to join this system—potentially to lease it—when it becomes available. One possible disadvantage of using PRINCE is that Puerto Rico will have limited input in managing it.[45]

## Human Capital for the Public Telecommunications Infrastructure of Puerto Rico

Human capital—a healthy, educated, and trained workforce—is critical for the recovery efforts of all twelve sectors, including communications and IT. The recovery plan has several COAs aimed at developing the island's workforce.[46] The recovery plan also has one strategic initiative fully devoted to this topic.[47] *CIT 1 Land Mobile Radio System* is one of the communications and IT COAs that addresses workforce needs for the Puerto Rico communications and IT sector.

First, as part of this COA, we addressed the important need of maintaining a workforce that can ensure the readiness of the public safety telecommunications infrastructure. A senior official with PRDPS stated that 56 technical personnel—in addition to current telecom personnel—would be sufficient to properly maintain the current and the ungraded first-responder public safety telecommunications infrastructure. The costs for the additional 56 technical personnel are provided in Appendix C, and are based on the median salary for an engineer in San Juan, Puerto Rico.

---

[44] DHS, 2018.

[45] Moreover, in case of damage to PRINCE due to a hurricane or other disaster, its restoration will not be eligible for FEMA disaster relief funds. However, repair and maintenance of PRINCE will be the responsibility of the federal government. Thus, it can be expected that the federal government will provide the resources (including personnel) to restore it to full functionality as quickly as possible. Use of PRINCE by Puerto Rico will likely be less expensive than building and maintaining its own system.

[46] Examples in the municipality sector are *MUN 2 Create Regional Economic Development Plans* and *MUN 4 Build the Capacity of Municipalities to Apply for, Secure, and Manage Grants*. With economic development in mind, MUN 2 proposes that all municipal and regional economic plans assess available local workforces, provide additional education and training where necessary, and identify infrastructure needed to support industry locally. MUN 4 addresses the need for the municipalities to apply for and manage recovery funds, given that the number and size of grants received for rebuilding is likely to be much higher than Puerto Rico's grant-management workforce has ever experienced. See Central Office for Recovery, Reconstruction and Resiliency, 2018.

[47] Called *21st Century Workforce*, it intends to "develop and protect human capital to establish a world-class workforce, increase labor force flexibility, and create high-quality employment opportunities aligned with economic growth strategies." See Central Office for Recovery, Reconstruction and Resiliency, 2018.

Second, the Department of Public Safety indicated the need for workforce development in the form of four-year electrical engineering bachelor of science degrees for 20 of its personnel. Estimates of costs and further details are provided in Appendix C.

# 5. Partnering with the Private Sector for a Robust Telecommunications Infrastructure and Broadband Internet Deployment

## Overview

As described in Chapter 2, commercial communications were almost nonexistent in Puerto Rico immediately after Hurricane Maria. According to the data provided to the Telecommunications Bureau, as of September 24, 2017, Hurricane Maria had damaged 91 percent of the private telecommunications infrastructure (primarily antennas and fiber), which had a major impact on the government, retail stores, banks, pharmaceutical companies, food, transportation, and other businesses.[1] This was devastating to communications throughout Puerto Rico, since over 80 percent of residents rely on mobile phones as their primary means to communicate.[2] In large part, communications were lost because the hurricane destroyed the aerial fiber-optic cable that is an essential part of the telecommunications network. The aerial fiber cable in Puerto Rico is deployed predominantly above ground, on utility poles. It was reported that 80 percent of the approximately 350 linear miles of core above-ground fiber was destroyed and 85 to 90 percent of the approximately 1,500 linear miles of last-mile fiber was destroyed.[3] In addition to restoring and improving its commercial telecommunications networks, Puerto Rico needs to expand the deployment of broadband internet services. Currently, Puerto Rico lags far behind CONUS in broadband internet deployment.[4]

In the first two sections of this chapter, we will discuss COAs that address two main purposes: (1) achieving resilience to future disasters by strengthening the private telecommunications network in Puerto Rico; and (2) achieving deployment of broadband internet service throughout Puerto Rico. Making it easier for private telecommunications companies to do business in Puerto Rico and leveraging telecommunications-related federal programs to "build back better" will further these objectives.

---

[1] Torres Lopez, 2018.

[2] FEMA Hurricane Maria Communications Task Force, 2017.

[3] FEMA Hurricane Maria Communications Task Force, 2017. The core network serves to interconnect traffic from cellular, wireline, and broadband internet; last-mile fiber provides transmission from the core network to the retail end-user. According to a Puerto Rico homeland security official, 48,000 utility poles had been damaged by the hurricane; Puerto Rico Statewide/Territory-wide Interoperability Coordinator (SWIC), interview with the authors, Puerto Rico, March 1, 2018.

[4] For example, the Federal Communication Commission's data on broadband deployment indicate that over 98 percent of residents in rural areas of Puerto Rico lack a broadband connection to the internet. See FCC, "2016 FCC Broadband Report," GN Docket No. 15-191, FCC 16-6, January 29, 2016, Appendix D.

In the third section of the chapter we address COAs pertaining to the undersea network infrastructure of Puerto Rico. This infrastructure provides connectivity to CONUS and other regions of the world, and plays a major role during disaster-relief coordination.

## Achieve a Resilient Private Telecommunications Infrastructure

### Incentivize the Reduction of Aerial Fiber-Optic Cable

FEMA telecommunications experts determined that the resilience of the telecommunications network in Puerto Rico would be improved by reducing the amount of aerial fiber-optic cable.[5] There is another way to deploy fiber-optic cable—namely, in buried conduit. However, it is more expensive because of the cost of obtaining permits and ROW approvals from government agencies and municipalities (which can take up to a year) and the cost of trenching and laying the fiber conduit below ground.[6] Commercial telecommunications providers make decisions on their capital investments (such as extending their fiber-optic cable network) based on their expected return on investment (ROI). If a buried conduit network was made available to all telecommunications providers, it would change the ROI of deploying buried fiber and thus speed its deployment. That is because private companies would incur only the cost of pulling fiber through the conduit, rather than the combined expense of permitting and trenching to bury fiber. This might incentivize telecommunications providers to deploy buried fiber-optic cable throughout Puerto Rico, including geographic areas that have challenging terrain and rural and economically disadvantaged areas.

Telecommunications providers in Puerto Rico who met with the Telecommunications Bureau and HSOAC team members in May 2018 were enthusiastic about the opportunity to pull fiber through government-owned buried conduit. These providers agreed that, if the government of Puerto Rico trenched and laid empty conduit, it would defray the most expensive part of burying fiber-optic cable: the permitting and ROW approval process and the trenching costs (labor and equipment). The discussions during the May 2018 meeting led to the proposed COA, entitled *CIT 21 Government Owned Fiber-Optic Conduits to Reduce Aerial Fiber-Optic Cable and Incentivize Expansion of Broadband Infrastructure*. It consists of designing and deploying buried conduit for fiber-optic cable by the government of Puerto Rico in consultation with Puerto Rico telecommunications providers. The government would trench and lay empty conduit according to the design. It would own the conduit, but telecommunications providers would install and own their own fiber-optic cable. Thus, CIT 21 creates incentives for telecommunications providers to

---

[5] FEMA Hurricane Maria Communications Task Force, 2017.

[6] "Rights of way" refers to the legal right, established by usage or by government grant or permit, to pass along a specific route through grounds or property belonging to another.

restore lost aerial fiber-optic cable or extend their networks with buried fiber.[7] Another benefit to burying fiber would be the opportunity for joint trenching and installation of utilities, where appropriate, and coordination of a "dig once" approach.

A network of conduit provided by the government could circumvent the delays and expense imposed on telecommunications providers in Puerto Rico by the existing permitting and ROW approval process. A potential pitfall could arise, however, if funding from the government of Puerto Rico and federal assistance used to bury the conduit does not include maintenance costs, which would then be the responsibility of Puerto Rico. An option might be to require telecommunications providers to pay a reasonable user fee for using the conduit network or to create some form of cost-sharing mechanism. These funds could be escrowed for maintenance of the conduit network by the government of Puerto Rico. Telecommunications providers would probably want to have a service level agreement that would provide for immediate repairs to the conduit in case of damage or service interruption. Another potential concern is the possibility of disagreement among telecommunications providers if they are permitted to have input into the design of conduit network. For example, the network design may not be perceived as competitively neutral by all telecommunications providers. In any case, a network of conduit could be created in parallel to instituting regulatory reforms concerning permitting and ROW approval.

In addition to improving telecommunications resilience, CIT 21, as the latter part of its title implies, can also incentivize expansion of broadband infrastructure. That is because, if the government of Puerto Rico manages the time-consuming and expensive task of obtaining regulatory approvals for burying conduit, as well as the cost of trenching, it will significantly decrease the cost of broadband network expansion for commercial providers. Although aerial fiber-optic cable may also be used to extend the broadband network, giving commercial providers the option of pulling fiber-optic cable through buried conduit on a cost-effective basis will make the network more resilient to severe weather and improve network reliability. A government-owned buried conduit network could provide an attractive alternative to time-consuming and costly arrangements with the owners of utility poles about the use of their poles to deploy aerial fiber-optic cable in areas of Puerto Rico that are not densely populated.

As we shall explain later in this chapter, broadband expansion is an important part of the Puerto Rico recovery plan. However, we determined that it was most important to emphasize the need to reduce aerial fiber-optic cable to create a more resilient network, and that is why this

---

[7] The government of Puerto Rico could require telecommunications service providers to bury new fiber-optic cable in conduit via a governance change instead of by providing an incentive. For example, a requirement could be set by the Telecommunications Bureau (or new Public Service Regulatory Board) after a Notice and Comment administrative proceeding, mandating that all new fiber installations have to be in buried conduit. However, that would be very onerous for the private sector, due to the current price differential between installing aerial fiber-optic cable and burying fiber in conduit. Such a mandatory requirement could have the negative consequence of delaying telecommunications providers' restoration efforts and plans to extend their networks.

COA is presented in this section of Chapter 5. The creation of a more resilient network is a top priority for the government of Puerto Rico.

### Perform a Site Structural Analysis of All Towers Used for Government Telecommunications

Telecommunication towers are susceptible to damage by severe weather, such as hurricanes. For these reasons, it is important to perform a structural review of the towers that provide government emergency and other services. This review should assess whether all towers used for emergency communications meet the Puerto Rico tower code concerning structural loading. We include this COA in the section of our report pertaining to private telecommunications infrastructure because in Puerto Rico these towers are primarily privately owned. The recovery plan proposes a COA entitled *CIT 12 Perform Site Structural Analysis for All Government Telecom Towers (both Public and Privately Owned)*.

The first step of this COA is to review the existing tower code to identify the structural requirements for towers. The second step is to review the tower code enforcement authority of the Puerto Rico Planning Board (PRPB) and identify any other Puerto Rico agencies that are involved in permitting towers or enforcing the tower code.

The third step consists of performing the site structural analysis. To accomplish the analysis, an engineering team would need to inspect the designated towers. Such an analysis should be conducted as swiftly as possible, to assure the resilience of towers used for emergency services.

The requirement to perform the structural analysis could be introduced in a tower code–enforcement activity or in a notice and comment administrative proceeding. As of the time of the writing of this report, an administrative proceeding that included towers was underway at the PRPB. Decisions made in that administrative proceeding should be considered before implementing CIT 12.[8]

Failure to implement this COA and conduct the necessary repairs could result in loss of communications among government, emergency services agencies, and sectors other than communications and IT due to the next severe storm. Hurricane Maria demonstrated how loss of backhaul communications from cell towers resulted in PSAPs being unable to relay calls to police, fire, or EMS.[9] Additional information on these towers will help the government more effectively prepare for future extreme weather events. The blue-ribbon panel (CIT 25) or the Telecommunications Bureau would request tower companies to provide information on the sites that broadcast WEAs.

---

[8] Tower owners, not the government of Puerto Rico, are financially responsible for tower repairs and for bringing their towers into compliance with the code if they are found to be noncompliant.

[9] This information was documented in FEMA Hurricane Maria Task Force, 2017.

# Spur the Deployment of Broadband Internet Service Throughout Puerto Rico

## *The Importance of Making Broadband Internet Service Widely Available*

Broadband access can be linked to many benefits that the government of Puerto Rico wishes to achieve to "build back better." The President's Council of Economic Advisors' Issue Brief of March 2016 stated that broadband access has a significant impact on economic growth, wages, medical care, and education, among other benefits. The brief further stated that "addressing the digital divide is critical to ensuring that all Americans can take advantage of the many well-documented socio-economic benefits afforded by internet connections. These benefits are most evident when consumers have access to the internet at speeds fast enough to be considered broadband; these speeds are required to facilitate full interaction with advanced online platforms."[10]

The brief cited research on the economic impact of broadband connectivity. Before the widespread availability of streaming audio and video, broadband internet accounted for approximately $28 billion of the U.S. gross domestic product.[11] Nearly half of that total was due to households upgrading from dial-up broadband service. By 2009, broadband internet accounted for approximately $32 billion per year in net consumer benefits. These findings are consistent with studies in other countries. Moreover, growth is particularly concentrated in industries that are more IT-intensive.[12] In addition, insofar as developing internet skills allows a person to participate more fully in the economy, doing so may positively affect a person's wages.[13]

The brief also explained benefits in medical outcomes. Broadband has made medical care and medical information more convenient and accessible. Broadband-enabled virtual visits with trained medical professionals can improve patient outcomes at lower cost and with a lower risk of infection than conventional care provided in person.[14] Telemedicine is particularly valuable for rural patients who may lack access to medical care, as telemedicine allows them to receive medical diagnoses and patient care from specialists who are located elsewhere.[15] Broadband can

---

[10] Council of Economic Advisors, "The Digital Divide and Economic Benefits of Broadband Access," Issue Brief, March 2016.

[11] Shane Greenstein and Ryan McDevitt, "The Broadband Bonus: Estimating Broadband Internet's Economic Value," *Telecommunications Policy*, Vol. 35, No. 7, August 2011, pp. 617–632.

[12] Jed Kolko, "Broadband and Local Growth," *Journal of Urban Economics*, Vol. 71, No. 1, January 2012, pp. 100–113.

[13] Ernest P. Goss and Joseph M. Phillips, "How Information Technology Affects Wages: Evidence Using Internet Usage as a Proxy for IT Skills," *Journal of Labor Research*, Vol. 23, No. 2, 2002, pp. 463–474.

[14] S. M. Finkelstein, S. M. Speedie, and S. Potthoff, "Home Telehealth Improves Clinical Outcomes at Lower Cost for Home Health Care," *Telemedicine Journal and e-Health*, Vol. 12, No. 2, April 2006, pp. 128–136.

[15] GAO, "Telehealth Use in Medicare and Medicaid," Statement of A. Nicole Clowers, Managing Director, Health care, GAO-17-760T, Washington, D.C.: July 20, 2017b.

also be used to more accurately track disease epidemics. Various studies have demonstrated how large datasets from search engines and social media can be exploited in this way.[16]

The brief cited other benefits of broadband connectivity, including "supporting entrepreneurship and small businesses, promoting energy efficiency and energy savings, improving government performance, and enhancing public safety, among others. In addition, broadband has become a critical tool that job seekers use to search and apply for jobs." These are all benefits that will assist Puerto Rico in "building back better."

Finally, if the majority of students in Puerto Rico become proficient at using the internet, their ability to participate in Puerto Rico's economy would significantly improve. A digitally trained cohort of young people in the workforce could have a major impact on transportation, energy, telecom, emergency services, local integrated services, entrepreneurship, housing, health care, education, human capital, the visitor economy, social services, and investment in Puerto Rico.

### Improve Governance to Facilitate Doing Business in Puerto Rico and Spur the Deployment of Broadband

This section will explain COAs aimed at expanding the deployment of broadband internet services throughout Puerto Rico. The proposed COAs will take advantage of changes already made by the government of Puerto Rico to improve the governance of utilities, including telecommunications. The new agency was described in Chapter 1. This agency could work with a blue-ribbon panel of experts to evaluate and implement alternative methods to deploy broadband internet service throughout Puerto Rico (CIT 25). It could also consolidate and streamline the permitting and ROW processes for towers and the deployment of fiber-optic cable by telecommunications providers (CIT 13).

#### Create a Comprehensive Plan for Deployment of Broadband Internet in Puerto Rico

A comprehensive plan is required to deploy broadband internet effectively, using the existing fiber resources and leveraging available federal funding for broadband internet infrastructure. A COA entitled *CIT 25 Evaluate and Implement Alternative Methods to Deploy Broadband Internet Service Throughout Puerto Rico* proposes the formation of a high-profile team (a blue-ribbon panel) to prepare this plan.

A team of experts from the private telecommunications sector, or from the Telecommunications Bureau, will not be sufficient to gain the broad political and industry support that would be needed to implement a comprehensive plan. The benefit of a team that includes nationally recognized experts, as well as senior industry members, government officials,

---

[16] Jeremy Ginsberg et al., "Detecting Influenza Epidemics Using Search E-Query Data," *Nature*, Vol. 457, February 19, 2009, pp. 1012–1014; David A. Broniatowski, Michael J. Paul, and Mark Dredze, "National and Local Influenza Surveillance Through Twitter: An Analysis of the 2012–2103 Influenza Epidemic," *PLoS ONE*, Vol. 8, No. 12, December 9, 2013, e83672.

and civil society representatives, including disability community advocates, would be that the plan might gain support from many stakeholders and actually be implemented. The blue-ribbon panel might be convened by the Governor and assisted by an advisory board and an outside contractor. For example, the FCC could serve on an advisory board to the blue-ribbon panel. The blue-ribbon panel could work with carriers and regulators to gather additional information that is not currently available but would be important to analysis and decisionmaking, subject to reasonable protections. A blue-ribbon panel would have the additional benefit of being able to adjudicate among competing priorities and interests, especially among different government agencies, municipalities, and telecommunications providers, which may have different perspectives on the best path to recovery.

This COA would be a crucial first step to efficiently deploying broadband internet to education, health care, social services, the visitor economy, as well as other sectors, especially the emergency services sector.

### Streamline the Permitting and the Right-of-Way Processes

Streamlining the ROW process would make it easier and less expensive for telecommunications providers to rebuild their telecommunications network as well as to extend their broadband internet facilities in Puerto Rico. The lengthy ROW and permitting process in Puerto Rico has been identified by the New Fiscal Plan as one of the important factors that make it difficult to do business in Puerto Rico.[17] Specifically, the New Fiscal Plan aims to "Reduce unnecessary regulatory burdens to reduce the drag of government on the private sector."[18] In addition to telecommunications, the need to streamline ROW and permitting processes includes other sectors, especially transportation, energy, and water. A central ROW and permitting approval authority, with uniform, streamlined approval processes for telecommunications was endorsed by all Puerto Rico telecommunications providers who attended a May 11, 2018, meeting held at the Telecommunications Bureau. Similarly, in a June 13, 2018, meeting with the FCC Hurricane Recovery Task Force, the members of this task force identified the current ROW and permitting processes in Puerto Rico as a potential hindrance to the restoration and improvement of telecommunications infrastructure in Puerto Rico.[19] Based on the above findings, the recovery plan proposes a COA entitled *CIT 13 Streamline the Permitting and Rights of Way Processes for Towers and the Deployment of Fiber-Optic Cable.*

---

[17] "Rights of way" refers to the legal right, established by usage or by government grant or permit, to pass along a specific route through grounds or property belonging to another. Fiscal Agency and Financial Advisory Authority, "New Fiscal Plan for Puerto Rico," April 5, 2018.

[18] Fiscal Agency and Financial Advisory Authority, 2018.

[19] The FCC recognized the importance of reducing unnecessary government regulation to accelerate the deployment of broadband internet services in Report and Order, Declaratory Ruling, and Further Notice of Proposed Rulemaking in WC Docket No. 17-84, *In the Matter of Accelerating Wireline Broadband Deployment by Removing Barriers to Infrastructure Investment*, FCC-CIRC1711-04, November 29, 2017.

The Telecommunications Bureau has indicated that it controls access to many public ROWs, and this authority has been granted by regulation and statute and remains intact following its incorporation in the new Public Service Regulatory Board. For that reason, it would be appropriate (and more economical) to establish a centralized permitting and ROW office for telecommunications and internet services within the Telecommunications Bureau. We suggest an advisory status for an FCC expert, pending approval of such status by the Governor of Puerto Rico.

Some municipal stakeholders might resist a centralized Puerto Rico agency for telecommunications permitting and ROW approval, if they think it removes their control over disruption to streets in their municipality or their power to approve and exact revenues from telecommunications providers.[20] As an alternative, a uniform fee established by the new Telecommunications Bureau could be paid to municipalities. In informal discussions with members of the Telecommunications Bureau, we heard that the amounts of revenue paid to municipalities by telecommunications providers for permitting and ROW access are "not immense." A senior member suggested informally that one option to reduce the resistance of stakeholders in the municipalities to uniform and centralized permitting and ROW access would be to establish a reasonable, uniform payment that would be made by telecommunications providers to municipalities for use of municipal ROWs. This might incentivize municipalities to support a centralized authority.

### Leverage Federal Programs to "Build Back Better"

Two COAs address how Puerto Rico could strengthen existing programs by taking advantage of federal funding opportunities. The first is the E-Rate program, with funding supplied by the FCC, which supports telecommunications services (including broadband internet services) for qualified schools and libraries. The second program, which is currently funded by the Telecommunications Bureau, supports 58 municipal wi-fi hotspots throughout Puerto-Rico.

### Extend the E-Rate Program to Schools and Libraries Throughout Puerto Rico

One of our COAs addresses how to leverage the FCC's E-Rate program, which is currently used in Puerto Rico, to further promote the use of broadband internet in schools and libraries. This COA is entitled *CIT 22 Use Federal Programs to Spur Deployment of Broadband Internet to Schools and Libraries Island-Wide.* Through discussions with E-Rate program experts at the Telecommunications Bureau, the FCC, USAC, which administers the E-Rate program, and the Puerto Rico Department of Education, we determined that additional E-Rate funding could

---

[20] The Telecommunications Act of 1996 bars ROWs from being used to prevent competition. Some states have created systems that are intended to provide revenues to municipalities in a way that compensates them from a loss of revenue and authority over aspects of ROW pricing. Texas did this after a series of lawsuits brought against Dallas by Teligent Inc., GTE, and AT&T.

complement the use of the E-Rate program throughout Puerto Rico. CIT 22 proposes that a small task force would work with USAC, the Puerto Rico Department of Education, and the Telecommunications Bureau to develop a program to obtain broadband internet services for schools and libraries in all 78 municipalities.

This COA should build on the new pilot project established by Puerto Rico legislation for the use of E-Rate funding in approximately 20 schools in central Puerto Rico,[21] as well as the Puerto Rico Bridge Initiative funded by NTIA. The Puerto Rico Department of Education, assisted by the Telecommunications Bureau, has been designated by the new Puerto Rico legislation to administer the pilot project.[22] These agencies could use experience gained from the pilot project to develop a program throughout Puerto Rico. If additional assistance was needed on a short-term basis to assist the Puerto Rico Department of Education and Telecommunications Bureau with the program, a contractor could provide it. As part of CIT 22, a task force that includes employees of the Telecommunications Bureau and the Puerto Rico Department of Education could assist Puerto Rico schools and libraries with applications to receive funding for broadband internet services through the E-Rate program, as well as with the extensive follow-up compliance work required. This program would extend over at least a two-year period to reach schools and libraries in 78 municipalities. The task force could also assist with developing and implementing an evaluation method that the Department of Education, Telecommunications Bureau, and schools could use to assess the impact of the program.

The FCC's website addresses the equipment and services the E-Rate program will cover and the sliding scale of financial support that is available to schools and libraries:

> Eligible schools, school districts and libraries may apply individually or as part of a consortium. Funding may be requested under two categories of service: category one services to a school or library (telecommunications, telecommunications services and Internet access), and category two services that deliver Internet access within schools and libraries (internal connections, basic maintenance of internal connections, and managed internal broadband services). Discounts for support depend on the level of poverty and whether the school or library is located in an urban or rural area. The discounts range from 20 percent to 90 percent of the costs of eligible services. E-rate program funding is based on demand up to an annual Commission-established cap of \$3.9 billion.[23]

Potentially, all Puerto Rico schools could apply for E-Rate funding as a consortium. The E-Rate program is structured so that schools and libraries contribute a portion of the cost of services. The amount is determined by a sliding scale of economic need. Many schools and libraries in the central and rural parts of Puerto Rico may qualify for 90 percent funding under the program. In Puerto Rico, some schools cannot afford to pay teachers for an entire school year

---

[21] Resolucion Conjunta [Joint Resolution] 40-2018, R.C. de la C. 256, Puerto Rico Legislative Action, March 20, 2018.

[22] Resolucion Conjunta, 2018.

[23] FCC, "E-Rate: Universal Service Program for Schools and Libraries," February 9, 2018.

and so have no funds available for the E-Rate program. Therefore, we anticipate that additional non-FCC funding will be required for some schools and libraries to cover 100 percent of the expenses. Additional funding sources may include Department of Housing and Urban Development (HUD) Community Development Block Grants (CDBG), the U.S. Department of Agriculture (USDA), contributions by the Puerto Rico Department of Education, and nongovernment sources, such as charitable foundations.

According to new legislation in Puerto Rico, the Department of Education will decide which schools in Puerto Rico should be closed and which schools in central Puerto Rico should participate in the pilot project with the E-Rate program. The Department of Education will need to select the schools in the target municipalities that will participate in the pilot program. The Department of Education will also need to determine which schools should receive E-Rate funding in a broader program. Therefore, the Puerto Rico Department of Education should have oversight of this COA, assisted by the Telecommunications Bureau, as stated in the new legislation.

Finally, although the FCC has taken steps to simplify the E-Rate application and compliance process, it remains time-consuming and complicated. Some states, such as Florida, have hired a full-time E-Rate coordinator to manage the E-Rate process for schools. Potentially, a small team at the new Telecommunications Bureau (the former PRTRB) could serve as a central source of support for schools and libraries that engage in the E-Rate program.

### Leverage Federal Programs to Support Municipal Wi-Fi Throughout Puerto Rico

Currently, over 80 percent of Puerto Rico residents rely on their cellular telephones for business and personal communications.[24] Government-sponsored wi-fi in public spaces could provide a backup option when mobile service is not available.[25] Perhaps most important, wi-fi hotspots could serve as a way for the government of Puerto Rico to reach a majority of residents in the event of severe weather conditions or other disasters, when cellular communications may be unavailable. Such information could be provided through a main internet site, such as status.pr.[26] In addition, for those without residential broadband or those who need to access the internet while away from home, wi-fi-enabled public buildings, town squares, and parks could provide a place for browsing news, studying, accessing government services, or job information. For tourists and business travelers, access to government-sponsored wi-fi is a convenience that is increasingly expected throughout the world and a low-cost way for Puerto Rico to support its "visitor economy" through greater connectivity options.

---

[24] This information was documented in FEMA Hurricane Maria Task Force, 2017.

[25] A wi-fi hotspot is a wireless access point that provides internet access to network devices such as laptops or smartphones, typically in public locations such as cafes, libraries, airports, and hotels.

[26] As we discussed in Chapter 4, the government of Puerto Rico already provides some information to its citizens via status.pr. In that chapter we also described a course of action, *CIT 23 Data Collection and Standardization for Disaster Preparedness*, that has the goal of improving status.pr.

Wi-fi hotspots in public buildings and public areas can also increase adoption of government digital offerings, because citizens can browse information and complete forms online from a mobile device, rather than waiting in line. This COA, which is entitled *CIT 19 Municipal Hotspots*, would incorporate and expand the existing 58 municipal hotspots sponsored by the Telecommunications Bureau. Currently, resources from the Telecommunications Bureau sponsor installation of equipment and municipalities provide for recurring costs for service to a telecommunications provider. The wi-fi hotspots would be deployed in the following ways:

- Town centers: Each of the 78 municipalities in Puerto Rico has a center, in most cases next to the municipal administration buildings and a church. These "plazas" are shared gathering and recreational places for all residents and visitors. Free public wi-fi in all town centers would create an opportunity to reach residents in a place that is welcoming and accessible to all. The town center project might be implemented as a single project for all town centers in Puerto Rico, to take advantage of bulk-buying and contracting for the purchase of equipment and installation for all of Puerto Rico. It would also be important to establish partnerships with churches, municipalities, and the buildings surrounding municipal plazas to arrange for the placement of repeaters and routers in positions that ensure maximum coverage.
- Municipal and public buildings: Puerto Rico might set a goal that all public buildings will provide government-sponsored public wi-fi. This could expedite government processes by encouraging digital applications and forms and providing information resources online.

Many successful digital transformation efforts have started with extending broadband access to the population by connecting schools, ensuring affordable home access, and making wi-fi available in public areas, libraries, parks, and public buildings. Two examples are:

- Minneapolis: One of the first cities to offer ubiquitous public wi-fi, Minneapolis introduced 117 "Wireless Minneapolis" hotspots in 2010 in "places where people already gather and use computers, and places where free wireless access would encourage people to gather, including parks, plazas, schools, and businesses."[27]
- Vancouver: As a part of the city's Digital Strategy, approximately 550 locations in the downtown core and surrounding areas have free access to 10 Mbps wi-fi, with no data usage limit and no personal information required to access the network. Wi-fi-enabled locations include all 9 city-owned housing sites, all libraries, 27 community centers, 4 outdoor pools, 4 civic facilities, 3 public golf courses, 3 theaters, 2 marinas, and the City Hall campus. The service "enable[s] visitors to use their smartphones and devices to discover attractions, activities, and restaurants, for wayfinding, and to navigate Vancouver's transportation options." It also "enable[s] visitors to use social media on a real-time basis to share their Vancouver experiences with the world" and "is an important development for the Vancouver tourism industry as it moves us closer to our goal of being a 'smart' tourism destination."[28]

---

[27] City of Minneapolis, "Wireless Minneapolis," website, June 6, 2019.

[28] City of Vancouver, "Vancouver's Digital Strategy," website, 2019.

For additional examples of cities that have launched municipal wi-fi hotspots, please see Appendix C: *CIT 19 Municipal Hotspots*.

Reliable, affordable access to wi-fi would offer residents of Puerto Rico more ways to participate in every sector of society. When all citizens can access the internet, they can engage in planning activities, share their needs and views on proposed municipal or government actions, and feel more included in their communities. Moreover, internet access is one of the most important drivers for economic recovery because it provides information and opportunity. Although Telecommunications Bureau officials said that they have managed government-sponsored municipal hotspots since 1999, the Telecommunications Bureau could face resistance to expanding the program into areas currently covered solely by the private telecommunications market, a situation encountered by some municipalities in CONUS. However, a Telecommunications Bureau senior official explained that although their office pays for the initial installation of wi-fi equipment and internet service in the municipalities involved in the program, the municipalities pay the internet service provider directly after approximately two years. Thus, private telecommunications providers are paid for their services in rural or disadvantaged areas to which they might not otherwise provide service. In order to realize the full potential of affordable wi-fi access, individuals must also have access to digital devices and the digital skills to make use of the technology that connectivity makes possible.

Having addressed achieving a resilient private telecommunications infrastructure and spurring deployment of broadband internet service throughout Puerto Rico, we turn our attention to Puerto Rico's undersea cable infrastructure, another critical aspect to ensuring the resiliency of Puerto Rico's telecommunications infrastructure.

## Enhance Submarine Telecom Infrastructure

Puerto Rico's telecommunications network relies heavily on the availability of an undersea network infrastructure for connectivity to the CONUS and other regions of the world. The availability of this infrastructure has a substantial impact on Puerto Rico's economy and its ability to coordinate disaster response activities. As part of the global network, the subsea networked infrastructure around Puerto Rico also provides connectivity and alternative routes for other regions. As a core component of Puerto Rico's communications infrastructure, submarine or undersea fiber-optic cables support essential communications between Puerto Rico, CONUS, and other parts of the world whether these concern longer-term economic goals or urgent, disaster-related coordination. The next two sections describe COAs for expanding the capacity and availability of the existing undersea infrastructure.

*Expand Transoceanic Submarine Cable Network*

### Increase Availability and External Connectivity

Expanding the capacity and availability of communications relying on submarine infrastructure will help to meet Puerto Rico's goals for communications redundancy and demand for high-speed data links during critical periods, such as those that follow from disasters, either natural or manmade. Accordingly, *CIT 10 Transoceanic Submarine Cable* proposes to introduce new, very high bandwidth undersea cable(s) to Puerto Rico, situated away from San Juan: one landing point for the midterm, followed by additional ones in the long term to increase capacity and route options for communication.[29] The COA will also mitigate the known threats to existing landing stations and related infrastructure from disaster-level events.

Submarine cable–related infrastructure constitutes a primary, high-capacity communications link to and from the outside world. Both submarine cabling and related terrestrial infrastructure are vulnerable to storm damage.[30] Improvements to the undersea communications network topology would help create a highly resilient communications network with a reduced recovery time for network failures that may arise from disasters.[31] Further, redundant, high-capacity network channels to Puerto Rico will also improve the combined, overall communications capacity available within Puerto Rico and the quality of available services should a single route become impaired.

As a part of the risk-mitigation effort, the recovery effort calls for a risk evaluation for the existing as-is submarine cabling infrastructure and all related communications services to and from Puerto Rico. The preparation and implementation of a comprehensive plan to strengthen or make more resilient the submarine cabling infrastructure to Category 4 and 5 storm events are also required. This includes physical and cybersecurity (e.g., vulnerability to sabotage, communication intercept, or infiltration) of the cabling infrastructure in the vicinity of the shore and on-shore facilities, up to and including the landing station and required dependent operational infrastructure. Peak demands, operational activities, governance-related policies, and feedback from all stakeholders should be incorporated in the mitigation plan, along with

---

[29] As shown in Figure 1.4, all existing network routes to or from other parts of the world converge onto a single region of Puerto Rico, in and around San Juan. Having landing stations and associated infrastructure away from the San Juan area will allow communications with other parts of the world to be less vulnerable to disasters occurring in San Juan.

[30] Madory, 2017. Through onsite meetings in Puerto Rico and news reports, it was confirmed that outages due to storm damage impacted the submarine cabling infrastructure, which in turn resulted in outages. Several landing stations were flooded, suffering equipment and facilities-related damage. Outages were also reported in the region by other cable providers. See David Belson, "Internet Impacts of Hurricanes Harvey, Irma, and Maria," *Oracle + Dyn*, blog, September 25, 2017; and Nathália Guimarães, "Furacão Maria deixa conexão de internet lenta no Brasil," *LeiaJa*, September 9, 2017.

[31] We are referring to the network's ability to support alternate route options in the event of a disruption, not to the time it takes to physically assess and repair issues that caused the disruption and required the network to utilize new route options.

implementation of continuous monitoring and automation where feasible. New landing points and cable routes supporting fault-tolerant and resilient networking strategies (e.g., trunk-and-branch, elastic optical networks, and double-cable architectures) will also incorporate the risks and the impacts to economically sensitive areas and protected areas, as well as regulatory constraints, into the development and maintenance plans.

With a carefully considered, well-balanced near-term and forward-looking plan, several challenges can be mitigated. Coordination among stakeholders during discovery, implementation, and approaches to funding will contribute to a successful outcome when participants are actively engaged and share a common understanding through a well-structured process. If planning and construction are to span multiple hurricane seasons, stakeholders may need to form long-range operating principles for deployment of the new infrastructure. The plan will be well-balanced if single parties do not hold undue or unfair influence over one another (e.g., single-vendor dependencies pose a risk to Puerto Rico, the acceptability of which will need to be determined). Similarly, a broad consideration of the benefits desired for the region should account for the development impact to economically sensitive regions within or around Puerto Rico.

## Looking Forward

The effort described above will directly support digital transformation initiatives that rely on state-of-the-art available communication links and high-speed connectivity to regions outside of Puerto Rico. These communication links will enable opportunities to build and sustain digital governmental services for citizens, while leveraging cloud-based solutions for redundancy and computing services that are inherently resilient to disasters. They can also provide the physical communications bridge for highly coordinated emergency operations and planning for response activities required in the event of disasters on the scale experienced during hurricanes Irma and Maria in 2017.

As broadband internet access increases, the demands for this type of connectivity will grow and translate into needs for greater performance of the physical infrastructure for information exchange and communication.

The COA may also boost the economic value of the network infrastructure for Puerto Rico's economy and enhance the overall utility of Puerto Rico's submarine cable systems for U.S. and global communications. The additional capacity and capability to serve business entities could be leveraged as an asset to boost opportunities for commerce or the location of new industry.

New infrastructure for submarine networks and increased operational costs to sustain such networks will place additional demands on Puerto Rico utilities, the government of Puerto Rico, and the people of Puerto Rico people for power, funding, and workforce, respectively. At the same time, a robust communications infrastructure would mitigate the extent of challenges faced by response-and-recovery operations.

By reliably linking Puerto Rico to additional external networks, an improved submarine cabling infrastructure could have a very broad, positive cascading impact on numerous sectors, such as for the health and social services, water, and energy needs during a disaster, and the economic stability and well-being of Puerto Rico in the longer term.

### Incorporate an Undersea Cable Ring

#### Strengthen Continuity of Island-Wide Communications

Complementary to the inter-region submarine cable(s) described in *CIT 10 Transoceanic Submarine Cable*, the focus of *CIT 15 Undersea Cable Ring* is to strengthen the ability to communicate within and around Puerto Rico over high-speed and high-capacity links through a network infrastructure for a communications ring system.

The undersea network infrastructure should be developed to incorporate a point-to-point communications ring system encircling Puerto Rico. This ring network will connect landing points (present and future) around Puerto Rico to improve the availability of routes to or from Puerto Rico in the event of natural disasters in the long term. The ring system would support a resilient network with a reduced recovery time arising from subsea network failures by providing alternate route options and maintaining regional connectivity when parts of the subsea infrastructure are unavailable.[32] It would also boost both the value of Puerto Rico's network infrastructure for the Puerto Rico economy overall and the utility of Puerto Rico's submarine cable systems for U.S. and global communications.

Onsite meetings in Puerto Rico and news reports confirmed that storm damage to the submarine cabling infrastructure resulted in outages.[33] Landing stations are presently concentrated in the northeastern section of Puerto Rico. There, several landing stations were flooded, suffering equipment and facilities-related damage. Outages were also reported in the region by other cable providers.[34]

In the event of an outage, a point-to-point ring system provides alternate routes between points around Puerto Rico. While a pure ring structure can support one failure, a point-to-point topology allows regional high-bandwidth communications over available segments to continue despite multiple failures, which could be valuable in the context of a disaster, when communication is critical for life-saving activities and coordination for support.

#### Provide Complementary Network Services

The undersea ring system could also complement Puerto Rico's long-range goals. It could be leveraged as an economic attractor for industry, boosting the value and reliability of Puerto Rico's network infrastructure for the Puerto Rico economy by satisfying high-bandwidth needs

---

[32] We are referring to the network's ability to support alternate route options in the event of a disruption, not to the time it takes to physically assess and repair issues that caused the disruption and required the network to utilize new route options.

[33] Madory, 2017; Belson, 2017.

[34] Guimarães, 2017.

sought by industry, as well as the utility of Puerto Rico's submarine cable systems for U.S. and global communications. Similarly, the ring system can be leveraged in the delivery of broadband access to regions where it is generally difficult to extend terrestrial communication networks directly from interior points.

# 6. Information Technology for Critical Infrastructure and for the Digital Transformation of Puerto Rico

## Overview

For near, mid- or long-term digital transformation initiatives under consideration by Puerto Rico, key aspects of the technology infrastructure need to be developed and made available. In the first half of this chapter we describe COAs that pertain to computing infrastructure, data storage and data exchange standards, and the use of IT to make government information systems and functions more effective. The second half focuses on digital transformation through the use of IT.

The first set of COAs strives to improve the resilience of governmental functions while also laying a foundation for Puerto Rico's future. The core components reflect the essential goals for growth, stability, and capacity to drive and support new initiatives for communications and IT systems. Physical facilities will provide the computing capacity to support a variety of information and communications initiatives. Standards for digital interoperability will directly inform a modular, open approach for systems and applications design that is modern and cost-effective and acts as an economic driver in the long term. Development of information systems on a reliable, highly-scalable, extensible software architecture will enable governmental entities to more effectively respond, execute, and adapt their functions to address citizens' needs over time.

An open, modular, and standards-based common approach for information systems will improve the government's ability to deliver timely services to citizens and the private sector. The efficiencies and experience gained and shared through these efforts might incentivize economic activity in the private sector. The same systems may leverage workforce development initiatives to preserve essential trade skills developed in Puerto Rico. Success with these initiatives could inform publicly supported initiatives in health care, housing, and transportation, as well.

The second set of COAs focuses on using IT to achieve the "digital transformation" of Puerto Rico. "Digital transformation" is "the use of technology to radically improve the performance or reach of enterprises,"[1] with an emphasis on improving *performance*, not on technology for technology's sake.[2] This is especially true in the public sector, where "what separates digital leaders from the rest is a clear digital strategy combined with a culture and leadership poised to drive the transformation."[3] The government of Puerto Rico has embraced a strategy of digital

---

[1] George Westerman, Didier Bonnet, and Andrew McAfee, "The Nine Elements of Digital Transformation," *MIT Sloan Management Review*, January 7, 2014.

[2] Jeffrey Morgan, "Digital Transformation in the Public Sector," *CIO Magazine*, January 11, 2018.

[3] William D. Eggers and Joel Bellman, *The Journey to Government's Digital Transformation*, Deloitte University Press, 2015.

transformation in order to modernize government processes, improve citizen services, respond to fiscal pressures, and better coordinate post-Maria recovery efforts and preparations for the possibility of future disasters.

## Information Technology Infrastructure

### Computing Infrastructure

At the junction of digital transformation and disaster response (emergency services, public safety, and emergency operations) is the computing infrastructure relied on to support governmental goals and its ability to function smoothly, even when a disaster strikes. In this section we describe two COAs that are critical for the computing infrastructure of Puerto Rico. The first, *CIT 17 Puerto Rico Data Center*, addresses the need for a state-of-the-art, standards-compliant data center. The second, *CIT 2 Puerto Rico GIS Resource and Data Platform*, addresses the need for a comprehensive, real-time, and readily available information system that provides geolocated information to governmental public safety officials, emergency response teams, and community planning agencies.

### CIT 17 Puerto Rico Data Center

*CIT 17 Puerto Rico Data Center* satisfies the critical need of establishing a data center that will expand the government of Puerto Rico's capacity and independent ability to perform essential governmental functions and deliver essential governmental services both during disasters and in the forward-looking context of digital transformation.

CIT 17 will establish a robust, disaster-proof (Tier 3 or Tier 4),[4] cloud-enabled data center in Puerto Rico for government information systems,[5] initially targeting a small- to medium-sized capacity.[6] An independent, commonwealth-owned, and disaster-resilient data center can enable highly reliable governmental IT services for tracking, supporting, and coordinating response and recovery needs both within Puerto Rico and with organizations external to Puerto Rico, while preserving the integrity of all essential information systems. At present, there is no disaster-

---

[4] The ANSI/TIA-942 standard defines rating levels or tiers for data centers, as Rated-1: Basic Site Infrastructure, Rated-2: Redundant Capacity Component Site Infrastructure, Rated-3: Concurrently Maintainable Site Infrastructure, and Rated-4: Fault Tolerant Site Infrastructure. Telecommunications Industry Association, "About Data Centers," ANSI/TIA-942 Quality Standard for Data Centers, TIA-942.org, website, undated.

[5] Here, we use "cloud-enabled data center" to refer to a broad collection of systems, services, and tools that deliver highly scalable, highly reliable managed services for data and applications, utilizing virtualization and state-of-the-art engineering design principles for interoperability to prepare, support, and provide a computing environment for infrastructure-, platform-, and software-as-a-service within which web-based applications are typically deployed today.

[6] The U.S. Chamber of Commerce Technology Section defines small data centers as having a rack yield of 11–200, and medium-sized data centers as ranging from 201–800 racks. See U.S. Chamber of Commerce Technology Section, "Data Centers: Jobs and Opportunities in Communities Nationwide," undated.

resilient or hardened facility that reliably and comprehensively preserves important governmental data or ensures the availability of corresponding information systems. This complicates disaster response efforts (involving coordination, situational awareness, and so on) and undermines the government's capacity to assess and respond to disaster-related needs through informed decisionmaking.

There is potential to dramatically improve the planning efforts for disaster response and recovery through the evolution of existing commonwealth information systems in a cloud-based environment. If the resources made available by the data center are sufficiently utilized by software systems, COG and the efficiency in governmental services for disaster operations may be dramatically improved.

Although this COA primarily describes building one data center, multiple data centers accommodate the ability for a failover mechanism and redundancy for services. For example, two cloud-optimized data centers located on different parts of Puerto Rico will make the infrastructure more resilient to future power grid or network failures. The capacity of each data center could be individually reduced from a single, integrated data center while supporting a greater total capacity for computing services. A multi-cloud solution is also possible, wherein IT systems, applications, and related services are duplicated across multiple, different cloud-providers (including outside of Puerto Rico) for greater resiliency of governmental systems. As they are complementary methods for introducing resiliency, the methods can be designed and applied all at once or over time, depending on the strategy preferred by the government of Puerto Rico.

When viewed as a commonwealth-owned, highly available, scalable, and evolvable infrastructure for information systems, a data center is potentially a key infrastructure component that would enable Puerto Rico to pursue digitally driven economic goals and support a broad range of digital initiatives. The communications capacity of a broadband infrastructure may be leveraged directly by information systems residing in the center for workforce development initiatives and health related information systems, for example. The data center would be platform for such initiatives in addition to supporting COG, related information needs, and electronic services for coordinated activities. This COA therefore seeks to address needs for a computing infrastructure in a way that could simultaneously support multiple initiatives by providing a common, resilient computing infrastructure and capabilities for related services.

Potential challenges include determining a suitable location for a data center's facility, acquiring use rights, the availability of a trained workforce, and excessive dependence on nonresident workers. Improperly assessing power needs may also have a dramatic effect on upfront costs, leading to underutilization of resources or excessive annual expenses.

Robust power generation and supply systems, availability of broadband, and reliable submarine cable infrastructure are also necessary to fully realize the data center's role for governmental functions and digital transformation goals.

CIT 2 Puerto Rico Geographic Information Systems Resource and Data Platform

A key example of a COA utilizing a platform built in a cloud-enabled data center is *CIT 2 Puerto Rico GIS Resource and Data Platform*. CIT 2 will enable public safety, emergency response, and community planning decision-support systems to leverage geographic information systems (GIS) capabilities where practical. To accomplish this objective, it will be necessary to methodically examine how to share and use GIS data and the capabilities of GIS systems to achieve goals for disaster response and recovery-related planning decisions in all state agencies and municipalities in Puerto Rico. If GIS data is collected and shared in a uniform way across public safety, emergency response, and community planning agencies, it will greatly improve the government's responsiveness in the event of a crisis. The time needed to create actionable strategies to deal with crises such as hurricanes and earthquakes will be reduced, as will rebuilding and restoration efforts following a crisis.

As part of the implementation of CIT 2, IT/GIS technical and system/application support across all sectors will be provided by dedicated GIS personnel. This team will participate in the creation, sustainment, and enhancement of Puerto Rico's GIS infrastructure, support GIS applications and service requests from other sectors, and manage access and storage of GIS data.

Concerning access to privately owned GIS data, the appropriate agreements for information sharing and security would need to be established so that the government of Puerto Rico has access to such data to address disaster-related events comprehensively. National standards applied for information systems can serve as a basis, and the Puerto Rico Communications Steering Committee established through *CIT 24 Establish Puerto Rico Communications Steering Committee* and *CIT 18 Data Storage and Data Exchange Standards for Critical Infrastructure* may play key roles, in this regard.

A GIS system that is fully capable of seamlessly supporting the government during a disaster will need to satisfy several goals for day-to-day operation:

- *plan for sustainability* through modeling and comparing plans for resiliency and prosperity
- *improve quality of life* by understanding the movement of people, goods, and services to inform development decisions and to ensure continuity of supply chains
- *engage the citizens of Puerto Rico* by using smart maps to gather input and crowdsource ideas with citizen and business leaders.

CIT 2 provides a uniform means for all agencies to access and contribute real-time GIS data through a centralized platform. For implementation, all stakeholders will be engaged through a self-governing body that will set standards for GIS products and protocols for data use within Puerto Rico.[7] These standards and protocols are recommended to be in line with the United

---

[7] As facilitated by the steering committee proposed by *CIT 24 Establish Puerto Rico Communications Steering Committee*, for example.

Nations Economic and Social Council newly adopted Resolution of the Strategic Framework on Geospatial Information and Services for Disasters.[8]

Specific COAs from other sectors that will make use of CIT 2 are *HOU 5 Collect, Integrate, and Map Housing Sector Data, NCR 30 Create an Accessible Data Repository of Natural and Cultural Resources, ENR 11 Design and Deploy Technologies to Improve Real-Time Information and Grid Control, CPCB 3 Capacity Building to Incorporate Hazard Risk Reduction into Planning and Design, WTR 18 Invest in Stormwater System Management*, and *TXN 2 Harden Vulnerable Transportation Infrastructure.*

The availability of GIS data across Puerto Rico public safety, emergency response, and communities will be important for monitoring the status of the telecommunications infrastructure and for determining transportation routes and strategies in the event of a crisis. Granular GIS information about the power grid, public safety communications, telecommunications infrastructure, access roads, population, and location of medical facilities will all speed the government's response in a crisis. GIS data could provide Puerto Rico with real-time information to advise public safety and emergency responders. Similarly, GIS data could assist communities with rebuilding efforts following a crisis. However, government agency administrators may not move forward on collaborating to share GIS data unless a cross-agency core group of GIS technical experts works together to create a coherent plan to do so.

### Information Exchange

### CIT 18 Data Storage and Data Exchange Standards for Critical Infrastructure

The goal of *CIT 18 Data Storage and Data Exchange Standards for Critical Infrastructure* is to improve Puerto Rico's ability to plan for and comprehensively support CI needs by establishing data storage and data exchange standards for CI.

This COA will create an online data storage and data exchange standards for up-to-date, cross-sector data about CI (government and private sector) using an open, modular, and standards-based approach for information exchange, interoperability and storage. This COA will formally evaluate and develop a standardized interface definition (or definitions) for data exchanges and a mechanism by which data are stored in a standard form. This COA will also consider opportunities to leverage existing data standards and systems, such as those of the National Information Exchange Model or the DHS Infrastructure Protection Gateway. Such systems can be directly applied for a commonwealth-developed service or used as a basis for a service designed to satisfy Puerto Rico's specific near- and long-term needs.

Concerning access to privately-owned CI data, the appropriate agreements for information sharing and security will be established so that the government of Puerto Rico has access to such data to address disaster-related events.

---

[8] United Nations Economic and Social Council, "Strategic Framework on Geospatial Information and Services for Disasters," June 20, 2018.

A key aspect of a data-oriented information system for CI is its ability to receive, provide, and exchange current, accurate information across multiple sectors and thus be an integrated resource for CI data.[9] A common standard to receive high-quality data from a broad variety of sources will be essential to forming a coherent set of referential data about CI for Puerto Rico's areas of concern and to support a variety of related governmental initiatives. A standardized interface definition for information exchange will also need to consider the scope for CI data exchange, the range and types of data collected, definitions for CI, information-sharing and safeguarding policies, the capability of participating entities to provide such data, and the effort to implement the interface to exchange data between systems.

This effort will help provide comprehensive, quantitative, and data-driven assessments of the CI assets in Puerto Rico. A quantitative analysis relying on such a comprehensive picture could, for example, support burying aerial fiber (*CIT 21 Government-Owned Fiber-Optic Conduits to Reduce Aerial Fiber-Optic Cable and Incentivize Expansion of Broadband Infrastructure*), which would increase resilience in the face of severe weather. Accurate situational awareness of infrastructure issues could also inform emergency response activities prior to or after a disaster. Private companies would also know when and where PREPA, Puerto Rico Aqueduct and Sewer Authority (PRASA), and Puerto Rico's Department of Transportation may be trenching. Governmental and private-sector entities providing support for, or involved with, CI assets would also benefit, as would initiatives for the development of information systems relying on detailed CI information, such as described in *CIT 2 Puerto Rico GIS Resource and Data Platform*.

CI data can help streamline planning, budgeting, and supply-chain logistics and optimize recovery-related activities in a coordinated fashion. These improvements would benefit multiple sectors, such as water, energy, health and social services, transportation, and housing. The success of the system that will provide access to that CI data will rely on the standards for interoperability, comprehensiveness of policies for information-sharing and safeguarding, and the approach to governance adopted for government information systems predicated on an integrated platform (which improves the availability of information services and governmental functions).[10] A poorly developed implementation approach may undermine the value of the system, as would a poorly incentivized or funded mechanism to support the initiative. A new system may lead to high initial costs and to interoperability issues, such as with FEMA information systems.

---

[9] Here, an emphasis is being placed on the technical methods, tools, and services prepared for or applied to the detailed technical management of critical infrastructure data and corresponding aspects required for its dissemination. Higher-level functions, systems, and processes might naturally make use of CI information for assessments, analysis, reporting, and so on.

[10] *CIT 24 Establish Puerto Rico Communications Steering Committee* seeks to establish a steering committee for addressing governance issues such as the ones raised by this COA. *CIT 18 Data Storage and Data Exchange Standards for Critical Infrastructure* may be instrumental for development of the GIS resource and data platform described in CIT 2.

Given a broad range of private- and public-sector entities that are potential sources for CI data, there is an inherent challenge regarding stakeholder engagement through implementation and ongoing sustainment of standards and agreements. Lack of inclusion or availability of key stakeholders able to make authoritative decisions regarding standard interfaces, interoperability requirements, and data needs or to reconcile and prioritize functionality may undermine the quality of data incorporated and the final performance (measured against governmental "business" needs) of the system. Similarly, new business processes may need to be developed or changes to existing ones may be required to ensure the exchange of current information.

Availability of a secure data center for governmental information systems (CIT 14) and formalized governance for informed, authoritative decisions (CIT 24) will be necessary for successful implementation of this COA.

## More Effective Government Information Systems and Functions

Currently in Puerto Rico, challenges exist with the delivery of government services, the availability of timely, accurate, and coherent system of records management, and the use of legacy information systems for infrastructure and information management.

In December 2017, the Governor of Puerto Rico signed Act 122, which "seeks the reduction of 118 agencies to 35 more efficient ones."[11] In April 2018, the Fiscal Board noted that with 116,500 employees across 114 Executive Branch government agencies, the size of Puerto Rico's government is "outsized compared to the actual service needs of the people of Puerto Rico" and an outlier compared with U.S. states.[12] The Fiscal Board agreed with the Governor about the need for agency consolidation and right-sizing "to deliver services in as efficient a manner as possible."[13] In any jurisdiction, fundamental agency restructuring would require a corresponding digital reform. In Puerto Rico, given the needs of the recovery and future disaster preparedness, strategic digital reform is essential.

Agency consolidation and reorganization provide an opportunity to establish a "whole-of-government," people-centered, digitally designed, and data-driven approach that will improve service and service delivery, be more cost-effective, better serve the public, and make better policy. And generally, all sectors are expected to benefit from a holistic data-driven governing approach that is coordinated, effective, integrated, and responsive to feedback.

This section introduces two closely related COAs that address these issues. They introduce service cultures for coordination-by-design, which include data-driven, outcomes-based whole-systems governing to continuously improve governmental services and their delivery and apply

---

[11] "Governor of Puerto Rico Makes New Government Bill into Law" (Act 122 of 2017, New Government of Puerto Rico Act), La Fortaleza, Oficina del Gobernador, San Juan, December 18, 2017.

[12] Financial Oversight and Management Board for Puerto Rico (FOMB), *New Commonwealth Fiscal Plan*, April 2018, pp. 65–66.

[13] FOMB, 2018, pp. 68–69.

resources more effectively. The data-driven, customer-oriented approach described is intended to better serve public needs and to directly inform policymaking. By drawing on quantitative feedback from all of the governmental services being provided, improvements to services will, in turn, be guided and data-driven for continuous adaptation in an informed, methodical fashion.

### CIT 14 Consolidated Government Information Systems

The COA *CIT 14 Consolidated Government Information Systems* will establish and implement an open, modular, standards-based platform for information systems, and it will consolidate government systems across Puerto Rico (all 78 municipalities and the government). Through software and associated information systems the platform will natively enable interoperability, consistent standards and policies for information and data management, and the scaling of the system overtime at reduced expense and effort.

Presently, the landscape for governmental information systems is populated by a heterogeneous mixture of legacy systems that do not have an adequate capacity to scale, evolve, or interoperate for governmental needs. A fully integrated approach that uses a common, standards-based platform for information systems could reduce operating costs for all municipalities and simultaneously enable highly reliable governmental functions, including the coordination of response and recovery activities within Puerto Rico and externally (for example, in supply-chain logistics).

Development of a common platform with cross-agency and cross-municipality collaboration may be challenging without adequate stakeholder engagement. Goals, requirements, and capabilities of the underlying system will need to match today's needs with future ones and be prepared in a way that clearly defines usage and tools of the system, while enabling enough flexibility for specific configurations to meet needs at the level of a municipality or agency.

A standardized platform based on modern, cloud-based techniques and/or services has the potential to dramatically reduce the implementation costs of modernization efforts and to improve the sustainability and long-term value of information systems. The common platform will be enabled by the computing infrastructure improvements of *CIT 17 Puerto Rico Data Center* and the development of standards for information exchange of *CIT 18 Data Storage and Data Exchange Standards for Critical Infrastructure*. A substantial fraction of the cost for this COA will consist of transferring the information contained in the multiple systems from Puerto Rico agencies and the municipalities to the common platform.

Transition from existing systems to new ones may encounter opposition or fundamental changes to business process and organizational roles. Such issues will need to be addressed in a way to facilitate progress while addressing stakeholder concerns (see *CIT 33 Government Digital Process Reform*). Access to data will require the buy-in of many government of Puerto Rico agencies. An additional important issue will be properly addressing data security.

An experienced, dedicated workforce will also be required for implementation. Lack of skilled, experienced individuals may undermine the activity in terms of the integrity of the new platform, its perceived value, and actual effectiveness in meeting goals. The availability and participation of informed, authoritative decisionmakers will also be critical for the success of the effort.

Robust availability of broadband communications networks to all municipalities is necessary for a successful implementation of this COA.

### CIT 33 Government Digital Process Reform

In close coordination with CIT 14, the COA *CIT 33 Government Digital Process Reform* will (1) adopt a systems approach to government technology, with an emphasis on human-centered digital process design and data standardization to drive policy decisions; (2) establish people-centered, digital design and data science teams within the government of Puerto Rico to tackle cross-cutting policy and operational issues, coordinating different projects with agencies—especially during the agency consolidation process to assure clear accountability; and (3) open up government services internally and externally through the use of standards-based APIs, where appropriate, and for feedback to drive continuous improvement.

CIT 14 and CIT 33 will work together. CIT 33 emphasizes organizational culture and process change, while CIT 14 emphasizes technical aspects of consolidating information systems to run on a "uniform" IT platform (e.g., adopting a common implementation framework and physical infrastructure design). Consolidating information systems and corresponding technical challenges (CIT 14) will complement the effort to revise governmental methods and policies to drive how systems can or should be utilized with the intent of improving government services provided to citizens (CIT 33).

Incorporating a digital and data-driven approach to the operation of multiple government agencies allows for an outcomes-focused reframing of government priorities, rather than simply attempting to reinstitute the processes of the past. While the transformation will require significant culture and change management, as well as systems redesign, it is an opportunity to do things differently and introduce a data-driven, systematic approach around which public servants and their agencies can rally. As described later in this chapter, the COA for *CIT 32 Digital Citizen Services* will complement CIT 14 and CIT 33 by focusing on citizen needs for digital interactions with governmental services and their experience when using them.

It should be noted that a design-centered approach that reexamines government processes and policy-related practices is favored to achieve goals more fully, because modernization without reevaluating existing processes for their incorporation into a digital era is more likely to preserve existing challenges and miss opportunities for process improvements and architectural changes that capture current needs for information flows.

Similar to *CIT 32 Digital Citizen Services* and *CIT 14 Consolidated Government Information Systems*, this COA could potentially lead to internal friction, "change fatigue," workforce reduction, shift or reduction in influence and authority of existing leadership, and perceived loss of control by existing agencies. It will be crucial to ensure access and safety for the most vulnerable and underserved, as well as protection of citizen privacy and strong cybersecurity.

## Information Technology for the Digital Transformation of Puerto Rico

### *Roadmap for the Digital Transformation of Puerto Rico*

The recovery plan proposes a COA to create a roadmap for the digital transformation of Puerto Rico: *CIT 16 Government Digital Reform Planning and Capacity Building.*[14] Creating such roadmap will require (1) setting achievable goals and metrics for success; (2) a rigorous assessment of needs, costs, feasibility, and cultural and legal issues to be addressed; (3) the establishment of a clear strategy that can be communicated and championed both inside government and with the public.[15]

PRITS will lead the effort under this COA to create the roadmap and to establish priorities, assess needs, costs, and feasibility for a government-wide digital transformation strategy. This effort will require strong collaboration with private industry.

The first step in this COA will be to increase the human capacity within PRITS to add expertise and staffing to create the roadmap for digital transformation of Puerto Rico, with emphasis on digital transformation of the government. Thus, SMEs on data science, data architecture, IT architecture, and cybersecurity will be added, as needed.[16] They will then be responsible for the following:

- engaging experts who have led similar transformation processes at the federal government or U.S. state level and incorporating their lessons learned and best practices in creating the roadmap
- convening stakeholders and establishing shared goals for the digital transformation
- assessing needs, ideas, priorities, and feasibility with input from government agencies, members of the public, businesses, and stakeholder organizations; feasibility study to include
  - anticipating issues to be addressed such as governance and differences in cultures and processes across government agencies
  - estimating scope and cost of retraining or re-skilling public sector workers for new digitized processes
  - assessing project time frames
  - evaluating their potential benefits
- evaluating successful models from other jurisdictions
- establishing metrics of success and key progress indicators (KPIs)

---

[14] The phrase term "capacity building" in the context of this COA refers to increasing the human capacity of PRITS to add the expertise and staffing required to create the roadmap.

[15] Implementing this COA will require close coordination with the steering committee proposed in CIT 24 as well as coordination with the efforts to make government information systems and functions more effective (CIT 33).

[16] Concerning increasing human capacity, a rigorous review of the new roles and responsibilities will be performed.

- investigating whether governmental approvals or executive or legislative change are required
- reviewing the budget and resources required for successful implementation
- conduct ongoing outreach, training, and other management strategies to ensure successful implementation
- provide ongoing metrics tied to KPIs.

SMEs within PRITS or in coordination through PRITS will recommend implementation approaches for the COAs proposed in this report under the heading "Information Technology for the Digital transformation of Puerto Rico." These COAs are expected to lead the digital transformation in the short term. Additional projects may be considered.

This COA provides a framework for Puerto Rico to benefit from best practices and avoid the pitfalls that have beset digital transformation efforts in other jurisdictions. It will build stakeholder buy-in; solicit ideas, needs, and issues from a broad set of participants; and provide a comprehensive strategy with associated metrics to improve chances of success. Some of the pitfalls that have beset digital transformation efforts in other jurisdictions include

- "lack of strategy," which was identified as leading barrier to early-stage organizations taking full advantage of digital trends[17]
- "culture," which was cited by 85 percent of public-sector leaders surveyed as a challenging aspect of managing the transition to digital[18]
- "lack of system-wide prioritization" and navigating agency silos, which were challenges for New Zealand's digital transformation.[19]

The COA seeks to mitigate pitfalls for other COAs related to digital transformation. However, accommodating other sectors will require costs to the sectors that engage in this effort. Moreover, there may be lack of support by some government agencies.

### Citizen-Oriented Digital Initiatives

### CIT 26 Wi-fi Hotspots in Public Housing and Digital Stewards Program

In Puerto Rico, many public housing residents lack options for internet access without expensive data plans. Government-sponsored wi-fi in public areas provides reliable access and also represents a potential, postdisaster priority connection point for communications to the residents of Puerto Rico. This COA seeks to directly engage citizens to manage this access within communities and to utilize it for workforce development, educational or professional development, and entrepreneurship within the community.

---

[17] Eggers and Bellman, 2015.

[18] Eggers and Bellman, 2015.

[19] Darryl Carpenter, "Overcoming the Challenges of Digital Transformation—Lessons Learned from the NZ Government," University of Melbourne Power of Collaboration series, February 6, 2018.

*CIT 26 Wi-fi Hotspots in Public Housing and Digital Stewards Program* will establish a Digital Stewards program in Puerto Rico that is modeled after successful initiatives in Detroit, Michigan, and Red Hook, New York.[20] These programs create a way to improve connectivity options in public housing and provide valuable digital skills and employment experience for people in those communities. In Puerto Rico, its goals will include teaching public housing residents how to use the internet from a technical perspective and for education and entrepreneurial purposes in the community, as well as about cybersecurity and privacy best practices. Digital Stewards will also install and service wi-fi hotspots in public housing, following previously successful models. Wi-fi access and the ability to access computers, tablets, and smartphones can both help decrease the "digital divide" and provide a priority point of connection and coordination after disasters.

The Digital Stewards program began in Detroit and was successfully implemented in Red Hook, New York, where, in 2017, 92 percent of Digital Stewards agreed or strongly agreed that the program helped them "learn skills that allow them to succeed in the workplace and to make a difference in their neighborhood."[21] Within six months of completing the program, 77 percent remained employed or were actively pursuing further education.

Following Hurricane Sandy in 2012, the Red Hook Initiative (RHI), the nonprofit that leads the Digital Stewards program, became a community hub for disaster response, providing a gathering place for residents to charge phones, fill out FEMA forms, and check in with family members.[22] RHI helped to organize volunteers, posted updates on social media, and increased the reach of its mesh wi-fi to serve more than 1,000 people per day.[23] The Red Hook community wi-fi program's contribution after Sandy was recognized at a White House–hosted FEMA roundtable on emergency response best practices.

Affordable access to wi-fi will have the following benefits, by sector:

- CPCB: When citizens can access the internet they can participate in planning activities, share their needs and views on proposed actions, and feel more included.
- economy: Internet access is one of the most important drivers for economic recovery, providing information and opportunity.
- health and social services: Internet access can tangibly improve health.[24] Many hospitals and health plans (including those serving Medicaid patients) are expanding use of

---

[20] Digital Stewards are community organizers and community stakeholders who are trained in wireless technologies so that they can build their own local network and share internet connections with their community. The program sponsors provide the equipment and the training required by the Digital Stewards. See also Allied Media, Detroit Community Technology Project, Digital Stewards Training Program, undated; Red Hook WiFi, "Mission: Resilience, Opportunity, Community and Social Justice," 2019.

[21] Red Hook Initiative, "Digital Stewards," website, 2019.

[22] Red Hook Initiative, "A Community Response to Hurricane Sandy, Red Hook Initiative Summary Report," 2013.

[23] Red Hook Initiative, 2013.

[24] Evan Sweeney, "AMIA Sees Internet Access as a Social Determinant of Health," *Fierce Healthcare*, May 24, 2017.

internet-based communications tools to improve their members' health. In addition, new sensor technology can send health indicators to medical professionals and thereby improve compliance with doctors' recommendations. Especially in disaster-prone areas, there is significant need to ensure that chronic patients (who, for example, need insulin, oxygen, or dialysis) can be easily located and reached in case of emergency.

- public buildings: Wi-fi in public housing can create opportunities for expanding digital security systems, digitizing updates and outreach to housing clients, and increasing the ability to monitor building condition, maintenance needs, and energy use.

Through this COA participants will maintain and promote wi-fi connectivity in their communities, coordinate tech-related projects and events, gain tech skills and knowledge, build employment experience, and act as liaisons for internet connectivity and maintenance of hotspots in their communities.

The COA also consists of creating wi-fi hotspots with routers and repeaters for public housing in Puerto Rico. Already, legislation authorizing funds for rebuilding damaged public housing requires that "any substantial rehabilitation, as defined by the code of federal regulations 24 CFR 5.100, or new construction of a building with more than four rental units must include installation of broadband infrastructure."[25] The COA will be implemented by either Vivenda (Puerto Rico Department of Housing) or a nonprofit partner. They will lead the creation of a Digital Stewards program in public housing developments throughout Puerto Rico.

For a successful implementation of this COA, it will be important to ensure awareness of online safety, security, and privacy and to establish policies for appropriate use of systems. Moreover, existing commercial internet providers should not be displaced, and issues surrounding ownership of equipment and potential theft or vandalism will also need to be addressed. The program owner for the Digital Stewards program will be responsible for addressing these risks by implementing it according to best practices and lessons learned from similar efforts undertaken elsewhere.

It is also important to note that individuals will require access to digital devices and the digital skills to make use of the technology that connectivity makes possible. The Digital Stewards program will work to identify and address these types of needs based on the conditions in the residences where they are based. This could include establishing a publicly accessible "lab" for residents to use or refurbishing devices for use by residents. The access issue itself presents an opportunity for stewards to display ingenuity and entrepreneurship to address a community need.

---

[25] HUD, "Allocations, Common Application, Waivers, and Alternative Requirements for 2017 Disaster Community Development Block Grant Disaster Recovery Grantees," Docket FR-6066-N-01, *Federal Register*, Vol. 83, No. 8, February 9, 2018.

*CIT 27 Study Feasibility of Digital Identity* will study and evaluate secure methods to create a digital identity. It will also assess potential acceptance by Puerto Rico's citizens and business community of a secure, strong digital identity, based on resilient power and communications, to facilitate government and private sector transactions. The study will be designed to identify potential pitfalls, including concerns of privacy and security, as well as public perception.

The creation of a secure digital identity will facilitate digital transactions and reduce transaction costs and the potential for fraud and identity theft. The Office of CINO has described digital identity as "the key" to successful digital services implementation and has identified this as a top priority. The government of Puerto Rico has also identified identity verification as a key frustration for citizens and an "essential" element of the digital transformation of government.[26]

When viewed in this way for application in governmental functions, secure digital identities are a key component to digital transformation, facilitating financial transactions, contracts, and government services. A secure digital identity can increase accuracy and reduce costs associated with validation and access to government services, especially in disaster recovery when paper records can be inaccessible. By implementing a secure digital ID, Puerto Rico has an opportunity to lead in this new area.

A secure digital identity allows for secure login for all government services, prevents duplication, and allows for an "ask/update once" so that information is accurate and updated across government services. It also allows for more efficient business formation processes and transactions by eliminating the need for in-person authentication.

Any transaction between two or more parties requires identity verification. The World Economic Forum emphasizes the role of identity for establishing a trusted digital environment: "As the number of digital services, transactions and entities grows, it will be increasingly important to ensure the transactions take place in a secure and trusted network where each entity can be identified and authenticated."[27] Recently, two experts identified "a unique, uniform digital ID" as one of "three pillars of digital transformation."[28] Jurisdictions at the vanguard of digital transformation have prioritized digital identity through a variety of approaches ranging from (1) systems that are centralized and owned by the government,[29] (2) systems that are decentralized and managed jointly by the government and private sector, mainly banks,[30] and

---

[26] Puerto Rico Chief Innovation Officer, interview with authors, May 31, 2018.

[27] World Economic Forum, "On the Threshold of a Digital Identity Revolution," January 2018, p. 5.

[28] William D. Eggers and Steve Hurst, "Delivering the Digital State: What If State Government Services Worked Like Amazon?" *Deloitte Insights*, November 14, 2017.

[29] See Unique Identifier Authority of India, webpage, undated.

[30] E-Estonia, "We Have Built a Digital Society and So Can You," undated.

(3) systems that are established by the government and rely on nongovernment entities to provide validation.[31]

### CIT 32 Digital Citizen Services

This COA seeks to expand the scope of PRITS to include a focus on citizen-centered services and prioritize a "one-stop-shop" experience for accessing government services and information in an easy-to-use fashion. Citizens increasingly expect their experience with government services to be as easy, quick, and simple to understand as the user interfaces for consumer services. This COA will implement best practices for ensuring digital inclusion and accessibility, such as the ability to access government services from mobile devices, while streamlining experiences for effective online government services.

The expanded PRITS would work across agencies to complete projects with an examination of citizens' needs and experiences from start to finish, ensuring privacy, security, and agile thinking is implemented at the start of each project that develops iterative product releases where appropriate (some policy efforts may require alternative project methodologies), structuring budgets and contract deliverables to be outcomes-focused, and using data to drive decisions that improve the experiences of citizens engaging in government services.

The drive for updating the experience for citizens interacting with government is, for the most part, coming from citizens themselves. From booking an airline ticket, applying for a credit card, or paying bills, individuals are increasingly used to a seamless online experience that requires minimal wait with minimal complications.

While some government services have been gradually brought online in Puerto Rico, the experience from one government agency to the next is inconsistent. Some of the most important interactions, such as renewing a driver's license, still require an in-person application and can take hours. The same is true for many processes required for setting up a business or applying for government services. This has been exacerbated by budget-cutting moves that shut down district government offices and required people all over Puerto Rico to travel to the capital of San Juan in order to process documents or submit applications. These inefficient processes are not only frustrating for citizens and visitors, but also drive up the cost of government and contribute to greater data inconsistencies.

The COA proposes a PRITS-led embrace of these principles for the digitization of government services, led by PRITS product teams embedded within the agencies where they are working. This will require (1) learning from experts who have been a part of creating digital services in other jurisdictions; (2) communicating the new approach within government *and* valuing the expertise and experience of SMEs; (3) project teams undertaking new projects with

---

[31] Government of UK, "Taking GOV.UK Verify to the Next Stage," Government Digital Service, October 11, 2018.

adherence to the United States Digital Service (USDS) principles;[32] and (4) evaluating effectiveness, cultural acceptance, and success.

If successful, this COA could also increase government transparency and accountability and public trust; increase the use and adoption of digital services tailored to citizens' needs and experiences; streamline internal government processes; and reduce the number of people and resources required for rote government services, thus allowing for more attention to human interaction and other challenges to improve the public experience overall.

Digitized government services would impact all sectors by providing better data for planning and evaluation, transparency and accountability, and more efficient citizen interaction for services, as well as saving money, streamlining reporting, and reducing time-to-decision.

By embracing the principles of citizen-centric digital services and at the same time by getting off legacy systems that can impede delivery of better digital services—to be accomplished by *CIT 14 Consolidated Government Information Systems* and *CIT 33 Government Digital Process Reform*—Puerto Rico will be able to achieve long-term cost savings, increased public transparency, and better data for policy making. Digitizing government transactions can save time and resources on the front end for a citizen who no longer spends hours in line or mailing in paper forms and for a clerks manning a desk or processing mail. Digital services also enable standardization, which can fuel automation, real-time analytics, data-sharing, and process optimization.

An increasingly digitized government sector will also provide more opportunities for highly skilled graduates to remain in Puerto Rico after graduation and for the members of the diaspora who want to apply their skills at home.[33] Residents of Puerto Rico have been involved in the evolution of digital services in the United States for many years, as catalogued by former CIO, Giancarlo Gonzalez.[34] The first Chief Technology Officer for the United States, Todd Park, recruited technologists from Puerto Rico to join the first USDS cohort.[35]

This COA could potentially lead to internal friction, "change fatigue," workforce reduction, and perceived loss of control in existing agencies. It will be crucial to ensure protection of citizen privacy and strong cybersecurity.

---

[32] United States Digital Service, website, undated.

[33] Jim Glade, "Puerto Rico Turns to Tech and Entrepreneurialism to Revitalize the Economy," *TechCrunch*, January 15, 2017, writes that "the 2012–2013 Global Competitiveness Report from the World Economic Forum ranked Puerto Rico third in the availability of scientists and engineers. According to Lucy Crespo, . . . 'Puerto Rico graduates 22,000 STEM students and 60 to 70 percent leave the island.'" See also Lisette Alvarez, "As Others Pack, Some Millennials Commit to Puerto Rico," *New York Times*, August 5, 2017.

[34] Gonzalez has written extensively about ongoing efforts to digitize government services in Puerto Rico, and catalogued milestones and best practices through his *Dear Fiscal Board*, blog, 2009–present.

[35] Giancarlo Gonzalez, "PR—Former CIO Team Is Recruited by USDS," *Dear Fiscal Board*, blog, December 7, 2015.

## Health Initiatives

For effective delivery of health care following a disaster, access to patient data both within and outside of clinical facilities is critical. The COA *CIT 29 Health Care Connectivity to Strength Resilience and Disaster Preparedness* proposes a network to improve the resilience of clinic-to-clinic communications for improved coordination and service outcomes in health and medical domains following a disaster, such as the one incurred by Hurricane Maria.

This COA has two complementary objectives. First, it aims to provide robust, resilient, multimodal "mesh" communications connectivity to the 86 community clinics across Puerto Rico, using satellite and low-power radio and line-of-site technologies, to complement connectivity that is available through the telecommunications infrastructure or to provide redundancy when such infrastructure is damaged.[36] Each clinic will have a satellite uplink to the internet plus radio connectivity to other clinics for COOP and coordination during emergencies and provide the ability to connect to existing hospitals.

Second, this COA will use the increased connectivity and IT to ensure real-time access to clinical data—including mobile and telehealth—from many access points to improve clinical care delivery and to better adapt to disaster impacts. This COA will also take advantage of the increased connectivity and IT to support situational awareness, behavioral health, environmental monitoring, and social services, as well as other services' delivery when bandwidth permits and during hours when clinics are closed.

Efforts to implement this COA need to be coordinated with health authorities and with ongoing private-sector efforts by hospitals and insurance companies. This COA will leverage the network of health clinics spanning Puerto Rico and seek to improve care and emergency response capabilities. It will also seek to enable medical innovation and provide real-time clinical electronic health record access and telehealth, behavioral health, and other social services delivery. The proposed network will bolster access to local services and situational awareness due to a disaster, limited connectivity, or loss of power.

A mesh network, which is deployable in the near term with currently available technology, could also enable community health centers to function as local disaster response and recovery centers. Network components are also straightforward to deploy at facilities or to bring online after a disaster, if components such as satellite dishes are not utilized during normal operations. The communications facilitated by the mesh will enable more access to medication information, allow prioritization of and allocation of resources for dialysis patients, and provide remote monitoring of individuals who are unable to reach a clinic. If nodes in the clinic-mesh network are viewed as a group of facilities acting as access points spread throughout Puerto Rico, there are also potential opportunities for broader, health-related initiatives in a variety of areas, such as

---

[36] A mesh network is a type of network in which a device (node) both transmits its own data and serves as a relay for other nodes. In the event of a hardware failure—e.g., one node fails—many routes are available to continue the network communication process.

telemedicine within Puerto Rico or with CONUS counterparts, medical and health-related population studies, food-safety and agricultural monitoring, and health-related community engagement or education.

Mesh network capacity would enhance access to other government services (benefiting municipalities). This COA would also leverage clinic connectivity for resilience, education, and emergency response (benefiting CPCB).

### Forward-Looking Economy and Human Capital

A broad, coordinated push to provide technology access and digital and coding skills-training will help develop a digitally literate employment pool for recruiting or expanding tech-reliant industries, consistent with the idea of a "human cloud"—a skilled digital workforce that can work with companies around the world from Puerto Rico. This initiative reinforces the message that Puerto Rico is "open for business" and intends to welcome new technologies and a digital workforce to support economic growth through innovation. According to the Information Technology and Innovation Foundation, signs of an emerging innovation economy include:

- entrepreneurs taking risks to start new ventures
- companies funding breakthrough research
- regional clusters forming to foster innovation
- research institutions transferring knowledge to companies through patents
- policies fostering widespread adoption of new technologies and broader digital transformation of society.[37]

*CIT 28 Innovation Economy/Human Capital Initiative* will provide people in Puerto Rico with the skills to work and participate in an increasingly digital society through skills-training in schools, access to technology, and cultivating a culture of entrepreneurship.[38] This COA will create a public-private initiative to provide digital skills training, entrepreneurship programs, and access to new technologies for people throughout Puerto Rico through a network of innovation hubs and entrepreneur centers, training partnerships with schools, and outreach via mobile labs to rural and underserved areas. The initiative will follow successful models such as those implemented in Tennessee, Oklahoma, and Atlanta, to name a few.

This initiative is to be led by PRITS, in cooperation with the Department of Economic Development and Commerce [Departamento de Desarrollo Económico y Comercio], as a public-private partnership. The initiative itself should create a nonprofit entity, with an appointed board that includes representatives from government, the private sector, the investment community,

---

[37] Information Technology and Innovation Foundation, "Innovation Economics: The Economic Doctrine for the 21st Century," undated.

[38] "Innovation economy" can be understood as establishing economic growth through innovation initiatives.

and academia, and the appointment of a CEO and provisions of staff to make the initiative a success.[39]

The Puerto Rico Innovation Economy/Human Capital Initiative will examine successful models in other parts of the country and establish an action plan for Puerto Rico, which may include a network of innovation hubs, innovation partnership with schools for digital skills and coding courses, and mobile innovation labs. A network of innovation hubs across Puerto Rico would provide residents with the opportunity to learn to use new digital technologies, access to training, tools, and co-work space, and foster the creation and scaling of new technology businesses. Following other successful models, the hubs would include makerspace and co-work space and would regularly host events such as entrepreneurship trainings, mentorship, and pitch competitions.[40] Puerto Rico already has several successful private co-work and makerspaces that could be candidates for joining the innovation hubs network. The key difference between existing private hubs and those that choose to participate in the innovation network will be that the latter will be eligible for matching resources tied to metrics measuring training, outreach, and technology access for the public; new businesses created or incubated as a result of the hubs; and an opportunity to partner with local schools for digital skills and coding training programs.

Innovation partnership with schools would sponsor a digital-skills-training platform that will allow students to sign on, learn skills, and earn badges at their own pace. Such an arrangement would involve

- teams based at innovation hubs partnering with local schools to support teachers who serve as leads for the coding program
- curriculum that provides for a school-wide ranking of skills attained by the students, which can in turn be used this to recognize and reward student coding champions with prizes such as donated devices, awards ceremony, and/or scholarships[41]
- ultimately extend availability of the digital learning platform to all residents of Puerto Rico to provide the digital skills required for jobs in technical fields.

Mobile units that bring training and innovation opportunities to rural areas, public housing developments, schools, festivals, and town centers would include

- "biz buses" that offering training in areas such as business formation, accounting, internet protocol, social media, marketing, and creating a website
- "innovation labs" that allow participants to explore and use new technologies, including computer numerical control (CNC) routers, laser etchers, 3D printers, robots, drones, science, technology, engineering, and mathematics (STEM) kits, and virtual reality sets.

---

[39] Board members for a similar program in Tennessee are appointed by the governor and state legislature. See Launch Tennessee, "Launch Tennessee Announces New Board Members," March 22, 2018.

[40] A makerspace is a place where people come to work openly, creatively, and together on technology projects.

[41] The "Dev Catalyst" has been running successfully in Tennessee since 2013. Dev Catalyst, website, undated.

"TN Driving Innovation," "a program to spark curiosity and inspire Tennesseans to realize the specific roles that they can play in the future economy," was successfully launched in Tennessee.[42]

Innovation hubs, partnerships with schools, and mobile labs provide a point of entry for people throughout Puerto Rico to learn about and develop new technologies, start high-growth businesses, and enter the digital workforce. As a result, this COA will have positive impacts on multiple sectors, including community planning and capacity building, health and social services, and economic.

The initiative will create a framework for opening the opportunities of the innovation economy to communities in Puerto Rico through access to physical locations (innovation hubs) that will serve as centers for activities that demonstrate new technologies and provide training and mentorship to encourage the creation of new products and businesses.

The partnership with schools will introduce students and teachers to a new, experiential way to learn digital skills. This will help increasing the number of skilled workers in Puerto Rico and will provide greater opportunity for residents to "invent" new technologies, start their own businesses, and contribute to a modern, digital Puerto Rico.

Mobile innovation labs will address the growing disparity between those who are exposed to and understand emerging technologies and those without access by allowing people of all ages and backgrounds to be introduced to new digital technologies in their own communities.

The initiative will require that physical equipment be protected from theft and extreme weather. The initiative will also require dynamic leadership and relationships with stakeholders in Puerto Rico and in the tech community worldwide.

It will also require a considerable amount of coordination across school districts and classrooms. There may be cases in which some schools do not have teachers who feel comfortable enough with technology to be the lead for the project (this can be addressed through training and organizational support).

Finally, participation in the initiative will require connectivity and access to a desktop or laptop computer. Its success will depend heavily on other COAs that offer wi-fi, expanded access to computers, and the ability to borrow computers from public libraries; see *CIT 19 Municipal Hotspots* and *CIT 26 Wi-fi Hotspots in Public Housing and Digital Stewards Program*.

### Digital Initiatives to Improve Resilience

Due to its location, Puerto Rico is vulnerable to natural disasters and needs innovative capacity-development approaches to enhance resilience and overcome human capital and investment constraints. Hurricane Maria alone caused over $100 billion in estimated damage.[43]

---

[42] "Tennessee Is Driving Innovation and We're Bringing It to You," undated.

[43] See Jeff Masters, "Hurricane Maria Damage Estimate of $102 Billion Surpassed only by Katrina," *Weather Underground*, November 22, 2017.

This section addresses two initiatives that take advantage of IT to improve the resilience of Puerto Rico. They are *CIT 30 Resiliency Innovation Network Leading to Development of Resiliency Industry* and *CIT 31 Resilience/e-Construction Learning Lab.*

### CIT 30 Resiliency Innovation Network Leading to Development of Resiliency Industry

This COA will create a Resiliency Innovation Network (RIN) across Puerto Rico to build on the existing Puerto Rico Science, Technology, and Research Trust (PRSTRT) and university facilities to teach, test, and refine existing resiliency products and services, as well as to develop new ones to enhance capability and stimulate new commercial ventures.

This COA will encourage established companies to set up new operations or expand into Puerto Rico, opening new export opportunities and import substitution. This COA could also empower the residents of Puerto Rico at a community and individual level to learn, tap into resources, and help one another. This real-world experience could give Puerto Rico a competitive "resiliency industry" advantage.

RIN's first goal is to create businesses that could enhance Puerto Rico's resilience. As human and technical talent grow, RIN will leverage research and development (R&D) for new resiliency products and services and encourage new ventures for their commercialization. This could transform Puerto Rico into a significant producer, consumer, and exporter of these products and services. PRSTRT, along with local universities, would lead the Resilient Puerto Rico Advisory Commission (funded by the Rockefeller, Open Society, and Ford Foundations), and Fomento. RIN would accomplish a variety of objectives, including:

- establishing research priorities based on the comparative advantage of Puerto Rico researchers in areas of resiliency innovation, such as telecommunications, energy, water, and so on, and working closely with the Rockefeller Foundation's 100 Resilient Cities to expand public-private partnerships and private industry offerings for increasing resiliency in communities and cities.[44]
- rolling out the Healthy Generations Project community resiliency model,[45] which trains community peers in such areas as understanding effects of trauma on individual and social well-being; building resiliency individually and community-wide; adapting to the new baseline ("new normal") and establishing benchmarks for growth and success; and developing communication strategies that lead to shared healing
- setting up makerspaces, hackerspaces, and business incubators at all 25 municipalities that host Puerto Rico universities outside of the San Juan metro area
- establishing two resiliency innovation labs, one at PRST's San Juan headquarters and another at PRSTRT's Guanajibo Research and Innovation Park in Mayaguez
- leveraging PRSTRT's existing resources, such as entrepreneurial programs, corporate ties to encourage established companies to collaborate and invest in resiliency innovations,

---

[44] 100 Resilient Cities, website, undated.

[45] Healthy Generations Project, "About HGP," webpage, undated.

government ties to Fomento for tax incentives and business credits; and the Technology Transfer Office to provide IP protection and negotiate licensing agreements

- establishing a Resiliency Center of Education and Innovation (RCOEI) to institutionalize progress.

Ultimately, the aim of RIN would be to develop a resiliency industry in Puerto Rico that includes "resiliency maturity models,"[46] ties to the insurance and re-insurance industry,[47] and ways to engage volunteers.[48] Puerto Rico leadership in this emerging area is possible, based on real-world needs and experiences, and would provide competitive advantages and potentially long-term benefits.

This COA will have an impact on other sectors. First, the RIN will develop innovation and entrepreneurship to expand business and community capacity and promote Puerto Rico educational institutions (CPCB). Second, this will attract investment while RIN's distributed network opens rural and municipal areas to investors (municipalities, economic). Third, RIN will support economic objectives (economic).

## CIT 31 Resilience/e-Construction Learning Lab

The purpose of this COA is to leverage state-of-the-art resilient e-construction approaches to accelerate socioeconomic development. This COA will establish a Resilience/e-Construction Learning Lab for a one-year pilot project to digitize assessment, permitting, and reporting processes in one Puerto Rico municipality and present findings to inform the feasibility of an e-Permitting and e-Construction ecosystem throughout Puerto Rico.

A streamlined paperless construction administration delivery process will facilitate all legacy and new construction documentation and digital management in a secure environment. It will also save money by decreasing paper use, printing, and document storage costs, and it will save time by decreasing communication delays and transmittal time, all while increasing transparency and facilitating tax collection.

This COA would help accelerate the adoption of efficient ways to build affordable and resilient homes and structures (housing). It would help in facilitating the development of roads and bridges (transportation) and in ensuring public safety and the continuity of essential government functions. It would also streamline paperless construction administration, increase transparency, facilitate tax collection (municipalities).

This COA requires coordination to organize a multidisciplinary team. The risks are minimized by conducting an initial pilot project to present findings and cost-benefit analysis before expansion.

---

[46] Derived from Carnegie Mellon's cyber capability maturity models. See for example, Richard Caralli, Julia Allen, Pamela Curtis, David White, and Lisa Young, *CERT Resilience Management Model, Version 1.0, Technical Report CMU/SEI-2010-TR-012*, Pittsburgh, Penn.: Software Engineering Institute, Carnegie Mellon University, 2010; and CMMI Institute, "What Is Cyber Resilience? A Step Beyond Compliance," undated.

[47] As in CAMICO's model for CPA-related insurance and reinsurance. CAMICO, website, undated.

[48] As in the California Health Medical Reserve Corps (CH-MRC), website, undated.

The new Resilient e-Construction Learning Lab will be staffed with multidisciplinary teams consisting of an IT Lead, a government process SME, a finance SME, an architect/ master planner, architect support, including a civil engineer, and an analyst focused on successes/best practices outside of Puerto Rico. The team will run a pilot program in collaboration with the Department of Housing, the Department of Transportation, academic institutions (especially those with architecture and business programs), and a municipality. The COA will establish one physical Resilient e-Construction Learning Lab in partnership with two academic institutions and two government agencies and a municipality, which will engage in a one-year investigation and assessment, including a practical on-the-ground project. Pilot deliverable should include:

- engaging in a three-month assessment to gather input from government, industry, academia, and the public to identify strengths, weaknesses, threats, and opportunities
- identifying priority problem sets and providing recommendations aligned to achieving e-resilient construction, to include legacy issues (e.g., lack of documentation, rigorous identification system for title and property rights, ROWs, housing and building ownership in the case of multiple shared owners, existing infrastructure repair, renovation, reinforcement, and disaster-readiness) and new projects
- establishing reporting metrics (time, costs, efficiency, safety, and so on) and monitoring progress
- developing a paperless e-construction administration process by facilitating all construction documentation and digital management in a secure environment
- streamlining paperless permits for roads and buildings, while increasing transparency
- conducting market research, evaluating promising technical solutions, determining best solution and costs for adoption, and providing rational justification of need
- delivering quarterly progress reports and an annual final report
- developing a five-year resilient e-construction master plan that embraces protocols and government mandates for a self-sufficient "Smart Island."

Findings from the first-year pilot will inform continuation work for the remaining ten years of implementation of this COA. The cost estimates presented in Appendix C assume that for subsequent years the COA will require the same level of yearly effort. Once a successful pilot is completed, these costs can be refined.

The establishment of a new Resilient e-Construction Learning Lab falls under Vivienda (Puerto Rico Department of Housing), with support from the Department of Transportation, the communications and IT sector, and academia, in collaboration with a municipality. The motivated teams from the resilient e-Construction Learning Lab will be mobile-ready to embed with the agencies or the areas where they are focused—whether in agency offices or out in the community.

# 7. Portfolios, Funding, Implementation, and Concluding Remarks

## Overview

The 33 COAs of the Puerto Rico Recovery Plan pertaining to the communications and IT sector were presented to the government of Puerto Rico in June 2018. They were organized into the three portfolios of the strategic objective *Modernize the Telecommunications System.* Each portfolio was composed of a certain number of COAs and had an associated cost.

This final chapter of the report presents the three portfolios, provides a brief description of the potential sources of funding for the communications and IT COAs, and addresses general considerations for the successful implementation of the COAs. We end the chapter and the report with some concluding remarks.

## Communications/IT Portfolios

In Chapter 3 we introduced the concept of portfolios as the link between the top-down and the bottom-up paths. In this chapter we discuss the portfolios of the strategic objective *Modernize the Telecommunications System.*[1]

The three portfolios presented to the government of Puerto Rico are as follows:

*Portfolio 1. Resilient Communications.* Implement a state-of-the-art, survivable, resilient communications infrastructure for COG functions and for the provision of public safety to the residents of Puerto Rico. This infrastructure will include a robust data center to store and manage data on critical public infrastructure and with connections to redundant core (cloud) centers.

*Portfolio 2. Driver of Economic Growth.* Partner with the private sector to implement a state-of-the-art, survivable, resilient telecommunications infrastructure to provide commercial telecommunication services, including voice and data services to the residents of Puerto Rico. This infrastructure will support affordable access to broadband internet services in order to spur economic growth. This portfolio also includes the *Resilient Communications* COAs.

*Portfolio 3. Smart and Resilient Island Information Technology and Communications.* Make use of IT to further the economic and social vitality of Puerto Rico; foster government and private-sector innovation; increase economic opportunity; and help improve the quality of life for the residents of Puerto Rico. This third portfolio is the most expansive, and also includes the *Resilient Communications* and *Driver of Economic Growth* COAs. This satisfies all three communications and IT needs identified earlier in the report. Portfolio 3 includes all 33 COAs

---

[1] Strategic objectives other than *Modernize the Telecommunications System* also included many of the communications and IT COAs, and those strategic objectives were presented to the government of Puerto Rico in separate meetings. The cost of the overall plan was the sum of the costs of all the portfolios selected by the government of Puerto Rico, while making sure that the cost of every COA was counted only once.

described in this report at 100 percent level of implementation, at a total estimated cost of $3.2 billion. This most comprehensive portfolio was included in the recovery plan.[2]

Appendix C provides the costs associated with each COA and explains how the cost estimates were developed.

## Potential Sources of Funding

### Community Development Block Grant Disaster Recovery Funds

An important source of potential funding for the Puerto Rico Recovery Plan is Community Development Block Grant Disaster Recovery (CDBG-DR) assistance. HUD is the federal oversight agency for these funds. CDBG-DR assistance can fund a broad range of recovery activities to help communities recover from federally declared disasters. CDBG-DR funding can also be used as the nonfederal match requirement for other federal grants, which can potentially access additional federal funding. CDBG-DR initiatives "must demonstrate benefit to individuals and communities by meeting one of the program's three National Objectives for all money spent on projects. These are: (1) benefiting low-and moderate-income persons, (2) aiding in the prevention or elimination of slums or blight, or (3) meeting a need having particular urgency (urgent need)."[3] As part of CDBG-DR funding, sizable investments in infrastructure and economic recovery, such as repairs to telecommunications, are allocated in order to restore the supporting infrastructure that makes housing recovery feasible and sustainable in the long term.

### Public Assistance Grant Program Funds

FEMA's Public Assistance (PA) Grant Program reimburses communities for actions taken in the immediate response to and during recovery from a disaster. In Chapter 2 we explained that FEMA PA Category A and B funds were allocated to response efforts in the communications and IT sector for restoration of the public service communications networks and for preparation for the hurricane season. Categories A and B obligations in the communications and IT sector amounted to $1.77 million as of the end of June 2018.[4] We expect that Category A and B funding will be available for the implementation of COAs in the Puerto Rico recovery plan. PA

---

[2] The structure of the portfolios does not necessarily align with the chapters of this report. Therefore, each chapter may discuss COAs in different portfolios.

[3] Government of Puerto Rico, *Puerto Rico Disaster Recovery Action Plan: For the Use of CDBG-DR Funds in Response to 2017 Hurricanes Irma and Maria*, July 29, 2018.

[4] FEMA Comms/IT sector PowerPoint presentation, *Incident Complex – Puerto Rico: Rico: Recovery Sectors Solutions Weekly Meeting*, Puerto Rico, July 3, 2018. This reference states that this is the total dollar amount for Category A and B Project Worksheets of the Communications/IT sector as of June 29, 2018. This dollar figure has been rounded. Categories A and B in FEMA terminology refer to "emergency work" or "work that must be performed to reduce or eliminate an immediate threat to life, protect public health and safety, and to protect improved property that is significantly threatened due to disasters or emergencies declared by the President." See FEMA website for a detailed explanation on FEMA public assistance program, which includes Categories A and B work and expenditures, see FEMA, 2010.

funding is available to states and territories, local and tribal governments, and certain nonprofit organizations. PA funding has a nonfederal match requirement; that is, a percentage share of the total cost must be provided from other sources. For Puerto Rico, the nonfederal match is 10 percent.

### Hazardous Mitigation Grant Program Funds

FEMA's Hazard Mitigation Grant Program (HMGP) is designed to help communities implement hazard mitigation measures to reduce the risk of damage, hardship, loss, or suffering from future disasters.[5] States and territories, local and tribal governments, and certain nonprofit organizations are eligible to apply for HMGP funding. HMGP funds are available to implement a range of mitigation projects in accordance with state, tribal, and local priorities. HMGP recipients are responsible for prioritizing, selecting, and administering hazard mitigation projects. Eligible activities may include mitigation reconstruction, structural retrofitting of existing buildings, nonstructural retrofitting of existing buildings and facilities, structure elevation, postdisaster code enforcement, and generators. Specific activities in the communications and IT portfolios that may be eligible for HMGP funding are included in one COA (*CIT 5 Implement Public Safety/Government Communications Backup Power*), which would implement standardized backup power systems for public tower sites used by local government and emergency responders and allow for the COG operations and emergency response.

### Emergency Management Performance Grant Program Funds

FEMA's Emergency Management Performance Grant (EMPG) program contributes to the implementation of the National Preparedness System by supporting the building, sustainment, and delivery of core capabilities in prevention, protection, mitigation, response, and recovery.[6] A resilient communications and IT sector improves the ability of local authorities to respond quickly to emergencies, mitigate the loss of life and property by lessening the impact of future disasters, and recover through timely restoration and revitalization of infrastructure, housing, and a sustainable economy. EMPG funding is available to state, territorial, local, and tribal governments. The government of Puerto Rico's administrative or emergency management agency must apply for EMPG program funds on behalf of territorial or local emergency management agencies. Eligible activities that may be funded under the EMPG include logistics and distribution management planning and evacuation plan/annex development—including risk management, risk and disaster resilience assessment, threats and hazards identification, and operational coordination. These activities address core functions (such as communications and

---

[5] FEMA, Hazard Mitigation Grant Program, website, September 19, 2018.

[6] FEMA, Emergency Management Performance Grant Program, website, June 7, 2018.

warning) that may require specific actions be taken during emergency response. Specific activities in the communications and IT portfolios that may be eligible for EMPG funding are in four COAs, which include upgrading and modernizing Puerto Rico's EOC (*CIT 6 Modernize the Emergency Operations Center*), establishing an alternative EOC (*CIT 7 Establish an Alternate Emergency Operations Center*), acquiring an MEOC (*CIT 8 Mobile EOC Vehicle*), and procuring a mobile emergency communications capability (i.e., deployable telecommunications equipment) for first responders (*CIT 11 Perform Site Structural Analysis for All Government Telecom Towers [Both Public and Privately Owned]*).

### Other Sources of Funding

In addition to the sources of funding described above, we identified the following as potential funding sources for communications and IT COAs: FCC, National Science Foundation, USDA, U.S. Department of Commerce, DoD, U.S. Department of Education, U.S. Department of Health and Human Services (HHS), U.S. Department of Transportation, U.S. Department of Veterans Affairs, the government of Puerto Rico, the private sector, and other nongovernment sources of funding. Potential sources of funding for each recovery action are indicated in the appendices of this report.

The potential sources of funding listed in the recovery plan are primarily at the federal agency level. However, to obtain funding, the government of Puerto Rico and other stakeholders will have to apply for specific federal grants and programs. The recovery plan did not explicitly seek to identify specific programs as potential funders due to uncertainty about eligibility and the likely success of future applicants. During the early stages of planning, in the development of COAs, program directors were also unable to comment on potential funding without sufficient detail on recovery actions that will materialize in the implementation process. However, the paragraphs below describe some programs details specific to the communications and IT sector.

On May 29, 2018, the FCC issued an order and notice of proposed rulemaking approving approximately $900 million in additional funding to rebuild, improve, and expand voice and broadband networks in Puerto Rico and the U.S. Virgin Islands.[7] Through the Uniendo a Puerto Rico Fund, the FCC intended to make available up to $750 million of funding to carriers in Puerto Rico, including an immediate $51.2 million for restoration efforts in 2018. Some of these funds, as well as other potential sources of funding, are administered by USAC. USAC is an independent, not-for-profit corporation designated as the administrator of universal service programs with the guidance of policy created by the FCC. USAC allocates funding through four programs focused on rural, underserved, and difficult-to-reach areas where broadband and connectivity needs are critical: High Cost Program, Lifeline Program, School and Libraries (E-Rate) Program, and Rural Health Care Program. Puerto Rico already participates in several of these programs and additional funding may be available for recovery and other needs. For

---

[7] FCC, Order and Notice of Proposed Rulemaking, FCC 18-57, May 29, 2018f.

example, *CIT 22 Use Federal Programs to Spur Deployment of Broadband Internet Island-Wide* describes Puerto Rico's past experience and additional opportunities for expansion of broadband internet services to schools and libraries with the E-Rate Program.

There are several other federal programs available for funding telecommunications and broadband services for health care providers. USDA's Distance Learning and Telemedicine Grant Program provides loans, grants, and advanced telecommunications technologies for education and health care opportunities in rural areas. USDA's Community Facility Grants Program also provides grants for essential community facilities. Both grant programs are available to municipalities as well as nonprofit organizations. HHS's Telehealth Grant Programs provide competitive grants to fund projects that use telemedicine to provide health care services for medically underserved populations. The Patient Protection and Affordable Care Act (Public Law 111-148) also established the Prevention and Public Health Fund, which has invested in a wide range of community and preventative health initiatives, including research surveillance, public health infrastructure, and public health workforce and training. The U.S. Department of Veterans Affairs has also invested in telemedicine solutions by providing rural broadband access to veterans.

The U.S. Department of Commerce's Economic Development Administration (EDA) provides assistance to rural and urban areas experiencing high unemployment, low income, or other severe economic distress. EDA oversees the Public Works and Economic Adjustment Assistance programs, which can support a wide range of construction and nonconstruction projects. The Public Works program helps distressed communities revitalize, expand, and upgrade their physical infrastructure, including technology-based facilities, skill-training facilities, and business and industrial parks. Under the Economic Adjustment Assistance program, EDA administers the Revolving Loan Fund Program, which provides small businesses with capital to start or expand their business as well as funds for planning or construction grants or market and environmental studies. Under the Bipartisan Budget Act of 2018 (Public Law 115-123), Congress appropriated an additional $600 million in Economic Adjustment Assistance funds for areas affected by Hurricanes Harvey, Irma, and Maria and by wildfires and other natural disasters. The U.S. Department of Education also offers programs such as Technology and Media Services for Individuals with Disabilities, which support the use of technology and education media activities that are available to state and local education agencies, institutions of higher education, and nonprofit organizations.

In addition to the programs described above, several communications and IT COAs may be able to benefit from cost-sharing with other sectors and other federal agencies. Specifically, joint-trenching or shared conduit (i.e., housing fiber-optic cable with other utilities, such as gas, water, or electric lines) may provide opportunities to share costs across different federal programs as well as to reduce overall costs. Thus, for example, trenching and laying fiber-optic cable could be done in tandem with road and highway repair to achieve a "dig once" strategy and realize a cost savings for the U.S. Department of Transportation as well as other potential funding sources.

Finally, assuming funding is indeed secured through various federal programs, the government of Puerto Rico will have to seek additional funds from other sources. These sources will include Puerto Rico's own contribution to the recovery and potential support from (1) the private sector, including institutional investors and public-private partnerships, and (2) charitable foundations. In addition to the funding these sources could provide, they could also bring expertise, innovation, and volunteers to the recovery efforts—and many are already contributing. Puerto Rico will also make a substantial contribution from its own resources to support recovery. This will include providing personnel, supplies, technical and oversight services, and other critical contributions. Funding may also come from revenue-generating projects, such as user fees for toll roads, leases, or sale of excess broadband capacity to private companies.

## Implementation

The successful implementation of the recovery plan will require obtaining federal and nongovernment funding, responsible management and oversight of those funds, effective project management, and increased resilience against future disasters. See, for example, the Hurricane Sandy Rebuilding Task Force guidelines and recommendations in the Hurricane Sandy Rebuilding Strategy.[8]

Program and grant applicants, including the government of Puerto Rico, municipalities, and nonprofit organizations, and in some cases, schools and school districts, libraries, and hospitals, should coordinate efforts to align funding to investments that are precursors to or priorities for recovery. The government of Puerto Rico, municipalities, and other organizations participating in the recovery should work with federal agencies to have a clear understanding of program eligibility requirements to maximize funding available for recovery actions. In some cases, the government of Puerto Rico will be responsible for prioritizing, selecting, and administering funds. Certain federal grant programs have a nonfederal match requirement, which means that they may be inaccessible without other funding. Limited local resources will have to be allocated to local priorities, but decisions should also weigh the potential to match additional federal resources, such as PA and HMGP funds. CDBG-DR funds can be used as the nonfederal match for federal grants, but will also be sought after for other local priorities and needs. Private and other nongovernment funding sources may be available for some investments, but may not always align with local priorities and federal programs for which they could provide a nonfederal match. COAs to support infrastructure resilience may span different agency programs and, potentially, multiple jurisdictions. There may be opportunities for cost-sharing with other agency

---

[8] Hurricane Sandy Rebuilding Task Force, Hurricane Sandy Rebuilding Strategy, August 2013.

programs, such as combining road and highway repair with trenching and laying fiber-optic cable or using shared conduit to carry other utilities.

Responsible management and oversight of funds are needed to support an effective recovery. In December 2017, the DHS Office of Inspector General issued reports concerned with ensuring that grant recipients properly manage disaster funds, including compliance with federal procurement requirements.[9] In July 2015, the GAO reported that there was no comprehensive strategic approach for identifying, prioritizing, and implementing investments for disaster resilience.[10] This lack of an investment strategy increased the risk that the federal government and nonfederal partners would experience lower returns on investments or lost opportunities to strengthen key CI and lifelines. Because lessons learned have been identified from previous major disasters, such as Hurricanes Katrina, Rita, and Sandy, the government of Puerto Rico has an opportunity to ensure proper management in its recovery from Hurricanes Maria and Irma in addition to benefiting from Puerto Rico's recovery plan.[11]

## Concluding Remarks

Hurricanes Irma and Maria caused unprecedented, widespread, and devastating damage to Puerto Rico's public safety and commercial telecommunications networks and systems, which in turn significantly increased the risk to public safety,[12] hampered emergency response and recovery efforts, and contributed to the shutdown of Puerto Rico's economy.[13] Failures in public safety communications left citizens unable to access critical emergency services at a time when such services were most needed. After Hurricane Maria struck, most telephone lines were not operational, so 911 operators often could not call first-responder dispatch centers and required emergency response could not be provided.

Great strides toward repair and restoration of the telecommunications infrastructure have been made by the government of Puerto Rico, FEMA, and the private sector. In the private sector, as of early June 2018, telecommunications carriers had been able to bring up to operational status 99.8 percent of the cellular sites in Puerto Rico.[14] In the public sector, a plan to

---

[9] DHS, Office of Inspector General, "Management Alert—FEMA Faces Significant Challenges Ensuring Recipients Properly Manage Disaster Funds," OIG-18-33, December 17, 2017; DHS, Office of Inspector General, "Lessons Learned from Prior Reports on Disaster-related Procurement and Contracting," OIG-18-29, December 5, 2017.

[10] GAO, *Hurricane Sandy: An Investment Strategy Could Help the Federal Government Enhance National Resilience for Future Disasters*, GAO-15-515, July 30, 2015.

[11] Central Office for Recovery, Reconstruction and Resiliency, 2018.

[12] As of September 24, 2017, Hurricane Maria had damaged 91 percent of the private telecommunications infrastructure (primarily antennas and fiber). This had a major impact on the government, retail stores, banks, pharmaceutical companies, food, transportation, and other businesses.

[13] Governor of Puerto Rico, 2017.

[14] This statistic refers to 2,653 out of 2,659 cellular sites. Restoration of cellular sites is particularly important for Puerto Rico because over 80 percent of residents rely on mobile phones as their primary means of communication.

allow for continuity of communications for government leadership and CI stakeholders such as hospitals, public safety answering points, and emergency medical services was in place by the end of July 2018.

In spite of these efforts, the telecommunications infrastructure remains vulnerable to another disaster as catastrophic as Hurricane Maria. Capabilities need to be developed to achieve robust public communications networks and systems so that emergency services and government functions will be effective and responsive in the aftermath of a disaster, natural or manmade. Similarly, commercial telecommunications networks need to be made more resilient to future disasters by using state-of-the art technologies and encouraging greater collaboration between the government of Puerto Rico and private telecommunication carriers. In addition to achieving adequate resiliency, recovery efforts in the communications and IT sector require bringing many systems up to federal telecommunications standards, taking advantage of technological advances to upgrade systems and equipment, improving maintenance procedures for CI, and maintaining a cadre of qualified communications and IT personnel. The goal of the recovery plan is not just to "build back"—repair and restore what was destroyed or damaged—but to "build back better." Thus, the first objective of the COAs developed for the communications and IT sector under this plan is to "build back better" the telecommunications infrastructure, both public and private.

A great opportunity presented by the recovery plan is to expand broadband internet access to the population, which in Puerto Rico is lower than any U.S. state.[15] Broadband internet access can be linked to many of the benefits that the government of Puerto Rico wishes to achieve to "build back better." Broadband internet access has been shown to have a significant impact on economic growth, wages, medical care, and education and to foster prosperity and well-being. Moreover, the recovery plan offers the opportunity to address the "digital divide" and to ensure that all residents of Puerto Rico can take advantage of the many well-documented socioeconomic benefits afforded by broadband internet. Thus, the second objective for the COAs developed for the communications and IT sector under this plan is widespread access to broadband internet service throughout Puerto Rico.

Puerto Rico needs to build modern, state-of-the-art telecommunications networks and implement more resilient emergency communications systems that leverage cutting-edge technology. With a reliable telecommunications infrastructure in place and with widespread broadband internet access, IT can help accomplish the digital transformation of Puerto Rico and spur economic growth. IT will enable the government of Puerto Rico to implement citizen-centric digital services. Digitizing government transactions can save time and resources on the front end for citizens, who will no longer need to spend hours in line or mailing in paper forms.

---

[15] Ten percent of the U.S. population does not have access to the broadband threshold the FCC deems necessary for high-quality video, voice, and data applications. In Puerto Rico, 62 percent of the population (over 2.3 million people) does not have access to that service. The FCC's data on broadband deployment indicates that over 98 percent of residents in rural areas of Puerto Rico lack a broadband connection to the internet. See FCC, 2016.

Digital services also enable standardization, which can fuel automation, real-time analytics, data-sharing, and process optimization. Additional benefits of using IT will be increased public transparency and better data for policymaking. The efficiencies and experience gained and shared through these efforts could incentivize economic activity in the private sector. Thus, the third objective of the COAs developed for the communications and IT sector under this plan is to make use of IT for the betterment of the residents of Puerto Rico.

The government of Puerto Rico has embraced a strategy of digital transformation in order to modernize government processes and improve citizen services, respond to fiscal pressures, better coordinate post–Hurricane Maria recovery efforts, and prepare for future disasters. Proper implementation of the COAs proposed in this report, together with actions taken in other sectors,[16] will result in a resilient infrastructure and assist in realizing the recovery plan's vision for the Digital Transformation of Puerto Rico.

---

[16] Central Office for Recovery, Reconstruction and Resiliency, 2018.

# Appendix A. Illustration of the Methodology to Estimate the Costs of the Courses of Action for the Communications/IT Sector

This appendix describes the methodology used to estimate costs for the COAs of this sector.[1] It illustrates the approach with reference to a particular COA that was introduced in Chapter 5.

## Cost Methodology

To estimate the cost of COAs for the communications and IT sector, we relied on existing cost studies for similar recovery actions and investments, judgment-based estimates of SMEs, and input from Recovery Support Function teams, SBTs, and other relevant agencies. In general, we developed rough order-of-magnitude cost estimates intended to support high-level planning and inform decisionmaking. Costs were estimated for the period from FY 2018 through FY 2028. All costs were presented in 2018 dollars, and future costs were not discounted. To illustrate the cost estimation approach, we describe the cost methodology for *CIT 21 Government Owned Fiber-Optic Conduits to Reduce Aerial Fiber-Optic Cable*. This COA represents the highest-cost COA for the communications and IT sector. It aims to incentivize private companies to invest in the deployment of fiber in buried conduit to every municipality in Puerto Rico. Technical details and additional information on this COA are presented in Chapter 5 and in Appendix C.

## Cost of Government-Owned Fiber-Optic Conduits to Reduce Aerial Fiber-Optic Cable

To estimate the cost of constructing government owned fiber-optic conduits, we relied on unit cost estimates from the Federal Highway Administration's (FHWA) 2009 *Rural Interstate Corridor Communications Study*.[2] The study estimated rural installation costs of 48-singlemode fiber-optic (SMFO) cable along the Interstate 90 corridor (including parts of South Dakota, Minnesota, and Wisconsin) and the Interstate-20 corridor (including parts of Louisiana, Mississippi, and Alabama). We received additional input on unit costs from SMEs with experience in a variety of similar projects.[3]

In Table A.1, we report the unit costs derived from the FHWA study and SMEs. In cases where a range of costs was provided, we used the midpoint value unless otherwise noted. All

---

[1] For a more general discussion of the cost methodology for the recovery plan and not specific COAs, see RAND Corporation, undated.

[2] U.S. Department of Transportation, Federal Highway Administration, "Rural Interstate Corridor Communications Study: Report to States, February 2009.

[3] Technical director at Siemens Government Technologies, interview with the authors, June 5, 2018.

costs are inflation-adjusted to 2018 dollars. As shown in the table, for trenching we use directional boring costs of $8/foot for rural areas, $11/foot for urban areas, and $270/foot for mountain roads (the latter reflecting the high cost of construction in rocky terrain). We assume costs of approximately $1/foot for high-density polyethylene (HDPE) conduit plus $1,600 per handhole for every 1,500 feet of conduit. We also assume that regeneration buildings to house signal regeneration equipment will be required every 50 miles in flat terrain and every 35 miles in rocky terrain at a cost of approximately $340,000 per building. Finally, we assume a construction overhead rate of 48.5 percent for design, engineering, mobilization, administration, traffic control, and contingency.

**Table A.1. Installation Unit Costs**

| Service/Product | Average Cost ($2018) | Unit | Notes |
| --- | --- | --- | --- |
| Directional boring—rural areas | $7.70 | Foot | |
| Directional boring—urban areas | $10.66 | Foot | |
| Directional boring—rocky terrain | $270.00 | Foot | High-end estimate, based on input from SMEs |
| Conduit (2" HDPE) | $0.92 | Foot | |
| Handhole | $754.80 | Each | Assumes one handhole for every 1,500 feet of roadway |
| Handhole installation | $828.80 | Each | |
| SMFO cable—48 count | $0.83 | Foot | |
| Regeneration building | $343,000 | Each | Assumes one building every 50 miles in urban/rural areas and every 35 miles in rocky terrain |
| Construction overhead | 48.50% | Percent of construction cost | |

SOURCE: Unit costs, U.S. Department of Transportation, 2009.

To install conduits reaching every municipal center, the government of Puerto Rico could construct along the strategic highway network, defined in the long-range transportation plan as the "core roadway network linking all regions of the island," plus arterial and collector roads maintained by the Puerto Rico Department of Transportation and Public Works (DTOP).[4] This would take advantage of existing roadways and allow for potential cost-sharing with transportation projects, such as highway and road repair. We selected information from FHWA's Highway Performance Monitoring System (HPMS) to map all DTOP roadways in Puerto Rico and GIS software to estimate pathways for laying conduit.[5] In Table A.2, we report the length of highways and roads in Puerto Rico (in miles) by ownership and functional class. There are approximately 4,600 miles of DTOP roadways, including interstates and freeways, arterial roads,

---

[4] DTOP, "Puerto Rico 2040 Islandwide Long Range Transportation Plan," San Juan, Puerto Rico, 2013.

[5] Federal Highway Administration, Office of Highway Policy Information, "Highway Performance Monitoring System (HPMS)," 2018.

and collector roads.[6] We estimate that this network of roadways would functionally extend to every municipal center in Puerto Rico. Overlaying the DTOP map with topographical data for Puerto Rico, we estimate that approximately 25 percent of roadways are in mountainous regions.

**Table A.2. Miles of Roads and Highways in Puerto Rico**

| Total Miles of Roads and Highways: 16,650 | | | | |
|---|---|---|---|---|
| | Rural | | Urban | |
| **By Ownership** | **DTOP** | **Municipalities** | **DTOP** | **Municipalities** |
| Federal-aid | 545 | 2,060 | 2,710 | 220 |
| Non-federal-aid | 480 | 0 | 850 | 9,790 |
| **Total** | **1,025** | **2,060** | **3,560** | **10,010** |
| **By Functional Class** | | | | |
| Interstates and freeways | | 40 | 290 | |
| Arterial roads | | 275 | 1,410 | |
| Collector roads | | 490 | 1,245 | |
| Local roads | | 2,300 | 10,640 | |
| **Total** | | **3,100** | **13,550** | |

NOTE: Totals may not sum precisely due to rounding.

It may be redundant to install conduit along all roadways in each municipality because often two or more roadways intersect municipal centers. Therefore, we estimate that conduit would be required along 50 percent of roadways, or approximately 2,300 miles of DTOP roads in Puerto Rico (including the strategic highway network), to functionally extend to every municipal center. The government of Puerto Rico would need to engage with telecommunications providers, the Telecommunications Bureau, and essential stakeholders such as hospitals, clinics, and schools to determine the best location for the conduit. Based on the distance and unit costs reported above, we estimate the total cost of trenching and installing conduit would be $1.33 billion. This implies an average cost of $580,000 per mile, which is higher than the cost of rural interstate projects in FHWA's study (which ranged from $150,000 to $210,000 per mile), reflecting in part the higher cost of construction in rocky terrain and the mountainous geography of Puerto Rico.

Figure A.1 summarizes the key assumptions, calculations, and total estimated cost for this COA.

---

[6] For this estimate we counted DTOP roads and excluded municipal roads. DTOP roads account for 1,025 rural plus 3,560 urban, or 4,585 miles. We focus on DTOP roadways because (1) they comprise most of the strategic highway network in Puerto Rico and provide a dense network of roads through which nearly every municipality can be connected, (2) policymakers have to deal with only one entity (DTOP) rather than local roads owned by 78 municipalities, and (3) the DTOP roadways are likely eligible for federal funding, in some cases from both FHWA and FEMA.

**Figure A.1. Total Estimated Cost of *CIT 21 Government-Owned Fiber-Optic Conduits to Reduce Aerial Fiber-Optic Cable and Incentivize Expansion of Broadband Infrastructure***

## THE COMPLEX PROCESS OF ESTIMATING COSTS

## EXAMPLE

**INSTALL UNDERGROUND CONDUIT ALONG ROADWAYS TO BURY FIBER-OPTIC CABLE (CIT 21)**

To prevent future damage to fiber-optic cable, one proposal is to install underground conduit along roadways so that telecommunications providers can run their fiber-optic cable underground instead of on poles. To estimate the costs of this initiative, there are precedents to draw on: The U.S. Federal Highway Administration white paper Rural Interstate Corridor Communications Study estimates the costs of installing 48-SMFO cable along Interstate 90 (through South Dakota, Minnesota, and Wisconsin) and Interstate 20 (through Louisiana, Mississippi, and Alabama). We also corroborated our unit cost estimates with an expert with local knowledge of Puerto Rico roadways and extensive experience laying fiber-optic cable for the U.S. military in different types of terrain.

### Construction Costs

**TRENCHING**
Uses these directional boring costs as a proxy:
- **$8/foot for rural areas,**
- **$11/foot for urban areas, and**
- **$270/foot for mountain road**

**LAYING CONDUIT**
Assumes these costs:
- **$1/foot for laying conduit**
- **$1,600 per handhole installed every** 1,500 feet of roadway ($750 in materials and equipment, and $830 in installation costs)

**BUILDINGS TO HOUSE SIGNAL REGENERATION EQUIPMENT**
Assumes these costs:
- **Each building will cost $340,000**
- **Buildings will be required:**
  » Every 50 miles in flat terrain
  » Every 35 miles in rocky terrain

**CONSTRUCTION OVERHEAD**
Assumes an **overhead rate of 48.5%** for design, engineering, mobilization, administration, traffic control, and contingencies

### Amount of Conduit Needed

*How many miles of roads are there?*
**4,600 miles of roadways are operated by Department of Transportation and Public Works (DTOP)**
» We used Geographic Information System (GIS) data to assess that the roadway network reached every municipality

*Are the roads in tough terrain?*
**Approximately 25 percent of roadways are in mountainous regions**
» We arrived at this number using GIS data overlaying all DTOP roadways on a topographical map of Puerto Rico to estimate how many miles of roadway are in mountainous terrain.

*Is conduit needed along every roadway?*
**50 percent—about 2,300 miles of roads—will need cable**
» We arrived at this number because more than one road reaches most municipal centers, so laying cable along all of them is duplicative.

*What is the cost per mile?*
**Based on the unit costs above and accounting for the higher cost of construction in mountainous terrain, we estimate an average cost of $580,000 per mile.**

**In comparison, the costs of laying broadband along rural U.S. interstates (in generally flat terrain) ranged from $150,000 to $210,000 per mile.**

TOTAL ESTIMATED COST

# $1.3 billion

SOURCE: Central Office for Recovery, Reconstruction and Resiliency, 2018.

# Appendix B. Examples of Implementation Efforts, Opportunities, and Challenges After Delivery of the Recovery Plan

The implementation phase of the 33 COAs developed under the Puerto Rico recovery plan for the communications and IT sector will present opportunities for COA refinement and optimization, such as implementing several COAs concurrently to achieve efficiencies or combining their implementation when doing so makes sense. There will also be challenges to implementation.

The first section of this appendix explains an option to address the dependencies between upgrading and modernizing the primary EOC (CIT 6), establishing an alternate EOC (CIT 7), and procuring a mobile EOC (CIT 8) and the recommendation that these COAs need to be addressed together during the implementation phase. The second section provides examples of efforts to implement the communications and IT COAs presented in this report. The efforts presented here were ongoing as of November 2018. The third section describes COA implementation opportunities and challenges by using an important COA as an example; this COA is *CIT 21 Government-Owned Fiber-Optic Conduits to Reduce Aerial Fiber-Optic Cable and Incentivize Expansion of Broadband Infrastructure.*

## Emergency Centers

As indicated in Chapter 4, upgrading and modernizing the primary EOC (CIT 6), establishing an alternate EOC (CIT 7), and procuring mobile EOC (CIT 8) need to be addressed together during the implementation phase. This includes taking into consideration their estimated costs, the availability of an adequate site for the alternate EOC/PSAP (CIT 7), and the fact that the fastest to implement is the MEOC. We also indicated that an option combining these COAs had been discussed and that the participants in these discussions thought that option would be better addressed during the implementation phase of the recovery plan.[1] This option consisted of not implementing CIT 6—that is, to not modernize the current EOC that is colocated with PREMA headquarters. Instead, transfer the EOC functionality currently residing in the PREMA headquarters building to a new location (CIT 7). That location would be provisioned with standard-compliant and state-of-the art equipment, would meet all the requirements for a modern EOC, and would then function as Puerto Rico's primary EOC. Transferring PSAP and dispatch functionalities to the same center would accomplish the part of CIT 3 that pertains to consolidation of those two functions at the same location. Finally, the second or alternate EOC could reside in an MEOC (CIT 8).

---

[1] The participants were government of Puerto Rico officials, FEMA Comms/IT sector representatives, and members of the HSOAC team.

## Examples of Implementation Efforts

Implementation of the COAs proposed in this report is being led by the FEMA Comms/IT sector lead in close collaboration with the Telecommunications Bureau, PRDPS, and CINO.[2]

The following are examples of the COAs that, as of November 2018, were being pursued for implementation as part of the Economic and Disaster Recovery Plan for Puerto Rico in the communications and IT sector:

- *CIT 3 Upgrade and Enhance 911 Service, CIT 6 Upgrade and Modernize the Island's Emergency Operations Center*, and *CIT 7 Establish an Alternate Emergency Operations Center*: The government of Puerto Rico decided on the concurrent implementation of three of the COAs proposed in this report, which were discussed in Chapter 4. It decided to transfer the functions of the Puerto Rico (EOC) from its current location at PREMA headquarters to an alternate location (part of CIT 7). The EOC will be in a new building at a location that had been previously used as correctional facility of the government of Puerto Rico.[3] This new building will be upgraded with modern equipment and will meet EOC standards, including meeting FEMA guidelines (CIT 6). Moreover, a new PSAP will be installed at that same location and will consolidate PSAP and dispatch (part of CIT 3).[4]
- *CIT 8 Mobile EOC Vehicle*: The government of Puerto Rico is considering acquiring an MEOC.[5] This will be the implementation of another COA proposed in this report (CIT 8). Puerto Rico authorities are considering alternatives other than buying a NIMS Type-2 vehicle.[6]
- *CIT 15 Undersea Cable Ring*: The government of Puerto Rico is examining the design and construction of an undersea telecommunications ring. Initial inquiries with submarine cable vendors indicate that the design could be completed in six months and that building the submarine ring could be finished 12 months after the design was completed.[7] This effort is part of implementing another COA proposed in this report (CIT 15).

## Example of Implementation Opportunities and Challenges: Incentivize the Reduction of Aerial Fiber-Optic Cable

In this section, we provide an example of COA implementation opportunities and challenges. We selected *CIT 21 Government-Owned Fiber-Optic Conduits to Reduce Aerial Fiber-Optic Cable and Incentivize Expansion of Broadband Infrastructure* for several reasons. First, if

---

[2] Implementation of COAs that focus on IT for CI and for the digital transformation of Puerto Rico is being overseen by CINO and CIO.

[3] Senior official of the Puerto Rico Department of Public Safety, telephone communication with the authors, November 12, 2018.

[4] Senior Official of the Puerto Rico Department of Public Safety, telephone communication with the authors, October 25, 2018.

[5] FEMA Comms/IT sector, weekly Puerto Rico communications and IT leadership teleconference, October 9, 2018.

[6] See, for example, "Calling All TITANs: Nissan and the American Red Cross Mobilize Purpose-Driven Campaign with the Ultimate Service TITAN," *Nissan News*, October 4, 2018.

[7] FEMA Comms/IT sector, weekly Puerto Rico communications and IT leadership teleconference, October 9, 2018.

implemented, it has the potential to bring about the deployment of resilient telecommunications networks based on buried fiber conduit throughout Puerto Rico, including rural and mountainous areas. Second, this is a cross-cutting COA that identifies interdependencies between communications and IT and other sectors, since the trenches to be dug under this COA can be also used by other sectors, notably, transportation, energy, and water. Third, it will require an effective and long-term partnership between the government of Puerto Rico, the private sector, and other stakeholders (e.g., the FCC and civil society stakeholders). Fourth, it is the communications and IT COA that demands the largest investment of resources. As we shall explain, one implementation opportunity will consist of using CIT 25 to address several implementation challenges that we anticipate CIT 21 will face. CIT 25, the "blue-ribbon panel" COA, is entitled *Evaluate and Implement Alternative Methods to Deploy Broadband Internet Service Throughout Puerto Rico*. A second implementation opportunity will be to leverage ongoing efforts such as FirstNet.

In this section we outline some potential implementation challenges and opportunities for CIT 21. The challenges are of two types: challenges to obtain all the information required to successfully and efficiently implement this COA, and challenges to achieve the desired end result, which is that carriers rebuild and expand their networks throughout Puerto Rico utilizing buried fiber conduit.

### Some Potential Challenges

First, it is not clear how competition among carriers and how each carrier's individual market strategy will affect CIT 21 implementation, or even what role competition and market strategies play when the carriers are rebuilding or expanding their networks. The individual carrier's strategy depends on its market share, business plans, geographic markets, and the level of competition and/or differentiation with other carriers. A successful implementation of CIT 21 will require understanding of carriers' market shares, investment and market strategies, and whether these are different pre- and post-Maria.[8] It will require an understanding of the structure of the telecommunications market in Puerto Rico, including competition and investment, and an understanding of what the market can and cannot do. Finally, it will require an assessment of any possible unintended consequences.

Second, it will be important to consider what steps a carrier may take to achieve the desired effect—namely, deploy a resilient network throughout Puerto Rico—while taking into account the differences among carriers. A "one-size-fits-all" approach may not be effective, since not every carrier may be motivated by the same incentives, and certain carriers may already have plans and enough motivation to behave in the desirable way. Moreover, in spite of initial positive response to CIT 21 by the carriers' representatives during a meeting at the Telecommunications

---

[8] Most of this information tends to be proprietary or commercially sensitive and cannot be included in a report with wide-dissemination such as this one.

Bureau,[9] it is not clear whether they will commit to this COA. Telecom carriers operating in Puerto Rico might be enthusiastic about CIT 21, but before committing to it they will likely want to compare the potential capital and operating expenditures to other choices.

Third, CIT 21 suggests "carrots" to incentivize carriers to utilize buried fiber conduit when rebuilding and/or expanding their networks. Another tool potentially available to the government of Puerto Rico is the threat of using "sticks." Potential "sticks" are targeted regulations or the threat of additional regulations that carriers may want to avoid, and to do so they may be willing to make binding voluntary commitments.[10] Moreover, liability rules, eligibility for government contracts, favorable zoning relief, and whether generators should be required for approval of cell sites are other examples of "carrots" and "sticks" that could be used as leverage by the government of Puerto Rico to secure from carriers the commitment to develop a certain percentage of their networks as buried fiber-optic cable.

A fourth challenge is cost. Details on the sharing of ongoing costs will need to be clarified. For example, it will be necessary to determine whether ongoing costs will be supported from payment by or revenues from users and whether new participants will need to pay the whole cost of sharing the conduit network from its initiation if they are not currently engaged in the Puerto Rico market.

*How to Tackle the Challenges: First Implementation Opportunity*

An analysis of the challenges presented above, although very important, was beyond the scope of what HSOAC was requested to prepare for FEMA. The information on individual carriers' level of competition, business plans, and other relevant market information can be gathered (if it is not yet gathered) by the appropriate Puerto Rico agency, such as the Telecommunications Bureau (formerly PRTRB).[11] Moreover, the challenges outlined above could be explored in depth, possibly with proprietary business plans to be discussed and provided to that Puerto Rico agency. Analyzing these challenges as well as other issues required to successfully implement CIT 21 will be also an appropriate task for that agency. As part of this task the agency may consider expanding its already strong relationships and partnerships with the FCC to advocate for any policy changes or flexibility that it may find necessary.

An implementation opportunity is that if CIT 25 is implemented, these challenges could be addressed and the required analysis be performed by the blue-ribbon panel proposed by that COA. As explained in Chapter 5, CIT 25 proposes the establishment of a blue-ribbon panel by

---

[9] A meeting convened by the President of the Puerto Rico Telecommunications Regulatory Board on May 11, 2018, with representatives from cellular carriers and broadband internet providers.

[10] There is precedent for carriers taking action to avoid additional regulation. The Wireless Network Resilience Cooperative Framework (Wireless Framework) is a voluntary organization proposed and created in response to the FCC proposing regulation that the carriers did not want. The carriers were able to forestall adoption of that regulation by creating the Wireless Framework. See GAO, 2017.

[11] The carriers' market information to be gathered should be subject to reasonable protections.

the Governor of Puerto Rico. The panel is designed to include commercial carriers and government representatives from the Telecommunications Bureau, other relevant agencies, consumers, disability groups, and other important stakeholders. The panel is designed to have an advisory board, and it is apparent that a very significant amount of expertise will be required to fulfill its role. The blue-ribbon panel would need adequate time to assess these opportunities and challenges, but its work must inform the timely deployment of resilient telecommunications networks in Puerto Rico.

The decision on who will address these challenges and perform the analyses needs to be made by the Governor of Puerto Rico.

The Puerto Rico agency or the blue-ribbon panel could consider whether regulatory "sticks" should accompany the "carrots" of CIT 21 to ensure that all carriers use buried fiber cable to the extent possible, or it could require that any carrier's network expansion plans contain a binding commitment to use buried fiber cable in a designated amount of its network within a certain timeframe, which could be enforced through financial penalties. Also, a carrier's making use of CIT 21 conduit could be a condition of certain government funding or government contracts.

CIT 21 addresses and potentially resolves a fundamental bottleneck encountered by all carriers in Puerto Rico, namely, the extraordinary time, uncertainty, and expense of obtaining regulatory approvals to bury fiber-optic cable. The uncertainty, expense, and delay surrounding the permitting and ROW approval process have been major inhibitors for all carriers to rebuild using buried fiber. The members of the FCC's Puerto Rico Hurricane Task Force were very aware of the long-standing issues concerning permitting and ROW approval processes in Puerto Rico and indicated their interest in assisting the Telecommunications Bureau in resolving these issues. Members of the FCC's Puerto Rico Hurricane Task Force and other FCC partners may provide excellent targeted regulatory ideas or suggested contractual commitments for a successful implementation of CIT 21.

## Second Implementation Opportunity

There is also the opportunity for leveraging ongoing efforts such as FirstNet as well as 5G deployment. As explained in Chapter 4, an important aspect of the development of the public safety network in Puerto Rico (see *CIT 1 Land Mobile Radio System*) is closely monitoring and leveraging progress of the FirstNet network deployment by AT&T. FirstNet is intended be the first nationwide interoperable broadband network for first responders. AT&T's commercial interests include its public sector and potential FirstNet customers, as well as consumers and businesses. For the FirstNet network, AT&T will need to "harden" towers, antenna sites, and other elements of its network to provide the level of resilience and reliability required by the public safety community in an emergency. Thus, AT&T has special impetus to deploy the FirstNet network using buried fiber conduit. Moreover, AT&T will likely seek how to most efficiently invest so as to serve all its actual and potential classes of customers. Finally, AT&T's president of operations has said that he has three priorities: the company's FirstNet emergency responder network; its network upgrade to 5G; and the carrier's ongoing fiber build.[12] AT&T's network expansion plans (including FirstNet deployment) align with the purpose of CIT 21.

The use of CIT 21's conduit would resolve a complex regulatory problem for carriers. Therefore, CIT 21 could provide a significant benefit to AT&T when it is deploying its FirstNet network by providing substantial regulatory relief. CIT 21 can also provide more coordination on trenching and the availability of conduit and reduce the cost of deploying buried fiber for carriers. Carriers would no longer need to pay for obtaining permitting and ROW approvals, or for the labor and other costs of trenching and laying conduit. As previously indicated, the blue-ribbon panel might be an ideal forum to address these issues, since government regulators will be part of this panel.

### Issues to be Considered for a Successful Implementation of CIT 21

As stated, the challenges described above should be addressed by the Telecommunications Bureau or by the blue-ribbon panel and could make their work more effective. To this end we are proposing an outline of issues for consideration for a successful implementation of CIT 21, as follows:

- Could there be liability safe harbors of some kind if carriers bury fiber, but not if they do not?
- What can be done via contract or required as a condition for the government undertaking the trenching and conduit? What regulatory "carrots and sticks" would be appropriate?
- Even if trenching and conduit installation are handled by and at the expense of the government, how might the issue that carriers may not be ready to deploy fiber at the same time be handled?
- What about the competitive equities of AT&T, which has a network that includes previously buried fiber? And would FirstNet contribute to CIT 21 or share cost?
- What would the market potentially drive for deployment of buried fiber (for restoration and network expansion), absent the incentives of CIT 21? Is some sort of more extensive cost-sharing possible? Do the government and public safety organizations need to use conduit throughout Puerto Rico, so they would have to anticipate spending part of this money anyway?
- What is an appropriate user fee or cost-sharing arrangement so that the government will not have to shoulder the cost of maintenance and risk?
- Prior to initiating CIT 21, how can the government assure a high level of certainty about usage?

There are several additional considerations for the implementation of CIT 21 that could be addressed.

- To reduce risk to the government, carriers should provide assurances and specific commitments, so that policymakers gain a high level of confidence that buried fiber will

---

[12] Linda Hardesty, "AT&T Taps Synergies Between Its 5G and FirstNet Build-Outs," SDX Central, December 7, 2018.

be achieved broadly enough to make CIT 21 a success. The Telecommunications Bureau or the blue-ribbon panel could consider the specific types of assurances and specific commitments that will be required.

- The agency or the panel could consider the level of government assistance that will be needed to achieve the goal of burying fiber throughout Puerto Rico.

- CIT 21 envisions a user fee or other cost-sharing mechanism to obtain funding for maintenance and repair of the conduit network to prevent these financial burdens from falling on the government. The agency or the panel could consider the appropriate user fee or other cost-sharing mechanism.

- The agency or the panel could consider liability rules, eligibility for government contracts, favorable zoning relief, and whether generators should be required for approval of cell sites, among other incentives that could be conditional on having a certain percentage of buried fiber-optic cable.

# Appendix C. Full Description of the Courses of Action

This appendix presents the full descriptions of all 33 COAs developed to support the Puerto Rico recovery plan for the communications and IT sector. The numerical assignment of these COAs is random and does not indicate a specific prioritization. The ordering in this appendix also does not reflect prioritization. Moreover, the COAs are not always independent from each other. Their interrelationships are explained in Chapters 4, 5 and 6 of this report.

The format used here is included for consistency with the related sector reports supporting the recovery plan. For each proposed COA's associated cost, a rough order-of-magnitude cost estimate to support high-level planning and inform decisionmaking, in 2018 dollars is included. The estimates represent only the costs for which a specific payment is made by some source to carry out a specific action; the estimates do not include all of the costs to society that may be associated with recovery actions. Some cost estimates are more precise than others, given the quality of data available at the time, between February 2018 and July 2018. The estimated costs may vary depending on the scale of implementation or other choices made between technical, financial, or policy options.

Cost information should be regarded as preliminary because more final cost estimates would require specificity about the implementation of recovery options, or access to information such as proprietary estimates from private telecommunications companies. There is insufficient information to provide even rough order-of-magnitude estimates for all contingencies, so some actions may have incomplete cost estimates. More information about the cost-estimate methodology and how potential funding sources were identified can be found at the project website.[1]

COAs also indicate possible implementers as they were included in the recovery plan, identified during the time period from February 2018 to July 2018. These implementers should be considered preliminary because details about how the COAs would be implemented would not be known until there is additional clarity about available funding. In particular, we note that the Puerto Rico Telecommunications Regulatory Board was reorganized just *after* the recovery plan was published and that the new name is used in this appendix. The Board was renamed the Telecommunications Bureau of the Public Service Regulatory Board under legislation enacted on August 12, 2018. This reorganization is part of the Act for the Implementation of the Reorganization Plan of the Puerto Rico Public Service Regulatory Board, Chapter III, Telecommunications Bureau, No. 211-2018 (H. B. 1408).

---

[1] RAND Corporation, undated.

# CIT 1
## Land Mobile Radio System

### Sectors Impacted

Communications and Information Technology, Municipalities

### Issue/Problem Being Solved

Puerto Rico needs a state-of-the-art, survivable, resilient communications infrastructure to ensure the continuity of essential government functions and the provision of public safety services after a disaster. Beyond supporting effective disaster communications, better communications infrastructure is critical to improving first-responder operations more broadly.

### Description

This course of action (COA) would implement a state-of-the-art, resilient public land mobile radio (LMR) system in Puerto Rico.[1] This COA would first assess 2 potential solutions: (1) upgrading and consolidating the current public LMR systems and their supporting microwave networks and (2) making an arrangement to join the U.S. Department of Homeland Security's Puerto Rico and the U.S. Virgin Islands Interoperable Communications Network Engagement (PRINCE) system—the federal LMR system—when available. This COA would next put together a plan that aligns Puerto Rico's build-out in parallel with the PRINCE system build-out. This plan would have several options to choose from, depending on when PRINCE is available to Puerto Rico. The options may be different for the short and long terms, and they may be different for voice and nonvoice applications. Finally, this COA would implement the plan and regularly assess First Responder Network (FirstNet) as a backhaul service provider, complementary service, or potential replacement for the Puerto Rico's first-responder system.

Currently, Puerto Rico public LMR communications consist of five systems:

- the Puerto Rico Police Department P25 trunk system, which provides statewide radio coverage to police, fire, hospitals, and other entities (all using the same set of dispatchers)
- the Puerto Rico Police Department SmartZone, which allows communications among police teams within the greater San Juan metropolitan area
- the emergency medical services (EMS) system, which services the Medical Emergency Corps of Puerto Rico, responsible for premedical care and emergency transport to hospitals
- the Puerto Rico Emergency Management Agency (PREMA) system, which provides day-to-day mobile radio communications between municipalities and PREMA

---

[1] The FEMA Communications/IT solutions-based team suggested this course of action in an internal report dated June 30, 2018.

- the Interoperability System, which provides statewide communications with municipalities only in emergency or urgent situations such as tsunamis, earthquakes, and many others.

These systems use several microwave networks as a backbone. These systems and networks are critical for the proper operation of 911 emergency response in Puerto Rico.

The ideal solution would entail use of P25 by the full first-responder force, consolidating disparate microwave networks down to one microwave backbone that would allow the use of data, and aggregating the capacities of each into a more capable and resilient backbone to service all agencies' voice and data requirements. Final determination for the number of sites would require a more detailed study.

The FirstNet Authority, within the U.S. Department of Commerce, was authorized by the U.S. Congress in 2012. Its mission is to develop, build, and operate the nationwide broadband network for first responders, including police, fire, and emergency medical services. The network is to be used in disasters, emergencies, and daily public safety service. The Authority entered into a 25-year public-private partnership with AT&T to develop this network throughout the United States and provided initial funding. It should be monitored and regularly assessed as a backhaul service provider, complementary service, and potential replacement for Puerto Rico's first responders' system.

## Potential Benefits

Aligning Puerto Rico's build-out with the federal LMR system build-out would provide increased interoperability for all stakeholders within Puerto Rico and the U.S. Virgin Islands, including all federal and commonwealth users. This COA would also potentially reduce maintenance and logistics costs and facilitate repairs, restoration, and equipment upgrades.

The differences among the systems currently in use in Puerto Rico (frequency bands, modulation protocols, and microwave networks) make interoperability difficult to achieve. Integration would allow for interoperability and eliminate the need for some EMS first responders to communicate via radio with their hospitals using two radio system dispatch operators. The importance of achieving interoperability may drive the decision to build an integrated LMR network to support P25 radios sooner, instead of waiting for PRINCE or FirstNet. Furthermore, it will enable communication among first-responder groups (e.g., police and EMS). Integration will also lead to reduced maintenance and logistics costs and facilitate repairs, restoration, and equipment upgrades.

## Potential Spillover Impacts to Other Sectors

This COA would benefit municipalities. It would enable or facilitate communications among municipal first responders (police, fire, EMS), Puerto Rico Emergency Management Services, and non–first responders from commonwealth agencies during day-to-day emergency operations and during disasters.

## Potential Costs

Potential up-front costs: $62 million in estimated up-front costs[2]
Potential recurring costs: $77 million in estimated recurring costs over an 11-year period
Potential total costs: $140 million in total estimated costs[3]

The up-front costs include estimated costs for upgrading and consolidating the public LMR systems ($18 million), for provisioning P25 radios to first responders ($42 million), and for workforce development ($1.5 million). See details below.

The estimate for upgrading and consolidating the current public LMR systems is $14 million–$22 million, which includes antennas and associated electronic equipment.[4] The recovery plan used the average of this range for the estimated up-front cost.

The estimate for the P25 radios to provision Puerto Rico's first responder force is $42 million. This was calculated based on an estimate by an official of the Puerto Rico Department of Public Safety that about 7,000 radios should be sufficient for the full first responder force, which includes police, fire and EMS,[5] and on an estimate by a Federal Emergency Management Agency (FEMA) telecommunications expert that the cost of each required P25 compatible radio was approximately $6,000.

The costs above do not include upgrading the infrastructure that supports each public safety telecommunications site, which includes land (e.g., foundation reinforcements), towers, building structures, and access roads. Moreover, future analyses are required to ensure operations and maintenance funding estimates account for environmental challenges, which are estimated to be up to 10% of overall systems costs per year. Costs for the government of Puerto Rico to lease or use the federal LMR system and FirstNet are uncertain.

The recurring costs of $7 million per year are estimated based upon the personnel required to maintain the LMR system. These costs were calculated as follows: The Puerto Rico Department of Public Safety estimated that 56 technical personnel, in addition to their current telecom personnel, would be sufficient to properly maintain the system and satisfy police, fire, EMS, and the Puerto Rico Emergency Management Agency (PREMA) requirements; and the cost of these personnel is estimated based upon a median salary for an engineer in San Juan, using an annual cost of $124,600 per person as the fully loaded labor cost.

The Puerto Rico Department of Public Safety also indicated that telecom workforce development in the form of a 4-year electrical engineering degree for 20 of the additional engineers would be necessary, and this is included as workforce development in the up-front cost

---

[2] The up-front cost published in the recovery plan represents an estimated $65 million. That estimate is corrected here to $62 million.

[3] The potential costs published in the recovery plan are precise within 2 significant figures. If costs appear not to sum properly, it is due to rounding.

[4] FEMA Hurricane Maria Communications Task Force, 2017.

[5] This does not mean that there are only 7,000 first responders in the island. There are many more than that. However, many of them already have P25 radios.

estimate. The cost of workforce development is estimated based upon 20 graduates completing a bachelor's degree in electrical engineering at a Puerto Rico university (University of Puerto Rico at Mayagüez). The cost of the degree is assumed to be $72,200, for a total of $1.5 million for the 20 new engineers.

## Potential Funding Mechanisms

FEMA Public Assistance, Community Development Block Grant—Disaster Recovery

## Potential Implementers

Puerto Rico Department of Public Safety, Telecommunications Bureau,[6] Office of the Chief Information Officer

## Potential Pitfalls

In the short term, an upgraded and consolidated system may be more expensive than upgrading individual, separate systems. That is because the preferred system (recommended by the FEMA Communications and Information Technology solutions-based team) uses P25 radios. These radios offer technical advantages and are the ones used in the federal LMR system, but they are more expensive than standard UHF/VHF (ultra high frequency/very high frequency) radios currently used by some of the Puerto Rico LMR systems.

If Puerto Rico joins the U.S. Department of Homeland Security's PRINCE, it would have limited influence in managing the system. Moreover, in case of damage from a hurricane or other disaster, its restoration would not be eligible for FEMA disaster relief funds. However, it can be expected that the federal government would provide the resources (including personnel) to restore it to full functionality as quickly as possible.

P25 multiband radios are more expensive than UHF/VHF portable handheld radios: P25 multiband radios range from $6,000 to $8,000 each, whereas the cost of a typical UHF/VHF portable handheld radio does not exceed $1,000.

The FEMA Communications and Information Technology solutions-based team suggested using P25 multiband radios for several reasons, among them are that:

- the use of multiband radios significantly increases interoperability
- a large part of the Puerto Rico first responder force—including the majority of police and fire—already have some of those radios
- the current police microwave network that services both police and fire is based on the P25 standard

---

[6] The recovery plan lists the PRTRB as a potential implementer. The board's name was changed to the Telecommunications Bureau of the Public Service Regulatory Board in August 2018, after the publication of the recovery plan. Please note that Telecommunications Bureau has been named as the implementer throughout this appendix.

- P25 is a standards-based system produced through the joint efforts of the Association of Public Safety Communications Officials International (APCO), the National Association of State Telecommunications Directors, selected federal agencies and the U.S. Department of Homeland Security (DHS) Office of Emergency Communications (OEC), and standardized under the Telecommunications Industry Association (TIA)
- P25 radios are flexible and can communicate in analog mode with legacy radios and in either digital or analog mode with other P25 radios
- the P25 standard includes a requirement for protecting digital communications (voice and data) with encryption capability, which is optional by the user
- the radio systems manager is able to remotely change encryption keys
- the P25 standard exists in the public domain, allowing any manufacturer to produce a P25 compatible radio product.

Ensuring that new P25 radios and the old P25 radios (used by police and fire) are compatible would be a priority.

## Likely Precursors

Before alignment with the federal LMR system can take place, significant improvement and modification of existing LMR systems are required.

## References

FEMA Communications/IT solutions-based Team, "Puerto Rico Communications/IT Solutions-based Team Report," June 30, 2018.

FEMA Hurricane Maria Communications Task Force, *DR-4339-PR Consolidated Communications Restoration Plan*, October 2017.

# CIT 2
## Puerto Rico GIS Resource and Data Platform

## Sectors Impacted

Communications and Information Technology, Energy, Economic, Health and Social Services, Housing, Transportation, Municipalities, Education

## Issue/Problem Being Solved

A centralized system for all commonwealth agencies and municipalities that comprehensively incorporates geographic information system (GIS) data for public safety, emergency response, and community planning efforts and informs decisionmaking for seamless coordination in the event of a disaster is presently not available in Puerto Rico.

## Description

This course of action (COA) would adopt a centralized network that all agencies can access, and it would establish a self-governing body that would set standards for GIS products and protocols for data use. Formation of this body would be informed by inputs from technical experts in Puerto Rico to determine how GIS data can be best utilized.

The GIS personnel in this COA would provide information technology, GIS technical support, and system and application support across all sectors. This team would create, sustain, and enhance the GIS infrastructure that would be needed to support GIS applications in, and service requests from, other sectors. This team would also manage access and storage of GIS data.

Concerning access to privately owned GIS data, the government of Puerto Rico would ensure appropriate memoranda of understanding (MOUs) are signed with those companies so that it will have access to such data during an emergency. To reassure those companies that their data would be protected, the government of Puerto Rico would ensure that appropriate security measures that adhere to federal standards are in place.

Puerto Rico needs a centralized GIS system for all its state agencies and municipalities in order to accomplish the following objectives for day-to-day operations:

- guide growth—use maps and analytics to guide open space and housing-stock planning, redistribution, and development, which the Planning Board is endeavoring to do with Community Development Block Grant Program as Disaster Recovery (CDBG-DR) funds
- plan for sustainability—model and compare plans for resiliency and prosperity
- improve quality of life—understand the movement of people, goods, and services to inform development decisions and to ensure continuity of supply chains
- engage the citizens of Puerto Rico—use smart maps to gather input and crowdsource ideas with citizen and business leaders.

These goals are the foundation for a GIS system that is fully capable of seamlessly supporting Puerto Rico during a disaster.

To move toward an integrated GIS system, Puerto Rico would adopt a centralized network that all agencies can access and establish a self-governing body set standards for GIS products and protocols for data use.

As a first step toward establishing a centralized self-governing body, Puerto Rico would convene a half-day meeting of technical experts in Puerto Rico public safety, emergency management, and community planning, to scope out how GIS data, particularly Federal Emergency Management Agency (FEMA) GIS data, can help them do their jobs even better. The group of experts would identify which Puerto Rico agencies need to be part of a government-wide plan to share and use GIS data about Puerto Rico. These experts would outline the scope and topics for a subsequent one-day meeting of Puerto Rico agency administrators and their GIS staff about the use and sharing of GIS data concerning Puerto Rico. The first portion of the day would focus on a high-level view of how GIS could enhance emergency management and communication in a disaster situation. The second half of the day would be a more technical conversation for the GIS analysts. Topics may include the data layers to share, how data may be shared among agencies as well as protected, the quantity and maintenance of data that will be shared, and best practices for dealing with third-party contractors. One outcome of the administrators' meeting would be to establish a GIS technical working group that would meet monthly for 6 months and then quarterly, which is consistent with the United Nations Economic and Social Council newly adopted resolution on the Strategic Framework on Geospatial Information and Services for Disasters.[1] A deliverable of the technical working group would be a template for collecting and sharing GIS data that will be provided to each agency and the municipalities, as well as a schedule for follow-up trainings across Puerto Rico.

The GIS personnel in this COA will provide IT/GIS technical, system, and application support across all sectors. Specific COAs from other sectors that will make use of this infrastructure are *HOU 5 Collect, Integrate, and Map Housing Sector Data, NCR 30 Create an Accessible Data Repository of Natural and Cultural Resources, ENR 11 Design and Deploy Technologies to Improve Real-Time Information and Grid Control, CPCB 3 Capacity Building to Incorporate Hazard Risk Reduction into Planning and Design, WTR 18 Invest in Stormwater System Management*, and *TXN 2 Harden Vulnerable Transportation Infrastructure*. All of them will make use of GIS data from the point of view of an end-user.

### Potential Benefits

Collecting and sharing GIS data in a uniform way across public safety, emergency response, and community planning agencies would greatly improve the government of Puerto Rico's responsiveness in the event of a crisis. It would also reduce the time needed to create actionable

---

[1] United Nations Economic and Social Council, "Strategic Framework on Geospatial Information and Services for Disasters," June 20, 2018.

strategies to deal with crises, such as hurricanes and earthquakes, as well as rebuilding and restoration efforts following a crisis.

## Potential Spillover Impacts to Other Sectors

The availability of GIS data across public safety, emergency response, and community planning agencies would have a positive impact on the resilience of the infrastructure of utilities, especially telecommunications infrastructure. It would also be important for determining transportation routes and strategies in the event of a crisis. Granular information about the power grid, public safety communications, telecommunications infrastructure, access roads, population, and location of medical facilities would all speed up the government of Puerto Rico's response in a crisis. GIS data could also provide the government with real-time information to advise public safety and emergency responders. Similarly, GIS data could assist communities with rebuilding efforts following a crisis.

## Potential Costs

Potential up-front costs: $1 million in estimated up-front costs
Potential recurring costs: $41 million–$44 million in estimated recurring costs (11 years)
Potential total costs: $42 million–$45 million in total estimated costs

The cost estimates assume that most of the hardware costs for this COA will be included in *CIT 17 Puerto Rico Data Center* and *CIT 14 Consolidated Government Information Systems*. Moreover, costs for IT infrastructure support for this COA will also be included in CIT 17 and CIT 14. Therefore, we avoid double-counting these investments; however, we estimate that additional up-front hardware costs specific to GIS resources will be $1 million.

The recurring costs depend on how many GIS professionals are involved in the effort. These personnel costs are estimated to be $683,000 annually, which reflect fully loaded labor costs for Puerto Rico–based staff, including a system administrator, 3 GIS specialists, 1 GIS lead and an administrative assistant. The annual costs are assumed to be $124,600 for the first 3 specialties, and $62,300 for the fourth specialty. A more robust staff with 6 GIS specialists is estimated to cost $1 million per year. Personnel composition was provided by FEMA GIS experts. The recurring cost for GIS software and 20 licenses was assumed to be $3 million, based on the professional judgment of FEMA GIS experts. In total, the annual cost will range between $3.7 million and $4.1 million, depending on the number of GIS professionals involved in the effort.

## Potential Funding Mechanisms

FEMA, Federal Communications Commission

## Potential Implementers

Office of the Chief Information Officer, Office of the Chief Innovation Officer

## Potential Pitfalls

Government agency administrators might not collaborate to share GIS data unless a cross-agency core group of GIS technical experts works together to create a coherent plan to do so.

## Likely Precursors

Precursor COAs are *CIT 10 Transoceanic Submarine Cable*, *CIT 17 Puerto Rico Data Center*, *CIT 18 Data Storage and Data Exchange Standards for Critical Infrastructure*, *CIT 21 Government-Owned Fiber-Optic Conduits to Reduce Aerial Fiber-Optic Cable*, and *CIT 24 Establish Puerto Rico Communications Steering Committee.*

## Reference

United Nations Economic and Social Council, "Strategic Framework on Geospatial Information and Services for Disasters," June 20, 2018. As of April 14, 2019:
https://undocs.org/E/2018/L.15

# CIT 3
# Upgrade and Enhance 911 Service

## Sectors Impacted

Communications and Information Technology, Community Planning and Capacity Building, Municipalities

## Issue/Problem Being Solved

Puerto Rico needs a state-of-the-art, survivable, resilient communications infrastructure to ensure the continuity of essential government functions and the provision of public safety services after a disaster.

Before the hurricanes, calls to 911 were answered at the public safety answering point (PSAP) and relayed over the public switched telephone network (PSTN) to the appropriate public safety agency, which handles dispatch. Because the PSTN was damaged by the hurricanes, the PSAP was unable to forward calls to many of Puerto Rico's police, fire, and emergency medical service (EMS) agencies.[1] This is despite the fact that the PSAPs remained largely operational, functioning on backup generators when commercial power was lost.[2] There is no radio communication in the PSAP.[3]

## Description

This course of action (COA) would upgrade the current 911 network to an Emergency Services IP Network (ESINet), implement Next Gen 911 (NG911), consolidate dispatch at the PSAP, and coordinate with government of Puerto Rico agencies in the housing sector for the adoption of E911 address conversion of rural route addresses.

## Potential Benefits

Moving to NG911 brings new features, such as automatic location information/automatic numbering information (ALI/ANI) and the ability to share photo, video, and Global Positioning System (GPS) location with first responders, all of which should improve the general effectiveness of 911 services.[4] Consolidating PSAP and dispatch would also help drive down 911 response times by removing redundant intermediaries, and it would improve the resiliency of the system. Converting rural route addresses to street-style addresses would ensure that emergency services can quickly and easily locate individual properties.

---

[1] DHS OEC, 2018.

[2] FEMA Hurricane Maria Communications Task Force, 2017.

[3] FEMA Hurricane Maria Communications Task Force, 2017.

[4] As we said in Chapter 4 of this report, Puerto Rico already had text-to-911 capability prior to Maria.

In moving to an IP-based network and by consolidating dispatch at the PSAP, there would be less reliance on the PSTN. In addition, the 911 call-answering process would be simplified by eliminating the need to forward calls to a separate dispatch center.

## Potential Spillover Impacts to Other Sectors

This COA would help improve disaster coordination at the community and municipality levels; thus, the sectors affected are community planning and capacity building and municipalities.

Currently, all 911 calls are answered at the PSAP in San Juan before being relayed to police, fire and EMS agencies throughout Puerto Rico. In order to consolidate dispatch, coordination would be needed among the various agencies and municipalities involved.

## Potential Costs

Potential up-front costs: $2 million–$6 million in estimated up-front costs
Potential recurring costs: $1 million in estimated recurring costs (11 years)
Potential total costs: $3 million–$7 million in total estimated costs

The estimated up-front costs to modernize the PSAP, including implementing ESINet, range from $2 million[5] to $6 million.[6] The estimated recurring cost includes maintenance and operations costs.

## Potential Funding Mechanisms

U.S. Department of Commerce

## Potential Implementers

Puerto Rico 911 Service Governing Board

## Potential Pitfalls

Strong governance and leadership with respect to the management of telecommunications and emergency services in Puerto Rico are essential for the successful implementation of this COA.

The Department of Homeland Security (DHS) Office of Emergency Communications (OEC) Interoperable Communications Technical Assistance Program identified governance as a key issue with respect to the emergency and public safety communication in Puerto Rico.[7] Many emergency agencies own and maintain independent systems. Getting these various agencies to coordinate in order to consolidate dispatch may be challenging.

---

[5] FCC, "A Next Generation 911 Cost Study," September 2011.

[6] FEMA Hurricane Maria Communications Task Force, October 2017.

[7] FEMA Hurricane Maria Communications Task Force, October 2017.

### Likely Precursors

Precursor COAs are *CIT 5 Implement Public Safety/Government Communications Backup Power*, *CIT 24 Establish Puerto Rico Communications Steering Committee*, and *HOU 11 Improve the Address System*. Having a generator and backup power were essential to keeping the PSAP functioning after the hurricane.[8] Thus, backup power is essential for the continuation of operations when commercial power is unavailable, which is addressed by *CIT 5 Implement Public Safety/Government Communications Backup Power*. Given the potential pitfalls, a steering committee could help coordinate police, fire, and EMS, as addressed by *CIT 24 Establish Puerto Rico Communications Steering Committee*.

### References

DHS Office of Emergency Communications, Interoperable Communications Technical Assistance Program, "Puerto Rico Public Safety Communications Summary and Recommendations Report," February 2018.

Federal Communications Commission, "A Next Generation 911 Cost Study," September 2011. As of October 29, 2018: https://www.911.gov/pdf/FCC_Next_Generation_911_Cost_Study_2011.pdf

FEMA Hurricane Maria Communications Task Force, *DR-4339-PR Consolidated Communications Restoration Plan*, October 2017.

---

[8] FEMA Hurricane Maria Communications Task Force, 2017.

# CIT 4
## Rural Area Network Task Force

## Sectors Impacted

Communications and Information Technology, Health and Social Services, Community Planning and Capacity Building, Municipalities

## Issue/Problem Being Solved

Advance public safety and health care delivery to loosely connected communities (e.g., "rural areas") by providing comprehensive, real-time situational awareness on weather conditions, emergency guidance, and medical needs. This course of action (COA) seeks to accommodate the special circumstances of individuals residing in disconnected regions by understanding and engaging their needs directly. It further seeks to bridge a gap in information systems to take advantage of broadband when it becomes available to service the specific needs of individuals in these areas should a disaster occur.

## Description

This COA would establish a task force for the development of a communications network and an information system (or networks and systems), with a focus on the public safety and health care needs of people situated in rural or disconnected areas of Puerto Rico, especially those who are isolated, have limited mobility, or are elderly, as well as their caregivers. The information systems and communications network would include devices within homes or at designated local stations, rely on new infrastructure (e.g., micro rings, First Responder Network Authority [FirstNet], Rescue 21), be practical and efficient, operate in the immediate aftermath of a disaster, and provide key information (e.g., guidance, health issues) prior to and after a disaster for residents, emergency services, and medical providers alike.

In the event of an emergency or disaster, this information system would enable better quality and timeliness of information for greater coordination between medical facilities or health care providers and emergency services, to improve survival for the people of Puerto Rico in the wake of a disaster. The ability of individuals of all ages (from children to elderly) to communicate with service providers about their situation should be greatly improved compared with past experiences.

## Potential Benefits

This COA would be a first step toward improving survivability (avoiding loss of life) and the medical health of people for whom the current limited communications infrastructure constrains the quality or timeliness of required health or medical services.

## Potential Spillover Impacts to Other Sectors

Extending the communications infrastructure to poorly connected areas can improve the readiness, planning, and allocation of resources for services sourced from medical facilities and emergency responders. The government of Puerto Rico would have an improved, detailed awareness of disaster-related issues, which would better inform decisions and coordination activities at a governmental level.

## Potential Costs

Potential up-front costs: $400,000–$800,000 in estimated up-front costs[1]
Potential recurring costs: —
Potential total costs: $400,000–$800,000 in total estimated costs

The up-front cost is estimated based upon a range of full-time staffing of a task force of subject matter experts for 1 year with 6–12 staff, at a cost of $62,300 per staff member. Thus, the up-front costs are estimated to be between $400,000 and $800,000.

This COA will need to be executed in parallel with and rely on other Communications/ Information Technology COAs, as explained in Chapter 4 of this report. Detailed recommendations for an information system(s) and a communications network will correspondingly depend on the decisions made for implementing underlying systems. A greater range of interoperability may be required to accommodate flexibility in requirements for dependent systems. Costs for additional networks systems that may be recommended by the task force are not included.

## Potential Funding Mechanisms

Government of Puerto Rico, private sector

## Potential Implementers

Telecommunications Bureau, Puerto Rico Emergency Management Agency (PREMA)

## Potential Pitfalls

The proposed solutions could potentially focus more on the availability of a municipality- or facility-level network infrastructure than on acute public safety and health care information and communication needs of people situated remotely from quality medical and emergency services in the wake of a disaster. Although the availability of a network infrastructure is a major and critical component of communications, it only partly addresses the challenge of delivering public safety and health care services to rural regions in a way that maximizes their benefit to the

---

[1] These dollar figures are rounded off to one significant figure. The figures that are not rounded off are $375,000 for the 6-member task force and $750,000 for the 12-member task force.

population residing in those areas, especially in the wake of a disaster. In this regard, the goals of governmental acts cited for "insular and rural areas" are at risk of being undermined.

Should the scope of the effort expand too broadly, there is also a risk for duplicative information systems, which could have numerous negative implications for development costs arising from increased complexity, data quality and privacy issues, and interoperability conditions.

### Likely Precursors

The availability of reliable power and broadband (or corresponding communications backbone) throughout Puerto Rico is necessary for this COA. Precursors COAs are *CIT 23 Data Collection and Standardization for Disaster Preparedness and Emergency Response, CIT 24 Establish Puerto Rico Communications Steering Committee, CIT 29 Health Care Connectivity to Strengthen Resilience and Disaster Preparedness, CIT 6 Modernize the Emergency Operations Center*, and *CPCB 9 Coordinated Local Recovery Planning Process*.

# CIT 5
# Implement Public Safety/Government Communications Backup Power

## Sectors Impacted

Communications and Information Technology, Energy, Natural and Cultural Resources, Transportation

## Issue/Problem Being Solved

Puerto Rico's communications and information technology (IT) networks are dependent on the energy grid and lack resilience and redundancy. During Hurricanes Irma and Maria, the lack of sustained electrical power had a significant impact on communications and IT networks.

According to an October 2017 U.S. Department of Homeland Security (DHS) report, both public safety answering points (PSAPs) in Puerto Rico relied on generator power post–Hurricane Maria.[1] Due to the importance of PSAPs for 911, this demonstrates the importance of backup power sources. While PSAPs utilized backup power sources, the DHS report noted that 911 was limited or had no ability to relay call information to police, fire, and emergency medical service (EMS) stations across Puerto Rico. While the report does not detail the specific reasons, lack of power likely contributed to this challenge. Implementing backup power sources for public safety and government operations would increase resilience of communications and lessen the dependence on the electric grid following a disaster.

## Description

Lack of sustained electrical power had a significant impact on communications and IT networks during Hurricanes Irma and Maria. This course of action (COA) would invest in backup power sources, using standardized equipment where possible and appropriate, to provide the public safety and government communications networks with alternate power sources in the event of damage or destruction to energy infrastructure. A redundant portfolio of power sources would improve the resilience of public safety and government communications networks, thus helping to ensure government operations and emergency response in the event of a catastrophic loss of power across Puerto Rico.

Specifically, this COA would implement backup power systems for public tower sites; hospitals; police, fire, and EMS stations; municipal city halls; and government centers identified by the Puerto Rico Public Buildings Authority to allow for the continuity of government (COG) operations and emergency response. Having a diversified portfolio of power sources including but not limited to renewable energy sources such as solar, wind, and water could aid in ensuring resilience for public safety and government communications networks in the event of a

---

[1] FEMA Hurricane Maria Communications Task Force, 2017.

catastrophic loss of power across Puerto Rico. Careful consideration would be needed to determine the most appropriate mix of alternate sources depending on the region's needs. In addition, first responders could use small-suitcase, solar-based cell towers to provide redundancy and immediate restoration during natural disasters.

The Communications/ Information Technology solutions-based team (SBT) assumed backup power in certain instances (such as carriers to maintain 2–5 days of backup power at wireline facilities; mobile switching centers to have 72 hours of power; cell sites to have 4–8 hours of backup power).[2]

### Potential Benefits

Using backup power sources could increase the resilience and redundancy of Puerto Rico's communications networks. Renewable sources, including solar, wind, and water systems, have the potential to be self-sustaining, freeing resources for use by systems that must rely on nonrenewable sources. Use of standardized equipment where possible and appropriate could aid maintenance and repairs should Puerto Rico have access to relevant spare parts and components, according to Federal Emergency Management Agency (FEMA) officials.

Given that this COA would implement backup power solutions for public safety and government networks, emergency response and government operations could benefit from its implementation. By ensuring government agencies including municipalities and emergency services maintain power for telecommunications needs, essential functions could continue under emergency conditions.

### Potential Spillover Impacts to Other Sectors

The use of backup power systems could make Communications/Information Technology to a certain extent less dependent on the power grid (energy sector). As Puerto Rico makes improvements to the energy grid and reliability increases, there may be less dependence on backup power. However, these improvements may not be made in the near and midterm and therefore backup power would still be necessary.

Using fuel-based generators as backup power could affect transportation networks because of the need to distribute fuel. Fuel-based generators may also affect air quality or cause fuel spills that affect groundwater, flora, and fauna.

### Potential Costs

Potential up-front costs: $20 million in estimated up-front costs[3]

---

[2] FEMA Communications/IT solutions-based team, 2018.

[3] This is the average of the low-end and the high-end estimates in the description of potential costs provided in the text, and rounded off.

Potential recurring costs: $10 million in estimated recurring costs (11 years) [4]
Potential total costs: $30 million in total estimated costs

These estimates assume an average cost of $40,000 per backup system at a site.[5] The up-front costs are an average of a low-end estimate (452 sites) and a high-end estimate (537 sites). The low-end estimate has 30 public tower sites;[6] 68 hospitals;[7] 78 police, 93 fire, and 56 EMS stations;[8] 78 municipal city halls;[9] and 49 government centers identified by the Public Buildings Authority.[10] The upper-end estimate has 30 public tower sites; 68 hospitals; 78 police, 93 fire, and 56 EMS stations; 78 municipal city halls;[11] and 134 government agencies.

The recurring costs account for operations and maintenance at these sites is estimated at $1,900 per site.[12] Therefore, the low-end estimate is $18.1 million in up-front costs and $859,000 in annual costs for operations and maintenance. The high-end estimate is $21.5 million in up-front costs and $1 million in annual costs for operations and maintenance.

## Potential Funding Mechanisms

Hazard Mitigation Development Program, Community Development Block Grant Disaster Recovery, government of Puerto Rico

## Potential Implementers

Government of Puerto Rico

## Potential Pitfalls

Logistical and maintenance support for fuel-based generators may be more challenging in remote or mountainous areas. Investments in alternate/renewable energy sources, such as solar, wind, and water, may not be appropriate in some regions, too expensive, or less efficient than the current fuel-based generators, and they may require a more structured maintenance program to ensure effective operation. Additional assessments regarding the survivability of such technologies in view of future extreme weather events would be important. In addition, instances

---

[4] The average of the low-end and the upper-end estimates for annual costs is approximately $930,000.

[5] This estimate is based on PlugPower, undated.

[6] Government of Puerto Rico telecommunications official, discussion with the authors, April 10, 2018.

[7] Governor of Puerto Rico, 2017.

[8] FEMA Hurricane Maria Communications Task Force, 2017.

[9] Based on discussion with RAND Corporation staff to confirm each municipality has a city hall.

[10] Puerto Rico Public Buildings Authority representative, interview with the HSOAC Public Buildings team, March 29, 2018.

[11] Based on discussion with RAND Corporation staff to confirm each municipality has a city hall.

[12] PlugPower, undated.

of theft of generator fuel stocks in the aftermath of Hurricane Maria were noted by the Puerto Rico Telecommunications Regulatory Board. This would be another potential pitfall associated with fuel-based generators.

According to the Puerto Rico's Chief Innovation Officer, solar technology is very expensive to import to Puerto Rico.[13] This could be a limitation in expanding the use of solar technology as a backup power system.

## Likely Precursors

None

## References

FEMA Hurricane Maria Communications Task Force, *DR-4339-PR Consolidated Communications Restoration Plan*, October 2017.

Governor of Puerto Rico, *Build Back Better: Puerto Rico—Request for Federal Assistance for Disaster Recovery*, November 2017. As of April 24, 2018:
https://www.governor.ny.gov/sites/governor.ny.gov/files/atoms/files/Build_Back_Better_PR.pdf

PlugPower, *Comparing Backup Power Options for Communications, undated.* As of October 29, 2018:
https://www.plugpower.com/wp-content/uploads/2015/07/FCvGen_Stat_F1_101416.pdf

---

[13] Puerto Rico Chief Innovation Officer, interview with authors, May 11, 2018. We were unable to independently corroborate the CIO's statement on solar technology costs.

# CIT 6
## Modernize the Emergency Operations Center

### Sectors Impacted

Communications and Information Technology, Community Planning and Capacity Building, Municipalities

### Issue/Problem Being Solved

Puerto Rico needs a state-of-the-art, survivable, and resilient communications infrastructure to ensure the continuity of essential government functions and the provision of public safety services after a disaster.

In the May 2018 report by the Federal Emergency Management Agency (FEMA) solutions-based team (SBT), the current Emergency Operations Center (EOC) was identified as not meeting industry best practices or standards. Consequently, SBT recommends modernizing the systems and equipment in the EOC.[1]

### Description

This course of action (COA) upgrades and modernizes the EOC, which functions as a centralized location for the coordination and management of disasters, according to FEMA guidance.[2]

### Potential Benefits

An upgraded EOC, with improved data and communications systems, would better support response operations, incident management, and decisionmaking processes, and it would enhance emergency responders' ability to manage disaster response and recovery.

According to the SBT report, many systems in the EOC require modernization as the equipment is at or past end-of-life. Included in the report are several recommendations for upgrading and improving the current EOC. A modernized EOC should include video teleconference (VTC) facilities for communication and coordination. Critical government systems should be hosted in the EOC. A modernized EOC should include a monitoring center for surveillance systems, and it should employ emergency management software for real-time information sharing.[3]

---

[1] FEMA Communications/IT solutions-based team, 2018.

[2] FEMA, 2009.

[3] FEMA, 2009.

## Potential Spillover Impacts to Other Sectors

This COA would help improve disaster coordination at the community and municipality levels; thus, the other sectors affected are community planning and capacity building and municipalities.

## Potential Costs

Potential up-front costs: $250,000–$6.3 million in estimated up-front costs[4]
Potential recurring costs: $6.7 million in estimated recurring costs (11 years)[5]
Potential total costs: $7 million–$13 million in total estimated costs[6]

For the potential up-front costs, the unit cost estimates are derived from the U.S. Department of Transportation Costs Database.[7] The range of potential up-front costs assumes a low-end up-front cost of $250,000, which includes only hardware and software costs, and a high-end up-front cost of $6.3 million, which includes both the cost of a new facility and the hardware and software costs for the facility. Our high-end estimate assumes that a new facility will be required. If the current facility can meet the needs and requirements of an upgraded and fully modernized EOC, then the cost to implement this COA could be significantly less than the high-end estimate. The Federal Communications Commission (FCC) provides a checklist for assessing the hazards, vulnerability, and risks to an existing EOC. This checklist addresses the suitability of the facility, including its location and size, as well as its sustainability, including the availability of sources of backup power such as generators.

The potential recurring costs include facility operations and maintenance and additional labor. Operations and maintenance costs for the facility are estimated at $360,000 per year. This reflects an average annual operations and maintenance cost of approximately $6 per square foot.[8] The current Puerto Rico EOC is located inside the Puerto Rico Emergency Management Agency (PREMA) headquarters. For estimating purposes, the EOC facility is assumed to span between 50,000 and 70,000 square feet, and the average size of 60,000 square feet is used as the basis of

---

[4] The high-end estimate for the up-front costs did not capture the cost of the hardware and software for the new facility as represented in the recovery plan. The high-end estimate is corrected here.

[5] The recovery plan represents the potential recurring costs as $10 million over an 11-year period. The recurring cost estimate is updated here (1) using a labor cost consistent with other COAs and (2) assuming a smaller size EOC with reduced square footage as the basis for the operations and maintenances costs. The size of the facility was revisited after the publication of the recovery plan, resulting in this update.

[6] The recovery plan represents the total costs as a range of $11 million–$16 million. The total estimated costs are updated here to reflect the reduced recurring cost estimate.

[7] U.S. Department of Transportation, Office of the Assistant Secretary for Research and Technology, Intelligent Transportation Systems Joint Program Office, Costs Database, webpage, undated.

[8] In 2018, the average operations and maintenance cost for a U.S. municipal building was $5.66 per square foot.

estimate for the recurring cost. Also included in the recurring cost are the labor costs for 2 additional personnel at the existing EOC, estimated at $250,000 per year.[9]

## Potential Funding Mechanisms

FEMA Emergency Management Performance Grant, government of Puerto Rico

## Potential Implementers

Puerto Rico Emergency Management Agency, Telecommunications Bureau

## Potential Pitfalls

None

## Likely Precursors

*CIT 5 Implement Public Safety/Government Communications Backup Power* is a likely precursor because backup power is essential for the continuation of operations when commercial power is unavailable.

## References

Federal Emergency Management Agency, *Emergency Operations Center Assessment Checklist*, Washington, D.C., August 7, 2009.

FEMA Communications/IT Solutions-based Team, "Puerto Rico Communications/IT Solutions-based Team Report," June 30, 2018.

U.S. Department of Transportation, Office of the Assistant Secretary for Research and Technology, Intelligent Transportation Systems Joint Program Office, Costs Database, webpage, undated. As of June 20, 2018: https://www.itscosts.its.dot.gov/its/benecost.nsf/SubsystemCosts?ReadForm&Subsystem =Emergency+Response+Center+(ER)

---

[9] In the cost estimate published in the recovery plan, a lower personnel cost was used as the basis of estimate—only $100,000 per staff member as compared with $124,600 per staff member used in the current estimate.

# CIT 7
## Establish an Alternate Emergency Operations Center

## Sectors Impacted

Communications and Information Technology, Community Planning and Capacity Building, Municipalities

## Issue/Problem Being Solved

Puerto Rico needs a state-of-the-art, survivable, and resilient communications infrastructure to ensure the continuity of essential government functions and the provision of public safety services after a disaster.

In its May 2018 report, the Federal Emergency Management Agency (FEMA) Communications and Information Technology solutions-based team (SBT) recommend the establishment of an alternate Emergency Operations Center (EOC). The alternate EOC would function primarily as a backup facility for emergency management activities, but it would also include facilities and equipment to function as an alternate public safety answering point (PSAP) and serve as a continuity of government (COG) and continuity of operations (COOP) site. A key element of this course of action (COA) is to establish the alternate EOC outside of San Juan, where the two PSAPs and current EOC are located. This ensures an extra layer of redundancy by providing a backup location where emergency and other critical functions can be carried out when the sites in San Juan are degraded or disabled.[1]

## Description

This COA establishes an *alternate* EOC outside the San Juan area, and it establishes a PSAP in the alternate EOC to serve as a backup to the primary PSAPs in San Juan. The alternate EOC would be designed to serve as a COG and COOP site for non-EOC activities. The alternate EOC would not be located in vulnerable areas, such as floodplains or tsunami zones.

## Potential Benefits

An alternate EOC would serve as a backup location for emergency management activities should the primary EOC in San Juan become inoperable. The primary benefit of establishing an alternate EOC is to serve as a backup location for emergency management activities should the primary EOC in San Juan become inoperable. In designing and implementing an alternate EOC, several secondary benefits would be achieved. For example, this COA envisions establishing a PSAP in the alternate EOC to serve as a backup to the primary PSAPs in San Juan. Furthermore,

---

[1] FEMA Communications/IT Solutions-based Team, 2018.

the alternate EOC would be designed to serve as a COG and COOP site for non-EOC activities. The alternate EOC will not be located in vulnerable areas such as floodplains or tsunami zones.

The primary outcome of this COA is to provide an alternate location to support emergency and disaster management functions. However, the alternate EOC would be designed to support additional, non-EOC activities. In addition, the alternate EOC would improve the availability of the government of Puerto Rico's operational infrastructure by designating a site, outside San Juan, that has redundant, fault-tolerant telecommunications systems, including satellite backup.

## Potential Spillover Impacts to Other Sectors

This COA would help improve disaster coordination at the community and municipality levels; thus, the sectors affected are community planning and capacity building and municipalities. If the EOC is built to support COG and COOP activities, then there must be coordination with the municipal governments. We would expect they would be part of the requirements gathering process and involved in any training or exercises conducted at the EOC.

## Potential Costs

Potential up-front costs: $6.3 million in estimated up-front costs
Potential recurring costs: $6.7 million in estimated recurring costs (11 years)[2]
Potential total costs: $13 million in total estimated costs[3]

The unit cost estimates are derived from the U.S. Department of Transportation Costs Database.[4] The estimate includes capital costs of $6 million for an EOC facility,[5] as well as hardware plus software costs of $250,000 for the equipment that is to be installed at the facility. The cost depends on many factors, such as the location and size of the facility that is needed to support the missions and responsibilities normally carried out by the primary EOC. The alternate EOC must also be able to function as an alternate PSAP and COG/COOP site.

The potential recurring costs include facility operations and maintenance and additional labor. Operations and maintenance costs are estimated at $360,000 per year. This reflects an average annual operations and maintenance cost of approximately $6 per square foot for a 60,000 square foot facility.[6] For estimating costs, EOC facilities that span between 50,000 and

---

[2] The recovery plan represents the potential recurring costs as $10 million over an 11-year period. The recurring cost estimate is updated here (1) using a labor cost consistent with other COAs and (2) assuming a smaller size EOC with reduced square footage as the basis for the operations and maintenances costs. The size of the facility was revisited after the publication of the recovery plan, resulting in this update.

[3] The recovery plan represents the total costs as $17 million. The total estimated costs are updated here to reflect the reduced recurring cost estimate.

[4] U.S. Department of Transportation, Office of the Assistant Secretary for Research and Technology, undated.

[5] This is the same cost as the upper limit for upgrading the existing EOC in *CIT 6 Modernize the Emergency Operations Center*.

[6] In the recovery plan, a facility that was twice as large was used as the basis of estimate.

70,000 square feet were considered reasonable, with the average of that range used for the recurring cost estimated here. Labor costs for 2 additional personnel to staff the alternate EOC are estimated at $250,000 per year.[7]

## Potential Funding Mechanisms

FEMA Emergency Management Performance Grant, government of Puerto Rico

## Potential Implementers

Puerto Rico Emergency Management Agency, Telecommunications Bureau

## Potential Pitfalls

None

## Likely Precursors

Precursor COAs are *CIT 3 Upgrade and Enhance 911 Service, CIT 5 Implement Public Safety/Government Communications Backup Power*, and *CIT 6 Modernize the Emergency Operations Center*. Prior to establishing an alternate PSAP, the government of Puerto Rico should upgrade and enhance the current EOC in San Juan, bringing it up to industry standards and upgrading end-of-life systems. In addition, governance and infrastructure issues related to the consolidation of dispatch should be addressed prior to taking action on this COA. Backup power is essential for continuation of operations when commercial power is unavailable. Alternate resilient communications paths must be provided to ensure communications are available in case the primary path fails or is disrupted.

## References

FEMA Hurricane Maria Communications Task Force, *DR-4339-PR Consolidated Communications Restoration Plan*, October 2017.

U.S. Department of Transportation, Office of the Assistant Secretary for Research and Technology, Intelligent Transportation Systems Joint Program Office, Costs Database, webpage, undated. As of June 20, 2018: https://www.itscosts.its.dot.gov/its/benecost.nsf/SubsystemCosts?ReadForm&Subsystem =Emergency+Response+Center+(ER)

---

[7] In the cost estimate published in the recovery plan, a lower personnel cost was used as the basis of estimate—only $100,000 per staff member as compared with $124,600 per staff member used in the current estimate.

# CIT 8
## Mobile EOC Vehicle

### Sector Impacted

Communications and Information Technology, Community Planning and Capacity Building, Municipalities

### Issue/Problem Being Solved

Puerto Rico needs a state-of-the-art, survivable, and resilient communications infrastructure to ensure the continuity of essential government functions and the provision of public safety services after a disaster.

The May 2018 report by the Federal Emergency Management Agency (FEMA) solutions-based team (SBT) recommended that Puerto Rico procure a mobile Emergency Operations Center (MEOC) similar to the National Incident Management System Type-2 Mobile Command Center vehicles that arrived in Puerto Rico after the Hurricanes Maria and Irma.[1]

### Description

This course of action (COA) calls for an MEOC, which is a deployable asset that functions as a centralized location for the coordination of disaster recovery and first-responder activities when other infrastructure is disabled. The mobile Emergency Operations Center (EOC) would be able to operate independently from remote and austere locations.

### Potential Benefits

An MEOC is a useful tool when coordinating an on-site response to a local emergency or disaster. It provides independent communications over civilian and military frequencies, cellular or satellite. It can generate its own power. It may also contain computers to run incident management software and video teleconferencing (VTC) equipment. In addition to being used for emergency purposes, an MEOC can function as a command center to monitor special events.

According to an October 2017 U.S. Department of Homeland Security (DHS) report, the communications infrastructure in Puerto Rico was severely damaged by hurricanes Irma and Maria.[2] Only 5% of cell towers were operational. Similarly, most of the towers that provide backhaul communications for police, fire, and emergency medical services (EMS) were damaged, destroyed, or without power. In addition, most of the public switched telephone network (PSTN) lines connecting public safety agencies were nonfunctional and needed to be repaired or replaced. Together these communications failures severely affected the ability of first

---

[1] FEMA Communications/IT Solutions-based Team, 2018.

[2] FEMA Hurricane Maria Communications Task Force, 2017.

responders and emergency managers to communicate and coordinate disaster response activities. In order to provide some amount of survivable communications capability, the SBT recommended that Puerto Rico procure an MEOC. It will enable HF/VHF/UHF radio communications and transmission of voice and data over satellite.

## Potential Spillover Impacts to Other Sectors

This COA would help improve disaster coordination at the community and municipality levels; thus, the sectors affected are community planning and capacity building and municipalities.

## Potential Costs

Potential up-front costs: $1.1 million in estimated up-front costs
Potential recurring costs: $1.4 million in estimated recurring costs (11 years)
Potential total costs: $2.5 million in total estimated costs

The up-front cost estimate is based on a search of similar MEOCs used in cities in the United States. The estimated cost range is $250,000 to $2 million. The recovery plan used the midpoint of this range for the up-front cost. The recurring cost is estimated as $124,600 annually, for 2 staff employed at half time.

There are several ways to achieve mobile communications for emergency purposes in addition to an MEOC.[3] On the low end, the government of Puerto Rico could procure fly-away communications kits. Each kit would be equipped with computers and communications networking equipment to provide secure communications over 4G cellular or satellite networks. However, this COA recommends that the government of Puerto Rico procure a specialized vehicle for this purpose.

## Potential Funding Mechanisms

FEMA Emergency Management Performance Grant, government of Puerto Rico

## Potential Implementers

Puerto Rico Department of Public Safety

## Potential Pitfalls

None

## Likely Precursors

None

---

[3] FEMA Communications/IT Solutions-based Team, 2018.

## References

FEMA Communications/IT solutions-based team, "Puerto Rico Communications/IT Solutions-based Team Report," June 30, 2018.

FEMA Hurricane Maria Communications Task Force, *DR-4339-PR Consolidated Communications Restoration Plan*, October 2017.

# CIT 9
## Auxiliary Communications—Volunteer Radio Groups and Organizations

### Sector Impacted

Communications and Information Technology, Health and Social Services, Community Planning and Capacity Building, Municipalities

### Issue/Problem Being Solved

A workforce of available communications volunteers may be accessible to Puerto Rico, but it is not formally engaged for disaster services.

### Description

This course of action (COA) would enhance the capacity of disaster response services through a coordinated, structured engagement between Puerto Rico and uniformly trained, highly skilled, and certified communications volunteers or volunteer groups. This COA would cultivate a highly skilled, uniformly trained volunteer workforce to leverage auxiliary emergency communications (AUXCOMM) during disasters by engaging with volunteer radio groups and organizations. This COA would support, incentivize, and encourage volunteer radio operations (e.g., the government of Puerto Rico would set up backup stations, distribute radios, or have representatives attend AUXCOMM working group meetings), following on, or improving on, models implemented in disaster-prone regions.

Volunteer radio organizations represent an underutilized capability for communications. Although the COA focuses on auxiliary volunteer communications, experience that results from its implementation can be applied to formally develop other programs in Puerto Rico. From a governmental standpoint, this is important to consider when weighing the value (beyond cost).

An example approach might utilize an memorandum of understanding (MOU) or memorandum of agreement (MOA) to pre-establish KP-4 assets at emergency management offices across municipalities and in all Puerto Rico Emergency Management Agency (PREMA) Zones, PREMA HQ, Emergency Operations Center (EOC), Puerto Rico Police Department (PRPD) districts, the 12 FD fire districts, the 911 call center, and with emergency medical services (EMS) dispatch.[1] A provision for municipal or commonwealth government personnel to be trained, licensed, and equipped with KP-4 radio systems should also be considered.

---

[1] KP-4 is a call sign assigned by the FCC for amateur radio in Puerto Rico.

## Potential Benefits

This COA extends emergency operations, at minimal cost, by utilizing an experienced, intrinsically motivated group of volunteers.

## Potential Spillover Impacts to Other Sectors

Highly dispersed AUXCOMM could benefit multiple sectors, including housing, transportation, health and social services, water, and energy, by providing detailed information about a disaster-related event. Guidance might also be relayed from informed, authoritative sources via recognized volunteers to individuals residing in affected regions. Coordination with the community planning and capacity building sector may be beneficial for implementation of this COA.

## Potential Costs

Potential up-front costs: $100,000 in estimated up-front costs
Potential recurring costs: $1 million in estimated recurring costs (11 years)
Potential total costs: $1.1 million in total estimated costs[2]

Annual costs were estimated at $100,000 based on examples from state-based initiatives for auxiliary emergency communications and volunteer fire department personnel.[3] The exact range depends on the state's approach to volunteer initiatives and expectation for volunteer participation. In New York, for example, the annual expense for a volunteer firefighter ranges from $800 to $1,500. In 2016, Colorado allocated approximately $60,000 for volunteer radio groups into its budget. Example program expenses or costs may include credentialing services (e.g., background checks), training services and exercises, travel and other approved expenses incurred for the performance of duties, increased demand for supplies and repairs, incentives for participation (e.g., tax incentives), capacity-related enhancements to infrastructure to support increased communications needs, and those expenses or costs required by statute, local law, or ordinance.[4]

## Potential Funding Mechanisms

Government of Puerto Rico, nongovernment sources

---

[2] The total cost for this COA is $1.1 million. The recovery plan represents the estimated cost using a precision of one significant figure for the COA, and therefore the estimate total cost appears to be rounded.

[3] See also Veronica Rose, "Cost of Operating Different Types of Fire Departments," Connecticut General Assembly, Office of Legislative Research, 2014; John A. Peterson, "Auxiliary Emergency Communications: Recognition of Its Support to Public Safety," U.S. Department of Homeland Security, SAFECOM, July 11, 2016; Colorado ARES, "New Colorado Law Creates Auxiliary Emergency Communications Unit within the State's Division of Homeland Security and Emergency Management," Colorado Auxcomm, news release, June 6, 2016; Colorado General Assembly, "HB16-1040 Auxiliary Emergency Communications," 2016 Regular Session.

[4] Colorado Revised Statute § 24-33.5-705.5, "2016 Colorado Revised Statutes Title 24," Justia U.S. Law, undated.

## Potential Implementers

PREMA, Telecommunications Bureau, volunteer groups

## Potential Pitfalls

If volunteer groups become too large or experience high turnover, volunteers might not receive adequate training and preparation for disaster scenarios. The postdisaster availability and condition of an emergency communications infrastructure poses an operational uncertainty for AUXCOMM activities, which potentially undermines the program's utility. However, this uncertainty is shared across emergency services and not unique to AUXCOMM or availability of a prepared workforce.

## Likely Precursors

An emergency operations center with point(s) of contact for auxiliary or volunteer communications is necessary. Precursor COAs are *CIT 6 Modernize the Emergency Operations Center* and *CPCB 9 Coordinated Local Recovery Planning Process.*

## References

"2016 Colorado Revised Statutes Title 24—Government—State Principal Departments Article 33.5—Public Safety Part 7—Emergency Management § 24-33.5-705.5. Auxiliary emergency communications unit—powers and duties of unit and office of emergency management regarding auxiliary communications—definitions." As of April 29, 2019: https://law.justia.com/codes/colorado/2016/title-24/principal-departments/article-33.5/part-7/section-24-33.5-705.5/

Colorado ARES, "New Colorado Law Creates Auxiliary Emergency Communications Unit within the State's Division of Homeland Security and Emergency Management," Colorado Auxcomm, news release, June 6, 2016. As of April 29. 2019: http://www.coloradoares.org/wordpress/links/colorado-auxcomm/

Colorado General Assembly, "HB16-1040 Auxiliary Emergency Communications," 2016 Regular Session. As of April 29, 2019: https://leg.colorado.gov/bills/hb16-1040

Peterson, John A., "Auxiliary Emergency Communications: Recognition of Its Support to Public Safety," U.S. Department of Homeland Security, SAFECOM, July 11, 2016. As of April 29, 2019: https://www.dhs.gov/safecom/blog/2016/07/11/auxiliary-emergency-communications

Rose, Veronica, "Cost of Operating Different Types of Fire Departments," Connecticut General Assembly, Office of Legislative Research, 2014. As of April 29, 2019: https://www.cga.ct.gov/2014/rpt/2014-R-0147.htm

# CIT 10
## Transoceanic Submarine Cable

### Sectors Impacted

Communications and Information Technology, Economic, Municipalities, Energy

### Issue/Problem Being Solved

Expanding the capacity and availability of communications that depend on submarine infrastructure would help meet Puerto Rico's goals for communications redundancy.

### Description

This course of action (COA) would introduce new, very high bandwidth undersea cable(s) to Puerto Rico, situated away from San Juan: one landing point for the midterm, followed by additional ones in the long term to increase capacity and route options. This COA would also mitigate the known threats to existing landing stations and related infrastructure from disaster events. Alternative approaches considering the near and longer terms may be combined in realizing this COA.

Near-term approaches would evaluate the risk to existing as-is submarine cabling infrastructure and all related communications services to and from Puerto Rico; and prepare and implement a comprehensive plan to strengthen or make more resilient the submarine cabling infrastructure to Category 4 and 5 storm events. Specific steps would include:

- developing and implementing a security plan to mitigate the risk of accidental or storm related damage, or an intentional compromise (e.g., sabotage, communications intercept) to submarine cabling infrastructure in the vicinity of the shore and on-shore, up to and including the landing station and required, dependent operational infrastructure; include a comprehensive maintenance plan and develop the plan so that it can be implemented by multiple entities with minimal dependence on any single provider or organization
- developing and implementing a comprehensive risk mitigation plan for communication services that takes into account peak demands, operational activities, governance, feedback across all stakeholders, and continuous monitoring of the submarine cable–related infrastructure, using automation where feasible
- performing root cause analysis of issues pertaining to the recovery of internet connectivity and service delivery to Puerto Rico and any substantial dependencies, including public and private sector issues, and highlighting the issues, needs, and decisions being faced
- identifying alternative sources of power and power efficient systems for landing stations that improve the self-sufficiency of the stations, including renewable resources
- applying adaptive restoration and reallocation techniques for routes (e.g., elastic optical networking where appropriate)
- evaluating the fault tolerance of the submarine cabling infrastructure undersea and on-land to develop and deliver additional fault tolerant mechanisms to bolster the existing

infrastructure services, such as by forming mesh networks using multiple cables and landing points.

Long-term considerations include constructing a new landing station for a very high-bandwidth cable situated away from the northeastern side of Puerto Rico (e.g., near Ponce) with supporting fault tolerant undersea and shallow water network infrastructure. Specific steps would include:

- performing a thorough cable route and feasibility study for a new landing station located on the northwest, west, or southern side of Puerto Rico
- preparing a development and maintenance plans based on a comprehensive picture of risk, taking into account the impact of all threats, natural influences, economically sensitive areas, protected areas, regulatory constraints, political issues, and private- and public-sector interests
- constructing or utilizing off-shore branch points to improve availability of services at potentially lower costs
- constructing a landing station infrastructure that is hurricane hardened for Category 4 and 5 storms
- considering using trunk-and-branch, elastic optical networks, and double-cable architectures to improve resiliency and flexibility
- installing stubbed branching units to support future expansion
- developing a comprehensive view of operational, workforce, and maintenance needs and implementing a plan to support and monitor the submarine cabling infrastructure before, during, and after disaster events.

## Potential Benefits

Improvements to the undersea communications network's design would help create a highly resilient, commonwealth-scale communications network with a reduced recovery time for failures that may arise from disasters. Redundant, high-capacity network channels to Puerto Rico would also improve the overall communications capacity and quality of available services should a single route become impaired. This COA also helps to boost the value of Puerto Rico's network infrastructure for the economy, and it enhances the overall utility of Puerto Rico's submarine cable systems for U.S. and global communications.

Submarine cable–related infrastructure constitutes a primary, high-capacity communications link to Puerto Rico from the outside world. Both submarine cabling and related terrestrial infrastructure are vulnerable to storm damage.[1] Puerto Rico's landing stations are also concentrated on the northern side of Puerto Rico centered on San Juan. These landing stations are located in Punta Salinas, Miramar, San Juan, and Isla Verde support 14 submarine cables.

---

[1] Madory, 2017. Through onsite meetings in Puerto Rico and news reports, it was confirmed that storm damage impacted the submarine cabling infrastructure resulting in outages. Several landing stations were flooded, suffering equipment and facilities-related damage. Correspondingly, outages were also reported in the region by other cable providers. See, for example, Belson, 2017; and Guimarães, 2017.

Connectivity is provided to points distributed throughout the Caribbean, South America, and North America via these cables.

## Potential Spillover Impacts to Other Sectors

New infrastructure and increased operational costs would place additional demands on the utilities, the government, and people of Puerto Rico for power needs, funding, and the workforce, respectively. However, a robust communications infrastructure would mitigate the extent of challenges faced by response and recovery operations. A submarine cabling infrastructure would link Puerto Rico to external networks and embody the primary mechanism to connect Puerto Rico to the rest of the United States for all activities. Consequently, this COA could have a positive cascading effect across many sectors, such as health and social services, water, and Energy.

## Potential Costs

Potential up-front costs: $25 million–$130 million in estimated up-front costs (2 years)
Potential recurring costs: $42 million–$105 million in estimated recurring costs (11 years)
Potential total costs: $67 million–$235 million in total estimated costs

The capital expenditure is captured as potential up-front cost.[2] Potential recurring costs include operational expenditures, estimated on an annual basis. Upgrades and loan payments with interest are excluded from the estimate. Depending on assumptions, costs other than operational expenditures may constitute the bulk of the annual cost. Example of considerations to be made for a detailed implementation include the need for terrestrial infrastructure to operate the cable, the complexity of the route chosen for introduction of additional cabling, the level of fault-tolerance required, workforce needs, and connectivity to global networks.

## Potential Funding Mechanisms

Government of Puerto Rico, private sector, sale of capacity via indefeasible right of use or by lease

## Potential Implementers

Telecommunications Bureau, government of Puerto Rico agencies, private industry

---

[2] Costs were estimated based upon input from representatives of the government of Puerto Rico and Federal Emergency Management Agency (FEMA) solutions-based teams (SBTs). Elaine Stafford, "The Suboptic Guide: Chapter 1 Planning, Contracting, Constructing, Owning and Operating A Submarine Cable Network," Suboptic, 2013, was also referenced.

## Potential Pitfalls

Poor coordination among stakeholders during discovery, implementation, or funding could lead to higher than anticipated costs. Other potential pitfalls are irreparable loss to economically sensitive natural resources, unanticipated repairs to undersea cables, and single-vendor lock-in that does not allow Puerto Rico to exercise influence over key decisions. A development plan also requires a strategy for spanning multiple hurricane seasons.

If the implementation effort is insufficiently considered or not well coordinated during discovery and implementation among all potential stakeholders, the value of enhancements to existing infrastructure or development of new infrastructure may be diminished over the long term. Suboptimal design would increase the costs for the project over the long term.

The total cost is not well understood or poorly supported by funding mechanisms. Planning and construction could take between 18 months and 30 months, thus spanning multiple hurricane seasons. If this is not accounted for, costs could increase beyond planned amounts.

Areas of vulnerability for cabling infrastructure include those that are land-based, such as landing points, front- and backhauls, landing stations (damage, flooding, and power dependence), and transmission equipment situated at a point-of-presence.

Sea-to-shore transition is also vulnerable. For example, hurricane generated currents in shallow water can displace a cable, increasing tension on the cable body and resulting in physical damage to the cable leading to the landing point. Another undersea issue is that cables may snap during major storms or be cut as the result of marine traffic and ocean-bottom landslides. Cable integrity resulting from sabotage is also a threat. Any stress or physical strain applied to the fiber-optic lines can reduce the meantime between failures for the line, degrading the utility of the line for communications.

Moreover, Puerto Rico's submarine cable networks serve routes to South American and European networks. In the event of failure of other network gateways along the Eastern seaboard (e.g., due to natural disaster or arising from critical infrastructure [CI] attacks), Puerto Rico's infrastructure provides routing capabilities until network operations are restored. As such, there is a cascading risk to providing service from Puerto Rico to external communities.[3]

## Likely Precursors

Clear governance or policy for ownership rights

## References

Belson, David, "Internet Impacts of Hurricanes Harvey, Irma and Maria," Orcale + Dyn, blog, September 25, 2017. As of April 29, 2019:
https://dyn.com/blog/internet-impacts-of-hurricanes-harvey-irma-and-maria/

---

[3] See for example, submarine cable map, website, undated.

Guimarães, Nathália, "Furacão Maria deixa conexão de internet lenta no Brasil," LeiaJa, Technology, Internet, September 9, 2017. As of April 29, 2019:
http://www.leiaja.com/tecnologia/2017/09/22/furacao-maria-deixa-conexao-de-internet
-lenta-no-brasil/

Madory, Doug, "Puerto Rico's Slow Internet Recovery," Orcale + Dyn, blog, December 7, 2017. As of June 19, 2018:
https://dyn.com/blog/puerto-ricos-slow-internet-recovery/

Stafford, Elaine, "The Suboptic Guide: Chapter 1 Planning, Contracting, Constructing, Owning and Operating A Submarine Cable Network," Suboptic, 2013. As of April 29, 2019:
http://suboptic.org/wp-content/uploads/2014/10/The-Guide-1.pdf

Submarine cable map, website, undated. As of April 29, 2019:
https://www.submarinecablemap.com/#/landing-point/san-juan-puerto-rico-united-states

# CIT 11
## Procure a Mobile Emergency Communications Capability

### Sector Impacted

Communications and Information Technology, Municipalities

### Issue/Problem Being Solved

Puerto Rico needs a state-of-the-art, survivable, and resilient communications infrastructure to ensure the continuity of essential government functions and the provision of public safety services after a disaster. According to an October 2017 Department of Homeland Security (DHS) report, the communications infrastructure of Puerto Rico was severely damaged by hurricanes Irma and Maria.[1] Only 5% of cell towers were operational. Similarly, most of the towers that provide backhaul communications for police, fire, and emergency medical service (EMS) were damaged, destroyed, or without power. In addition, most of the public switched telephone network (PSTN) lines connecting public safety agencies were nonfunctional and need to be repaired or replaced. Together these communications failures severely affected the continuity of government (COG) and the ability of first responders and emergency managers to communicate and coordinate disaster response activities.

### Description

This course of action (COA) would develop the capability to quickly reestablish communications for emergency and government operations in the aftermath of a manmade or natural disaster that causes widespread, catastrophic damage to the telecommunications infrastructure. This system, consisting of deployable telecommunications equipment, is envisioned as a quick and temporary fix for the loss of primary communications capability.

This COA envisions the procurement of temporary, deployable assets that would be safely cached in Puerto Rico and quickly installed throughout Puerto Rico to restore voice and data communications for disaster response, emergency services, and government activities. This equipment would provide secure, interoperable, cross-agency communications. The solution may include but is not limited to utilizing civilian and military frequencies and a network of deployable nodes. The system would include portable power generation to ensure independent operations and/or remote deployment.

In addition to the procurement of systems, this COA envisions teams of technicians who specialize in the operations and maintenance of the equipment. The teams would coordinate with relevant agencies in training exercises for deploying and operating the system in the event of a

---

[1] FEMA Hurricane Maria Communications Task Force, 2017.

disaster. The team will also develop maintenance and deployment plans, taking into consideration the climate and terrain challenges of Puerto Rico.

## Potential Benefits

Reliable and interoperable communications are essential to providing effective and responsive disaster recovery, emergency services, and government operations.

## Potential Spillover Impacts to Other Sectors

Because this COA envisions providing temporary capability to support government operations, the limitations of the system must be communicated to the municipalities. This will ensure the municipalities understand what can and cannot be done with this system when the primary infrastructure fails. Similarly, the requirements of the municipalities must be captured and addressed when designing this system.

## Potential Costs

Potential up-front costs: $83 million–$165 million in estimated up-front costs
Potential recurring costs: $38.5 million–$58.3 million in estimated recurring costs (11 years)
Potential total costs: $122 million–$223 million in total estimated costs

In order to develop an estimate, the costs of a similar deployable network already implemented in several New Jersey counties (and called JerseyNet) were identified and extrapolated to the size of the network to meet the needs of Puerto Rico. Two cost bounds were estimated. For the lower bound, it is assumed that after a future disaster about 50% of Puerto Rico's telecom infrastructure remains operational and thus only 50% of the deployable network is needed. This deployable network is assumed to be interoperable with the telecom infrastructure that remains operational after the disaster. The lower end is $83 million in up-front costs and $3.5 million in annual recurring costs. The high-end estimate assumes that the full deployable network is needed. The high-end estimate is $165 million in up-front costs and $5.3 million in annual recurring costs.

JerseyNet was funded by a National Telecommunications and Information Administration (NTIA) grant.[2] The JerseyNet deployable network consisted of 42 mobile nodes.[3] The estimates are based on the cost of JerseyNet's terrestrial mobile nodes. What would be needed in Puerto Rico was extrapolated as follows: First, Puerto Rico was divided into (a) coastal, urban areas; (b) rural, mountainous areas; and (c) suburban areas with landscape of pastures and secondary forest; second, the number of mobile nodes required to cover these areas was estimated,

---

[2] U.S. Department of Commerce, undated. The NTIA grant was $39.64 million according this website; however, most of the funding went to the mobile nodes. Representative of NTIA, telephone communication with authors, undated.

[3] State of New Jersey, undated.

assuming that each node has a range of approximately 3 miles, with the result that 175 mobile nodes are required to cover Puerto Rico. The upper cost bound of $165 million was calculated by extrapolating the cost of JerseyNet (42 nodes) to Puerto Rico (175 nodes).

The estimated cost includes the technical teams needed to deploy, maintain, and store the mobile nodes and to operate and manage the network. An engineer from Puerto Rico was assumed to cost $124,600 per year and an engineer from the continental United States (CONUS) was assumed to cost $217,300 per year. A total of 60 technical personnel,[4] including 30 from Puerto Rico and 30 from CONUS, was assumed to be needed to operate and maintain the network for the 6 months of the Puerto Rico hurricane season. The annual cost was estimated as $5.1 million. It was also assumed that half of the personnel from CONUS would need to travel to Puerto Rico for a total of 4 weeks per year to help deploy the mobile nodes before the hurricane season starts and then store the nodes after the season ends. Initially, they would also be needed to train the local personnel. The travel cost per person was estimated to be $2,500 per week. Assuming 15 people would be needed for 4 weeks, the travel costs would be $150,000 per year. The total cost of personnel is thus $5.3 million per year. These costs are attributed to the upper-bound estimate. For the lower-bound estimate, it was assumed that only two-thirds of the personnel are needed to support 50% of the deployable network—thus $3.5 million per year.

## Potential Funding Mechanisms

Federal Emergency Management Agency (FEMA) Emergency Management Performance Grant, government of Puerto Rico

## Potential Implementers

Puerto Rico Emergency Management Agency (PREMA), Office of the Chief Information Officer, Puerto Rico Department of Public Safety

## Potential Pitfalls

The FEMA Communications/Information Technology solutions-based team (SBT) has stated that most of the mobile communications capabilities outlined in this COA are redundant to other capabilities already being pursued by private, commonwealth, and federal efforts. However, the government of Puerto Rico believes that such capabilities would not be available for some time and that Puerto Rico needs a solution that can be implemented in the near term. Uncertainty in the delivery of long-term communication capabilities being pursued could lead to additional costs for Puerto Rico to maintain the near-term solution beyond the anticipated need.

---

[4] This is close to the total number of technical personnel available in the Puerto Rico Department of Public Safety (56), as of June 2018.

## Likely Precursors

None

## References

FEMA Hurricane Maria Communications Task Force, *DR-4339-PR Consolidated Communications Restoration Plan*, October 2017.

State of New Jersey, Office of Homeland Security and Preparedness, website, undated. As of July 25, 2018:
https://www.njhomelandsecurity.gov/jerseynet/

U.S. Department of Commerce, National Telecommunications and Information Administration, BroadbandUSA Grants Awarded, New Jersey, website, undated. As of April 29, 2019:
https://www2.ntia.doc.gov/new-jersey

# CIT 12
## Perform Site Structural Analysis for All Government Telecom Towers (Both Public and Privately Owned)

### Sectors Impacted

Communications and Information Technology, Health and Social Services, Housing, Transportation, Municipalities

### Issue/Problem Being Solved

Increase the resilience of telecommunications infrastructure and emergency communications.

### Description

This course of action (COA) would survey all telecommunications towers and sites that have been identified by the Federal Emergency Management Agency (FEMA) Communications and Information Technology solutions-based team (SBT) and the U.S. Army Corps of Engineers as critical infrastructure for the provision of government emergency and other services. This COA would determine whether all towers used for emergency communications meet the Puerto Rico tower code for structural loading. The first step would be to review the structural requirements of the tower code and the enforcement powers of the Puerto Rico Planning Board (PRPB). The Telecommunications Bureau advised that there is a new proceeding at the PRPB (which controls the permitting of towers) that would address the structural overloading of towers. It is possible that the new proceeding at PRPB may make this COA unnecessary. However, it is likely that PRPB would require tower owners to pay for these structural inspections, which may not occur on a timely basis.

The COA would provide a review of the towers that are considered critical infrastructure to provide government emergency and other services. The COA would determine whether all towers used for emergency communications meet the Puerto Rico tower code concerning structural loading. The COA would be implemented by PRPB, which provides the permitting for towers. The first step would be to review the existing tower code to identify the structural requirements for towers, the tower code enforcement authority of PRPB, and any other Puerto Rico agencies that are involved in permitting towers or enforcing the tower code. The COA could be introduced in a tower code enforcement activity or in a notice and comment administrative proceeding such as the administrative proceeding that includes towers that is currently underway at PRPB. An engineer would need to inspect the designated towers (there may be 30 to 60 towers). The likely time scale to demonstrate the benefit of inspection is the next major hurricane or earthquake. The expected benefit will be towers that keep standing despite severe weather. The towers are located throughout Puerto Rico. The COA should serve to enforce the current tower code requirements and any new requirements determined by the PRPB.

## Potential Benefits

This COA would contribute to maintaining a resilient communications infrastructure and emergency communications. A resilient communications infrastructure is necessary to provide continuity of education, health care, social services, tourism supports, and the emergency services sector.

New tower codes that had been established in Puerto Rico several years ago were comprehensive; however, tower owners generally failed to comply with them.[1] This COA could focus attention on emergency communications towers that are structurally overloaded and thus more vulnerable to severe weather conditions. An additional benefit of this COA is that tower owners, not the government of Puerto Rico, will be financially responsible for bringing their towers into compliance with the code if their towers are found to be noncompliant.[2]

## Potential Spillover Impacts to Other Sectors

Failure to take this COA could result in loss of communications during a disaster, which would negatively affect all other sectors. Hurricane Maria demonstrated how loss of backhaul communications resulted in public safety answering points (PSAPs) being unable to relay calls to police, fire, or emergency medical service (EMS).[3]

## Potential Costs

Potential up-front costs: $1.5 million–$3 million in estimated up-front costs[4]
Potential recurring costs: —
Potential costs: $1.5 million–$3 million in total estimated costs [5]

A FEMA subject matter experts (SMEs) estimated that this work would require approximately $50,000 per tower.[6] An official of the government of Puerto Rico estimated the number of sites requiring this analysis at 30 to 60 towers.

---

[1] Senior member of management of a Puerto Rico cell tower company, interview with the authors, April 16, 2018.

[2] Senior official of Puerto Rico wireless telecommunications provider, telephone communication with authors, April 16, 2018.

[3] FEMA Hurricane Maria Communications Task Force, 2017.

[4] In the recovery plan, the total estimated cost is represented as $4 million. The estimated cost is being revised here by reducing the number of towers surveyed.

[5] In the recovery plan, the total estimated cost is represented as $4 million. The estimated cost is being revised here, based upon the change to the estimated up-front costs.

[6] The work will include the following: visual inspection by a team of structural engineers and other subject matter experts (this team will look for rusting, cracking or welds, bent members, and so on); tower mechanical structural analysis/modeling using mathematical engineering formulas for stress, along with the known behavior of different materials under stress; analysis of the soil and the ground of the sites where the towers are installed. The latter may be particularly important for locations in Puerto Rico that are susceptible to mudslides.

A very senior member of management at a large tower company in Puerto Rico explained that it would be difficult to estimate the overall cost for a structural review of towers that are used for public safety or critical communications, because each tower could be in very different condition.[7] Some towers are over 25 years old, which is very old for towers. Other towers are located in areas with adverse soil conditions, which can cause structural issues. According to the tower company expert, there are many variations in individual towers that would make cost estimates for a general inspection difficult.

### Potential Funding Mechanisms

FEMA Public Assistance, Community Development Block Grant Disaster Recovery

If a tower inspection reveals that towers fail to meet the existing Puerto Rico tower code, it would likely be the financial responsibility of the tower owner, not FEMA or the government of Puerto Rico, to shoulder costs to bring the towers into compliance.

### Potential Implementers

Puerto Rico Department of Public Safety, Telecommunications Bureau, Puerto Rico Planning Board[8]

### Potential Pitfalls

There could be a conflict between this COA and an administrative proceeding that is underway at PRPB. It is unclear what PRPB might order tower owners to do concerning tower inspections and structural analyses. Even if tower owners were ordered by PRPB to conduct tower inspections and structural analyses, it is unknown what the timeline would be for compliance and what PRPB's enforcement authority would be. However, it would be advisable to act in concert with the administrative proceeding at PRPB.

### Likely Precursors

The PRPB administrative proceeding that includes towers should be the first step in implementing this COA. PRPB should address requirements for tower owners to conduct tower inspections and structural analyses in the context of the Puerto Rico tower code.

### Reference

FEMA Hurricane Maria Communications Task Force, *DR-4339-PR Consolidated Communications Restoration Plan*, October 2017.

---

[7] Senior member of management of Puerto Rico cell tower company, interview with the authors, April 16, 2018.

[8] The Puerto Rico Planning Board was inadvertently not included in the recovery plan.

# CIT 13
## Streamline the Permitting and Rights of Way Processes for Towers and the Deployment of Fiber-Optic Cable

### Sectors Impacted

Communications and Information Technology, Energy, Economic, Health and Social Services, Housing, Transportation, Municipalities

### Issue/Problem Being Solved

This course of action (COA) would address the current long administrative lead times and expense of obtaining rights of way (ROWs) and other approvals needed to construct cellular towers and to trench and bury fiber-optic cable for telecommunications and broadband internet services. The expense in time and money of obtaining these approvals has limited the deployment of resilient telecommunications and broadband internet services throughout Puerto Rico. This COA is necessary to support objectives for digital transformation and a digital economy.

### Description

In collaboration with the new Puerto Rico Public Service Regulatory Board and the Telecommunications Bureau, this COA would establish and staff a central ROW and permitting approval authority, using uniform, streamlined approval processes. This authority may be part of the Public Service Regulatory Board or the Telecommunications Bureau. The central authority would supersede the powers of municipalities and other Puerto Rico agencies and departments to govern the ROWs and permitting approval processes for the deployment of fiber-optic cable and cellular towers. This COA was endorsed by all Puerto Rico telecommunications providers, as well as regulators, in a May 11, 2018, meeting held at the Puerto Rico Telecommunications Regulatory Board, the predecessor to the Telecommunications Bureau.

The Federal Communications Commission (FCC) has recognized the importance of reducing unnecessary government regulation to accelerate the deployment of broadband internet services in WC Docket 17-84:

> This *Report and Order*, *Declaratory Ruling*, and *Further Notice of Proposed Rulemaking* seeks to accelerate the deployment of next-generation networks and services by removing regulatory barriers to infrastructure investment; to speed the transition from legacy copper networks and services to next-generation fiber-based networks and services; and to eliminate Commission regulations that raise costs and slow broadband deployment.[1]

---

[1] FCC, 2017, para. 176.

The FCC has also recognized that unnecessarily burdensome government regulation may hinder rather than help recovery efforts after a natural disaster such as Hurricane Maria and has requested public comment about how and when the FCC might use its preemption authority in rebuilding and repairing communications infrastructure after natural disasters:

> We are committed to helping communities rebuild damaged or destroyed communications infrastructure after a natural disaster as quickly as possible. We recognize the important and complementary roles that local, state, and federal authorities play in facilitating swift recovery from disasters such as Hurricanes Harvey, Irma, and Maria. We are concerned that unnecessarily burdensome government regulation may hinder rather than help recovery efforts, and laws that are suited for the ordinary course may not be appropriate for disaster recovery situations. We seek comment on whether there are targeted circumstances in which we can and should use our authority to preempt state or local laws that inhibit restoration of communications infrastructure.[2]

### Potential Benefits

This COA is a crucial first step for the swift deployment of telecommunications services and broadband internet to the education, health and social services, and economic sectors, as well as the visitor economy and emergency services. The FCC recognized the importance of accelerating wireline broadband deployment by removing barriers to infrastructure investment in WC Docket No. 17-84. The FCC Fact Sheet for WC Docket No. 17-84 states:

> High-speed broadband is an increasingly important gateway to jobs, health care, education, information, and economic development. Access to high-speed broadband can create economic opportunity, enabling entrepreneurs to create businesses, immediately reach customers throughout the world, and revolutionize entire industries. But all too often, regulatory barriers increase the cost of deploying next-generation networks, resulting in higher prices, less competition, and worse service for consumers and small businesses, especially in rural America.[3]

### Potential Spillover Impacts to Other Sectors

The lengthy permitting process in Puerto Rico has been identified as one of the primary reasons that it is difficult to do business in Puerto Rico. This issue also affects other sectors, especially transportation, energy, and water.

The *New Fiscal Plan for Puerto Rico*, dated April 5, 2018, stated that an aspect of the new fiscal plan was to "reduce unnecessary regulatory burdens to reduce the drag of government on the private sector."[4]

---

[2] FCC, para. 176.

[3] FCC, 2017.

[4] Fiscal Agency and Financial Advisory Authority, 2018.

## Potential Costs

Potential up-front costs: $600,000 in estimated up-front costs[5]
Potential recurring costs: —
Potential total costs: $600,000 in total estimated costs

This estimated up-front cost was calculated based upon the cost of 5 professionals for a year, each with an annual salary of $124,600.

The cost of a central telecommunications ROW approval authority depends on whether it is established as a stand-alone agency or as an adjunct to an existing commonwealth agency. If it is established as an adjunct to an existing agency such as the Telecommunications Bureau, the cost would be primarily for any new staff that would be required.

## Potential Funding Mechanisms

Federal Communications Commission

Funding from the FCC might be available to create a new streamlined permitting and ROW capability within the Telecommunications Bureau. The FCC might provide the temporary loan of FCC staff to accomplish the changes needed for the permitting and ROW processes.

On September 27, 2018, the FCC released a Declaratory Ruling and Third Report and Order announcing its decision to accelerate the deployment of broadband internet services by removing several types of unnecessary government regulation. Specifically, as stated in the FCC's news release, the Ruling and Order "explains when a state or local regulation of wireless infrastructure deployment constitutes an effective prohibition of service prohibited by Sections 253 or 332(c)(7) of the Communications Act; the Order concludes that Section 253 and 332(c)(7) limit state and local governments to charging fees that are no greater than a reasonable approximation of objectively reasonable costs for processing applications and for managing deployments in the rights-of-way."[6] For this reason, the FCC might welcome the opportunity to engage in the streamlining of Puerto Rico's telecommunications permitting and ROW processes.[7]

## Potential Implementers

Telecommunications Bureau, Puerto Rico Department of Transportation and Public Works (Puerto Rico Highway and Transportation Authority), other government of Puerto Rico agencies, municipalities.

---

[5] The recovery plan represents the potential costs using one significant figure, and the estimated cost appears rounded when compared with the estimated cost described at $623,000.

[6] See Declaratory Ruling and Third Report and Order, *Accelerating Wireless Broadband Deployment by Removing Barriers to Infrastructure Investment*, WC Docket No. 17-84, FCC 18-133, rel. Sept. 27, 2018. See also, FCC News, *FCC Facilitates Deployment of Wireless Infrastructure for 5 G Connectivity*, September 26, 2018.

[7] See FEMA Hurricane Maria Communications Task Force, 2017.

## Potential Pitfalls

Some municipal mayors may resist a centralized commonwealth agency for permitting and ROWs if they think that it removes either their control over disruption to streets or their power to approve and exact revenues from telecommunications providers. As an alternative, a uniform fee established by the new central ROW and permitting authority could be paid by telecommunications and broadband internet providers to municipalities.

The following excerpt from a filing with the FCC in Docket No. 17-84 by Liberty Cablevision of Puerto Rico explains the challenges telecommunications providers face in dealing with Puerto Rico municipalities for permitting and ROW approvals:

> Liberty has a cable franchise covering the entire Commonwealth. Even so, for any particular project, Liberty still needs to deal with one or more Puerto Rico municipalities. In that context, Liberty encounters four main challenges. First, Liberty often encounters excessive fees that are unrelated to costs reasonably related to processing permit requests. Second, Liberty often encounters needless delays in obtaining the required approvals. Third, Liberty is often subject to burdens and obligations that are not imposed on others—particularly government-owned utilities who [sic] also use the public rights of way. Finally, there is a great deal of variation among the different municipalities' procedures, fees, and requirements.[8]

According to informal discussions with members of the Puerto Rico Telecommunications Regulatory Board (the predecessor of the Telecommunications Bureau), the amounts of revenue paid to municipalities by telecommunications providers for permitting and ROW access are, in their words, "not immense." A Board member suggested informally that one option to reduce the resistance of municipal mayors to uniform and centralized permitting and ROW access would be to establish a reasonable, uniform payment that would be made by telecommunications providers to municipalities for use of municipal ROW. This might incentivize municipalities to support a centralized authority.

## Likely Precursors

None. However, regulatory action to create a uniform, centralized telecommunications permitting and ROW authority, whether as a stand-alone agency or as part of the Puerto Rico Telecommunications Regulatory Board, would need to be accomplished to implement this COA. This COA might be implemented along with *CIT 21 Government-Owned Fiber-Optic Conduits to Reduce Aerial Fiber-Optic Cable and Incentivize Expansion of Broadband Infrastructure.* Following a meeting on June 13, 2018, the FCC provided model codes from other jurisdictions that could be modified for use by Puerto Rico.

---

[8] Liberty Cablevision of Puerto Rico LLC, Comments of Liberty Cablevision of Puerto Rico LLC, *In the Matter of Accelerating Wireline Broadband Deployment by Removing Barriers to Infrastructure Investment,* in WC Docket No. 17-84, June 15, 2017, p. 14.

## References

Comments of Liberty Cablevision of Puerto Rico LLC in WC Docket No. 17-84, In the Matter of Accelerating Wireline Broadband Deployment by Removing Barriers to Infrastructure Investment, June 15, 2017, p. 14.

Fiscal Agency and Financial Advisory Authority (Autoridad de Asesoría Financiera y Agencia Fiscal), *New Fiscal Plan for Puerto Rico*, April 5, 2018. As of August 23, 2018: http://www.aafaf.pr.gov/assets/newfiscalplanforpuerto-rico-2018-04-05.pdf

Report and Order, Declaratory Ruling, and Further Notice of Proposed Rulemaking in WC Docket No. 17-84, In the Matter of Accelerating Wireline Broadband Deployment by Removing Barriers to Infrastructure Investment, released November 29, 2017, FCC-CIRC1711-04.

# CIT 14
## Consolidated Government Information Systems

### Sectors Impacted

Communications and Information Technology, Economic, Municipalities

### Issue/Problem Being Solved

Currently in Puerto Rico, the delivery of government services is inconsistent; a timely, accurate, and coherent system of records management is lacking; and there is a reliance on legacy computing systems for infrastructure and information management.

### Description

In December 2017, the Governor of Puerto Rico signed Act 122, which seeks the reduction of 118 agencies to 35 more efficient ones. The Fiscal Board agreed with the Governor about the need for agency consolidation and right-sizing to deliver services in an efficient manner. This Course of Action (COA) recognizes that agency restructuring requires a corresponding digital reform.

This COA would establish and implement an open, modular, standards-based platform for information systems, and it would consolidate governmental systems across Puerto Rico. Through software and associated information systems, the platform would natively enable interoperability, consistent standards and policies for information and data management, and the scaling of the system overtime, at reduced expense and effort.

This COA will improve the continuity of government (COG) and the efficient delivery of its services to people of Puerto Rico and the private sector, while allowing for greater transparency and modern approaches to information management that keep citizens' best interests in mind.

If such a system is well designed, COG and the operational efficiency of governmental services for disaster operations may be dramatically improved. This COA can leverage broadband availability in municipalities and data-center services for implementing design decisions.

### Potential Benefits

Presently, the landscape for governmental information systems is populated by a heterogeneous mixture of legacy systems that are seen as not having an adequate capacity to scale, evolve, or interoperate for governmental needs. A fully integrated approach that uses a common, standards-based platform for information systems across Puerto Rico could reduce operating costs for all municipalities and simultaneously enable highly reliable governmental functions, including coordination of response and recovery activities within Puerto Rico and externally (e.g., in terms of supply-chain logistics).

This COA also seeks to improve the government's ability to function and provide citizens with reliable services in the event of a disaster by using high-quality, cross-sector data and information drawn from across the territory in a coherent way. If the opportunity is leveraged, the approach also is dual-purpose, in that it is supportive of and consistent with digital transformation initiatives that are also being undertaken.

A standardized platform based on modern, cloud-based techniques and/or services has the potential to dramatically reduce the implementation costs of modernization efforts and to improve the sustainability and long-term value of information systems.

There is potential to dramatically improve planning efforts and the availability of quality situational data for disaster response and recovery efforts through evolution of existing Puerto Rico information systems in a cloud-based environment.

## Potential Spillover Impacts to Other Sectors

An open, modular, and standards-based common platform for information systems improves the government's ability to deliver timely services to citizens and the private sector alike. The efficiencies and experience gained and shared through these efforts with the private sector might incentivize economic activity in the private sector. The same systems may leverage the products of workforce development initiatives to preserve essential trade skills developed in Puerto Rico. Success with this initiative could also inform publicly supported initiatives in health and social services, housing, and transportation.

## Potential Costs

Potential up-front costs: $152 million in estimated up-front costs
Potential recurring costs: $330 million in estimated recurring costs (11 years)
Potential total costs: $482 million in total estimated costs

The approach for implementation would strongly influence the costs for the project, as would the number of information systems in the government of Puerto Rico and municipalities participating in the consolidation. An initial deployment may focus on government of Puerto Rico services and information systems. Following this, municipalities could be added through a phased plan. A platform could be acquired through a competitive process or a new platform could be developed from the ground-up, but without careful consideration of the strategy the costs would not be well understood.

The costs were estimated for 134 agencies by the government of Puerto Rico. The up-front costs depend on the agency size. They include platform-related costs and implementation estimates for the consolidation effort. Platform costs include on-premise and cloud-based infrastructure and information systems such as on-premise systems and services, licensing fees, disaster recovery and business continuity, and colocation. Table C.1 summarizes the estimated cost based upon agency size. Implementation costs include data and application management, reengineering applications, planning and managing replatforming, and integration. Potential recurring costs are estimated as 40% of the initial platform costs per year, or $30 million annually.

**Table C.1. Up-Front Cost Estimates for 134 Agencies, Course of Action *Consolidated Government Information Systems***

| Estimated Up-Front Costs | Large Agency | Medium Agency | Small Agency |
| --- | --- | --- | --- |
| Platform costs per agency | $1.38 million | $0.83 million | $0.41 million |
| Implementation cost per agency | $1.39 million | $0.83 million | $0.42 million |
| Number of agencies (134 total) | 6 | 35 | 93 |
| **Total costs** | $16.6 million | $58.2 million | $77.2 million |

## Potential Funding Mechanisms

Community Development Block Grant Disaster Recovery, government of Puerto Rico

## Potential Implementers

Office of the Chief Information Officer, government of Puerto Rico agencies

A sizable, experienced, dedicated workforce would be required for implementation. Lack of availability of skilled, experienced individuals may undermine the activity in a variety of ways. The availability and participation of informed decisionmakers with the required authorities would also be critical for the success of the effort.

## Potential Pitfalls

Development of common standards with cross-municipality collaboration may be challenging without adequate stakeholder engagement. Goals, requirements, and capabilities of the underlying system would need to match today's needs with future ones and be prepared in a way that clearly defines usage and tools of the system while enabling enough flexibility for specific configurations to meet needs at the level of a municipality or agency.

Transition from existing systems to new ones may encounter opposition or fundamental changes to business process and organizational roles. Such issues would need to be addressed in a way to facilitate progress while addressing stakeholder concerns. Access to data would require the buy-in of many government of Puerto Rico agencies. An additional important issue would be properly addressing data security.

Given the number of information systems potentially involved in this effort, a sizable, experienced dedicated workforce will be required for implementation. Lack of availability of local or nonlocal skilled, experienced individuals may undermine the activity in a variety of ways. Participation by business analysts, developers, project managers (on the implementation side and governmental side), and highly skilled IT personnel for the computing infrastructure will all be needed. The availability and participation of informed decisionmakers with the required authorities will also be critical for the success of the effort. Together, these factors may place an undue burden on the local workforce, without corresponding workforce development (e.g., *ECN 2 Implement Workforce Development Programs*). Team structure and the detailed organizational approach (including vendor engagement) will be important to establishing a viable plan and will need to be carefully addressed at the start of the project and reevaluated over its course.

## Likely Precursors

Robust availability of on-island broadband to all municipalities is necessary. Precursors COAs are *CIT 17 Puerto Rico Data Center* and *CIT 21 Government-Owned Fiber-Optic Conduits to Reduce Aerial Fiber-Optic Cable and Incentivize Expansion of Broadband Infrastructure.*

# CIT 15
## Undersea Fiber Ring System

### Sectors Impacted

Communications and Information Technology, Economic, Energy, Municipalities

### Issue/Problem Being Solved

Increase the availability of Puerto Rico's submarine communications network through infrastructure expansion using a ring network, pursuant to Puerto Rico's goals for communications resiliency.

### Description

This course of action (COA) would evolve the undersea network infrastructure to incorporate a communications ring system around Puerto Rico. This ring network would connect landing points (present and future) around Puerto Rico and improve the availability of communication routes to or from Puerto Rico in the event of natural disasters in the long term.

Onsite meetings in Puerto Rico and news reports confirmed that hurricane damage to the submarine cabling infrastructure resulted in outages. Landing stations are presently concentrated in the north section of Puerto Rico. Several landing stations were flooded, suffering equipment and facilities-related damage.[1] Correspondingly, outages were also reported in the region by other cable providers.[2] A point-to-point ring system can improve resilience by increasing the number of route options available in the event of an outage. While a pure ring structure can support only one failure, a point-to-point topology allows regional high-bandwidth communications over available segments to continue despite multiple failures, which could be valuable in the context of a disaster where communication is critical for life-saving activities and coordination for support.

### Potential Benefits

This COA would support a highly resilient network with a greatly reduced recovery time arising from subsea network failures. It would also boost both the value of the Puerto Rico network infrastructure for the economy overall and the utility of Puerto Rico's submarine cable systems for U.S. and global communications.

---

[1] Madory, 2017; Belson, 2017.

[2] Guimarães, 2017.

## Potential Spillover Impacts to Other Sectors

New infrastructure and increased operational costs would place additional demands on Puerto Rico for power needs, funding, and the workforce. However, a robust communications infrastructure would mitigate the extent of challenges faced by response and recovery operations. A submarine cabling infrastructure would link Puerto Rico to external networks and embody the primary mechanism for communication with the United States for all activities. Consequently, this COA could have a positive cascading effect across many sectors, such as health and social services, water, and energy.

## Potential Costs

Potential up-front costs: $25 million–$130 million in estimated up-front costs (2 years)
Potential recurring costs: $42 million–$110 million in estimated recurring costs (11 years)
Potential total costs: $67 million–$240 million in total estimated costs

As with the basis for the estimate provided in *CIT 10 Transoceanic Submarine Cable*, the capital expenditure is captured as potential up-front cost. Potential recurring costs include operational expenditures, estimated on an annual basis. Upgrades and loan payments with interest are excluded from the estimate. Depending on assumptions, costs other than operational expenditures may constitute the bulk of the annual cost. Costs were estimated based upon input from representatives of government of Puerto Rico and Federal Emergency Management Agency (FEMA) solutions-based teams (SBTs).[3]

For an undersea ring system, the detailed approach may be able to leverage existing infrastructure, such as undersea branch points, to dramatically reduce the implementation effort while evolving both Puerto Rico's broadband capacity and redundancy of communications for the government of Puerto Rico. Similarly, the cost associated with cable length will be reduced to the cumulative length of segments required to connect cities. The total time to deploy and utilize the system will likely also be reduced from the 10-year window, as individual segments may be brought online at earlier points in time.

## Potential Funding Mechanisms

U.S. Department of Commerce Economic Development Administration, government of Puerto Rico, private sector, sale of capacity via indefeasible right of use or by lease

## Potential Implementers

Telecommunications Bureau, private industry

---

[3] The SubOptic Guide for a submarine cable network published by SubOptic was also referenced when developing the estimate; see Stafford, 013.

## Potential Pitfalls

Poor coordination among stakeholders during discovery, implementation, or funding could lead to higher than anticipated costs. Other potential pitfalls are irreparable loss to economically sensitive natural resources, unanticipated repairs to undersea cables, and single-vendor lock-in that does not allow Puerto Rico to exercise influence over key decisions. A development plan also requires a strategy for spanning multiple hurricane seasons.

## Likely Precursors

Precursors include clear governance or policy for ownership rights.

## References

Belson, David, "Internet Impacts of Hurricanes Harvey, Irma and Maria," Orcale + Dyn, blog, September 25, 2017. As of June 19, 2018: https://dyn.com/blog/internet-impacts-of-hurricanes-harvey-irma-and-maria/

Guimarães, Nathália, "Furacão Maria deixa conexão de internet lenta no Brasil," LeiaJa, Technology, Internet, September 9, 2017. As of June 19, 2018: http://www.leiaja.com/tecnologia/2017/09/22/furacao-maria-deixa-conexao-de-internet-lenta-no-brasil/

Madory, Doug, "Puerto Rico's Slow Internet Recovery," Orcale + Dyn, blog, December 7, 2017. As of June 19, 2018: https://dyn.com/blog/puerto-ricos-slow-internet-recovery/

Stafford, Elaine, "The Suboptic Guide: Chapter 1 Planning, Contracting, Constructing, Owning and Operating A Submarine Cable Network," Suboptic, 2013. As of April 29, 2019: http://suboptic.org/wp-content/uploads/2014/10/The-Guide-1.pdf

Submarine cable map, website, undated. As of April 29, 2019: https://www.submarinecablemap.com/#/landing-point/san-juan-puerto-rico-united-states

# Government Digital Reform Planning and Capacity Building

## Sector Impacted

Communications and Information Technology

## Issue/Problem Being Solved

The main issue to be addressed by this course of action (COA) is to establish a road map for the digital transformation of Puerto Rico. This will require (1) setting achievable goals and metrics for success; (2) a rigorous assessment of needs, costs, feasibility, and cultural and legal issues to be addressed; and (3) the establishment of a clear strategy that can be communicated and championed both inside the government and with the public.

The government of Puerto Rico has embraced a strategy of digital transformation in order to modernize government processes and improve citizen services, respond to fiscal pressures, and better coordinate post-Maria recovery efforts and preparation for future disasters. A transition of this scale must account for people, policy, processes, and technology and requires a shared vision, stakeholder buy-in, and a strategy that synchronizes efforts across silos.

## Description

This COA will create a road map to establish priorities and assess needs, costs, and feasibility for a government-wide digital transformation strategy. This COA will build stakeholder buy-in; solicit ideas, needs, and issues from a broad set of participants, including private industry; and provide a comprehensive strategy with associated metrics to improve chances of success.

The COA involves capacity-building and a government-wide assessment overseen by a chief innovation officer, to evaluate feasibility and set priorities for a government-wide digital transformation strategy, building on the Governor's agenda, the direction of the Fiscal Board, and the requirements attached to posthurricane recovery funds.

"Digital transformation" or "reform" is "the use of technology to radically improve performance or reach of enterprises,"[1] with an emphasis on improving *performance*, not on technology for technology's sake.[2] This is especially the case in the public sector, where "what separates digital leaders from the rest is a clear digital strategy combined with a culture and leadership poised to drive the transformation."[3]

Key objectives of this COA include:

---

[1] Westerman, Bonnet, and McAfee, 2014.

[2] Morgan, 2018.

[3] Eggers and Bellman, 2015.

- increasing human capacity by
  - adding expertise/staffing required for digital transformation (with emphasis on government), to address cybersecurity, data science, data architecture, and information technology (IT) architecture
  - engaging experts who have led similar transformation processes, at the federal government level or in U.S. states, to guide formation of a process with lessons learned and best practices
  - designating process for project teams
- convening stakeholders and establishing shared goals for the digital transformation
- conducting an assessment of the areas of greatest need and feasibility, with input from government agencies, members of the public, businesses, and stakeholder organizations
- conducting a feasibility study that will involve
  - refining cost estimates of proposed projects
  - anticipating issues to be addressed such as governance and differences in cultures and processes across government agencies
  - estimating scope and cost of retraining or re-skilling public sector workers for new digitized processes
  - assessing project time frames
  - evaluating their potential benefits
  - identifying metrics of success and evaluation process
- evaluating successful models from other jurisdictions
- establishing metrics of success and key progress indicators (KPIs)
- gaining necessary approvals within the government of Puerto Rico for proposed projects, finding out whether executive or legislative changes are required, and socializing proposals among stakeholders to increase potential for success
- establishing the budget and resources required for successful implementation
- determining what outreach, training, and other change management strategies are needed to ensure successful implementation
- providing ongoing metrics tied to KPIs.

Projects may include (but are not limited to):

- digital identity (e.g., *CIT 27 Study Feasibility of Digital Identity*)
- citizen-facing digital services improvements (e.g., *CIT 32 Digital Citizen Services*)
- citizen engagement portal
- government digital process reform projects (e.g., *CIT 33 Government Digital Process Reform*)
- eProcurement
- transparency portal
- data analytics.

The expected initial results from this include convenings of stakeholders and establishment of clear goals for the digital transformation; completed needs assessment and list of priority projects; establishment of success metrics, budget, and resources required; identification of policy changes or approvals required, of required trainings and outreach; first projects underway.

## Potential Benefits

This COA provides a framework for Puerto Rico to benefit from best practices and avoid the pitfalls that have beset digital transformation efforts in other jurisdictions. This COA intends to build stakeholder buy-in; to surface ideas, needs, and issues from a broad set of participants; and to provide a comprehensive strategy with associated metrics to improve chances of success. The pitfalls experienced in prior digital transformation efforts in other jurisdictions include:

- "lack of strategy," which has been identified as leading barrier to early-stage organizations taking full advantage of digital trends[4]
- "culture," which was cited by 85% of public-sector leaders surveyed as a challenging aspect of managing the transition to digital[5]
- "lack of system-wide prioritization" and navigating agency silos, which were challenges for New Zealand's digital transformation[6]
- "insufficient trust to share data," which was also a problem for New Zealand.[7]

## Potential Spillover Impacts to Other Sectors

This COA would result in better data and improved processes throughout the government of Puerto Rico and positively affect all sectors.

## Potential Costs

Potential up-front costs: $6.2 million in estimated up-front costs[8]
Potential recurring costs: $2.0 million in estimated recurring costs (11 years)[9]
Potential total costs: $8.2 million in total estimated costs[10]

The up-front cost estimate includes costs for staff to create the roadmap, undertake stakeholder outreach and planning process, work with small and medium enterprises and consultants on plan and metrics formation, and change-management training within commonwealth agencies. A multiyear task force would provide additional capacity to conduct needs assessments; convene with agencies; and prioritize projects, establish KPIs, and reporting mechanisms. The task force cost is estimated for 10 fulltime subject matter experts (SMEs) for

---

[4] See Eggers and Bellman, 2015, p. 5.

[5] See Eggers and Bellman, 2015, p. 15.

[6] Carpenter, 2018.

[7] Carpenter, 2018.

[8] The recovery plan represents the up-front costs as $14 million. The estimated up-front cost here is updated to reflect including training costs as recurring costs and to use labor costs consistent with the other COAs.

[9] The recovery plan inadvertently included training costs as up-front costs, and represents no recurring costs. The estimated recurring costs are updated here.

[10] The recovery plan represents the total estimated cost as $14 million. The total is updated here to reflect reduced estimated up-front costs.

5 years, which would cost approximately $6.2 million.[11] In addition, we estimate recurring costs for agency training would consist of approximately 1 week of training per year at an average cost of $1,200 per worker (based on an average administrative salary of $62,300 per year) plus the cost of 2 trainers at approximately $2,400 each (based on an average salary of $124,600 per year). For 150 workers,[12] the annual training cost would be approximately $185,000. Thus, the total recurring costs were estimated to be approximately $2.0 million over 11 years.

## Potential Funding Mechanisms

Community Development Block Grant Disaster Recovery, U.S. Department of Commerce Economic Development Administration

## Potential Implementers

Office of the Chief Innovation Officer, Office of the Chief Information Officer

## Potential Pitfalls

This COA seeks to mitigate pitfalls for and to other COAs. Accommodating other sectors would require costs to the sector that engages in this effort. Lack of support by government agencies would be a challenge to this COA.

## Likely Precursors

None. This COA is a starting point for several other COAs related to the digital transformation of Puerto Rico including *CIT 23 Data Collection and Standardization for Disaster Preparedness and Emergency Response*, *CIT 26 Establish Secure Digital Identity*, *CIT 32 Digital Citizen Services*, and *CIT 33 Government Digital Process Reform*.

## References

Carpenter, Darryl, "Overcoming the challenges of digital transformation – lessons learned from the NZ Government," University of Melbourne Power of Collaboration series, original February 6, 2018, updated September 19, 2018. As of April 29, 2019: https://melbourne-cshe.unimelb.edu.au/lh-martin-institute/insights/overcoming-the-challenges-of-digital-transformation-lessons-learned-from-the-nz-government

Eggers, William D., and Joel Bellman, *The Journey to Government's Digital Transformation*, Deloitte University Press, 2015. As of April 29, 2019:

---

[11] This dollar figure comes from multiplying the annual cost per member ($124,600) by the number of members (10) and the number of years (5).

[12] As of July 2018, Puerto Rico is expected to consolidate its 134 agencies into approximately 30 agencies. The number of staff needing training is based upon assuming 5 workers per agency.

https://www2.deloitte.com/content/dam/insights/us/articles/digital-transformation-in
-government/DUP_1081_Journey-to-govt-digital-future_MASTER.pdf

Morgan, Jeffrey, "Digital transformation in the public sector," *CIO Magazine*, January 11, 1018. As of April 29, 2019:
https://www.cio.com/article/3247305/digital-transformation-in-the-public-sector.html

Westerman, George, Didier Bonnet, and Andrew McAfee, "The Nine Elements of Digital Transformation," *MIT Sloan Management Review*, January 7, 2014. As of April 29, 2019:
https://sloanreview.mit.edu/article/the-nine-elements-of-digital-transformation/

# CIT 17
## Puerto Rico Data Center

### Sectors Impacted

Communications and Information Technology, Economic, Energy, Health and Social Services, Housing, Transportation, Water

### Issue/Problem Being Solved

Expand the government of Puerto Rico's capacity and independent ability to perform essential governmental functions and deliver essential governmental services throughout Puerto Rico.

### Description

This course of action (COA) would establish a robust, disaster-proof (Tier 3 or Tier 4),[1] and cloud-enabled data center in Puerto Rico for government of Puerto Rico information systems, initially targeting a medium-sized capacity. This COA would expand the government of Puerto Rico's capacity and ability to perform essential governmental functions and deliver essential governmental services efficiently by using a commonwealth-owned, highly available, scalable, and evolvable infrastructure for its information systems.

Two cloud-optimized data centers located on different parts of Puerto Rico would open up the design options from a power grid and networking perspective, while also accommodating the ability for a failover mechanism and redundancy of services. The capacity of each data center could be individually reduced from a single, integrated data center while supporting a greater total capacity for computing services.

### Potential Benefits

An independent, commonwealth-owned, and disaster-resilient data center can enable highly reliable governmental information technology (IT) services for tracking, supporting, and coordinating response and recovery needs both within Puerto Rico and externally while preserving the integrity of all essential information systems. At present, there is no disaster-resilient or hardened facility that reliably and comprehensively preserves important governmental data or ensures the availability of corresponding information systems. This complicates disaster response efforts (coordination, situational awareness, and so on) and undermines the government's ability to assess and respond to disaster-related needs. There is

---

[1] The ANSI/TIA-942 standard defines rating levels or tiers for data centers, as Rated-1: Basic Site Infrastructure, Rated-2: Redundant Capacity Component Site Infrastructure, Rated-3: Concurrently Maintainable Site Infrastructure, and Rated-4: Fault Tolerant Site Infrastructure. See Telecommunications Industry Association, undated.

potential to dramatically improve the planning efforts for disaster response and recovery through the evolution of existing commonwealth-owned information systems in a cloud-based environment.

## Potential Spillover Impacts to Other Sectors

This data center is potentially a key infrastructure component that would enable Puerto Rico to pursue digitally driven economic goals and support digital initiatives, leveraging the communications capacity of a broadband infrastructure for workforce development, health and medicine, and so on, in addition to supporting the continuity of government (COG) and related information needs or services. The COA also seeks to address needs for a computing infrastructure in a way that could support other initiatives by providing a common, resilient computing infrastructure and capabilities for related services.

## Potential Costs

Potential up-front costs: $7 million–$20 million in estimated up-front costs
Potential recurring costs: $61 million–$170 million in estimated recurring costs (11 years)[2]
Potential total costs: $68 million–$190 million in total estimated costs[3]

The range of up-front estimated costs scales depending on redundant power needs and workforce. Both up-front and annual costs will vary depending on engineering design for required capacity (cooling/power requirements, such as maximum available power versus consumed power), number of cabinets (a measure of computing units or servers), and staffing requirements. These values assume 150 cabinets for a Tier-4 data center, with an average power consumption of 60% available power. Potential additional costs may be incurred for failover to secondary cloud providers, depending on the approach taken for continuity of operations (COOP).

Up-front costs for a small- to midsize data center are $5 million–$15 million for initial capital and related expenditures. Annual costs are $4 million–$11 million. Allowing for uncertainty in initial capacity requirements (e.g., for all governmental information systems, including for all 78 municipalities) and corresponding staffing needs, and also increasing the rough order of magnitude estimates for a single data center by 35%, up-front costs may be from $7 million to $20 million and annual costs from $5.5 million to $15 million.[4] A strategy incorporating multiple data centers can use this approach as basis to estimate additional facilities, as costs are expected to be fairly consistent across sites.

---

[2] The estimated recurring costs are represented in the recovery plan using 2 significant figures.

[3] The estimated total costs are represented in the recovery plan using 2 significant figures.

[4] Rough order of magnitude costs were considered from various sources and discussions concerning data centers. Expedient's Data Center Build vs Buy Cost Calculator website was relied on to explore configuration options with improved granularity; Expedient, "The Complete Data Center Build vs Buy Calculator," website, undated.

The cost estimate for 2 data centers are projected to be $14 million to $40 million (up-front) and $11 million to $30 million (annually).

## Potential Funding Mechanisms

Government of Puerto Rico, leasing of excess capacity, nongovernment sources

## Potential Implementers

Office of the Chief Information Officer, government of Puerto Rico agencies

## Potential Pitfalls

Determining a suitable location for the data center's facility and acquiring use rights, the availability of workforce or excessive dependence on nonresident workers, and improperly assessed power needs could have a dramatic effect on up-front costs, as would underutilization of resources or excessive annual expenses.

## Likely Precursors

Robust power generation and supply systems, availability of broadband, and reliable submarine cable infrastructure are necessary. Precursors COAs are *CIT 10 Transoceanic Submarine Cable*, *CIT 21 Government-Owned Fiber-Optic Conduits to Reduce Aerial Fiber-Optic Cable and Incentivize Expansion of Broadband Infrastructure*, and *CIT 23 Data Collection and Standardization for Disaster Preparedness and Emergency Response*.

## References

Expedient, The Complete Data Center Build vs Buy Calculator, website, undated. As of April 29, 2019:
https://www.expedient.com/data-center-build-vs-buy-calculator/

Telecommunications Industry Association, About Data Centers, ANSI/TIA-942 Quality Standard for Data Centers, TIA-942.org, website, undated. As of September 12, 2019:
http://www.tia-942.org/content/162/289/About_Data_Centers

# CIT 18
## Data Storage and Data Exchange Standards for Critical Infrastructure

### Sectors Impacted

Communications and Information Technology, Economic, Energy, Health and Social Services, Housing, Transportation, Water

### Issue/Problem Being Solved

Improve Puerto Rico's ability to plan for and comprehensively support critical infrastructure needs.

### Description

This course of action (COA) would create an online data store and data exchange standards for up-to-date, cross-sector data about critical infrastructure (government and private sector) using an open, modular, and standards-based approach for information exchange, interoperability, and storage. This COA would formally evaluate and develop a standardized interface definition (or multiple definitions) for data exchanges and storing data in a standard form. This COA would also consider opportunities to leverage existing data standards and systems, such as ones offered by the National Information Exchange Model and the U.S. Department of Homeland Security Infrastructure Protection Gateway, either directly or as a basis for a commonwealth-developed service designed for and capable of satisfying Puerto Rico's near- and long-term needs.

A key aspect of a data-oriented information system for critical infrastructure is its ability to receive, provide, or exchange current, accurate information across multiple sectors in order to provide an integrated resource for critical infrastructure data. A common standard (or standards) to receive high-quality data from a broad variety of sources will be essential to forming a coherent set of referential data about critical infrastructure for areas of concern in Puerto Rico and to support a variety of related governmental initiatives. A standardized interface definition for information exchange will also need to consider the scope for critical infrastructure data exchange, the range and types of data collected, definitions for critical infrastructure, information-sharing and safeguarding policies, the capability of participating entities to provide such data, and the effort to implement the interface to exchange data between systems.

### Potential Benefits

The improved visibility into the critical infrastructure assets of Puerto Rico would support a comprehensive, quantitative picture of their availability. For example, a quantitative analysis may support burying aerial fiber, which would increase resilience to severe weather. Accurate, situational awareness of infrastructure issues may also inform emergency response activities prior to or after a disaster. Private companies can also know when and where Puerto Rico Electric Power Authority and Puerto Rico Aqueduct and Sewer Authority road repair will be trenching. Government and private-sector entities providing support for, or involved with, critical infrastructure assets would also benefit.

## Potential Spillover Impacts to Other Sectors

Coordinated reconstruction activities can be informed by critical infrastructure data to streamline planning, budgeting, and supply-chain logistics, as well as to optimize recovery-related activities. These improvements would naturally benefit multiple sectors, such as water, energy, health and social services, transportation, and housing.

## Potential Costs

Potential up-front costs: $1.8 million–$2.5 million in estimated up-front costs (2 years)
Potential recurring costs: $6.3 million–$13 million in estimated recurring costs (11 years)
Potential total costs: $8.1 million–$15 million in total estimated costs[1]

Costs will depend on the detailed approach, scope of the data being maintained, long-term plan for using critical infrastructure data, and staffing needs. For a cloud-based service, the total volume of data that are stored and the frequency of transactions made when requesting or exchanging data are potential cost drivers. Assuming that physical information technology (IT) infrastructure is sourced from the Puerto Rico Data Center (costed under CIT 17) and that the development effort will involve 15–20 team members, including informed, authoritative decisionmakers and key stakeholders, a new open, modular information system for critical infrastructure data will incur up-front costs for design and implementation, in addition to operational costs. Once operational, a smaller team (e.g., 8 members) can maintain the information system and support participating entities for integration efforts.

## Potential Funding Mechanisms

Private sector, government of Puerto Rico

## Potential Implementers

Office of the Chief Information Officer, government of Puerto Rico agencies

## Potential Pitfalls

Given a broad range of private- and public-sector entities that are potential sources for critical infrastructure data, there is an inherent challenges regarding engagement through implementation and ongoing sustainment. Lack of inclusion or availability of key stakeholders

---

[1] Potential costs are represented in the recovery plan using 2 significant figures.

able to make authoritative decisions about standard interfaces, interoperability requirements, and data needs or to reconcile and prioritize functionality may undermine the quality of data incorporated and the final performance (measured against governmental "business" needs) of the system. Similarly, new business processes may need to be developed or changes to existing ones may be required to ensure the exchange of current information.

## Likely Precursors

Availability of a secure data center for governmental information systems and formalized governance for informed, authoritative decisions are necessary. Precursor COAs are *CIT 17 Puerto Rico Data Center* and *CIT 24 Establish Puerto Rico Communications Steering Committee.*

# CIT 19
# Municipal Hotspots

## Sector Impacted

Communications and Information Technology, Community Planning and Capacity Building, Economic, Health and Social Services, Municipalities

## Issue/Problem Being Solved

Many residents and visitors to Puerto Rico lack options for reliable, affordable internet access. Government-sponsored wi-fi in town centers and public buildings will provide the government of Puerto Rico with a priority connection point for reaching a large number of residents in one place after a disaster. For residents, government-sponsored wi-fi in public spaces can provide a backup option when mobile service is not available. For those without residential broadband, or those who need to access the internet while away from home, wi-fi-enabled public buildings, town squares, and parks can provide a place for browsing news, studying, or accessing government services or job information. For tourists and business travelers, access to government-sponsored wi-fi is a convenience that is increasingly expected throughout the world, and a low-cost way for Puerto Rico to support its visitor economy through greater connectivity options.

Wi-fi hotspots in public buildings and public areas can also increase adoption of government digital offerings because citizens can browse information and complete forms online from a mobile device, rather than waiting in line.

## Description

This course of action (COA) would maximize public access to government-sponsored wi-fi in the main centers of public life, including municipal buildings, parks, libraries, and town squares across Puerto Rico. This COA would expand the existing 58 municipal hotspots sponsored by the Telecommunications Bureau (formerly, the Puerto Rico Telecommunications Regulatory Board), as feasible and appropriate. Currently, the Telecommunications Bureau sponsors installation of equipment, and municipalities provide for recurring costs for service to a telecommunications provider. The Telecommunications Bureau program has been working well for many years and is accepted by telecommunications providers. For that reason, the scope of this COA reflects the current Telecommunications Bureau program, which would also provide municipal hotspots for government outreach in the event of an emergency. There are 2 main kinds of sites to be considered:

- Town Centers: Each of the 78 municipalities in Puerto Rico has a center, in most cases next to the municipal administration buildings and a church. These "plazas" are shared gathering and recreational places for all residents and visitors. Government-sponsored public wi-fi in all town centers creates an opportunity to reach residents in a place that is welcoming and accessible to all. The town center project might be implemented as one project for all town centers in Puerto Rico to take advantage of bulk buying and contracting for the purchase of equipment and installation for all of Puerto Rico. It also would be important to establish partnerships with churches, municipalities, and surrounding buildings for the ability to place repeaters and routers in positions that ensure maximum coverage.

- Municipal and public buildings: Puerto Rico might set a goal that all public buildings provide government-sponsored public wi-fi by 2020. This could expedite government processes by encouraging digital applications and forms and providing information resources online.

## Potential Benefits

Government-sponsored public wi-fi would provide a priority postdisaster connection point for reaching a large number of residents in one place. Commonwealth or municipal governments would be able to provide information to citizens through a main internet page, such as status.pr.gov. Citizens would benefit from increased access to the internet, which may also help encourage a more digitally capable society.

For Puerto Rico, widely available public wi-fi would also:

- provide a backup option when mobile service is not available
- support the "visitor economy" through greater connectivity options for tourists and business travelers
- increase adoption of government digital offerings, as citizens can browse information and complete forms online from a mobile device, rather than waiting in line.

Many successful digital transformation efforts have started with connectivity—extending broadband access to the population by connecting schools, ensuring affordable home access, and making wi-fi available in public areas, libraries, parks, and public buildings. Examples include:

- Minneapolis: One of the first cities to offer ubiquitous public wi-fi, Minneapolis introduced 117 "Wireless Minneapolis" hotspots throughout the city in 2010 in places where people already gather and use computers and places where free wireless access would encourage people to gather, including parks, plazas, schools, and businesses.[1]
- Vancouver: As a part of the city's Digital Strategy, approximately 550 locations in the downtown core and surrounding areas have free access to 10 Mbps wi-fi, with no data usage limit and no personal information required to access the network. Wi-fi-enabled locations include all 9 city-owned housing sites, all libraries, 27 community centers, 4 outdoor pools, 4 civic facilities, 3 public golf courses, 3 theaters, 2 marinas, and the City Hall campus. The service "enable[s] visitors to use their smartphones and devices to discover attractions, activities, and restaurants, for wayfinding, and to navigate Vancouver's transportation options." It also "enable[s] visitors to use social media on a real-time basis to share their Vancouver experiences with the world"; it is also "an

---

[1] City of Minneapolis, 2019.

important development for the Vancouver tourism industry as it moves us closer to our goal of being a 'smart' tourism destination."[2]

- Berlin: Launched in 2016, "Free WiFi Berlin" provides free and unlimited wi-fi access through 650 hotspots (325 indoor and 325 outdoor) throughout Berlin, including tourist attractions such as the Red Town Hall, Brandenburg Gate, Gendarmenmarkt, Philharmonie, and Theatre of the West, as well as public buildings, including town halls, council offices, museums, schools, and libraries.[3] The city also supports the Freifunk ("free radio") initiative, a noncommercial open initiative of free radio networks that use mesh technology to interconnect multiple wireless local area networks (LAN) in 400 local communities, with over 41,000 access points.[4]
- Mexico: Google is launching 56 of its "Google Station" ad-supported wi-fi hotspots across Mexico, as an "entrypoint for the product in Latin America."[5] This follows the initial Google Station deployment in India, which reaches 8 million users a month.[6]
- Europe: Starting in March 2018, the European Union is offering 1,000 vouchers of €15,000 per year (with a total of €120 million committed overall) to help cities and towns set up wi-fi hotspots in public places such as parks, squares, and public buildings. The program is called "WiFi4EU." Applicants may contract with suppliers of their choice and must ensure installations are complete and functioning within 18 months, and make ad-free wi-fi available to all for at least three years.[7]

## Potential Spillover Impacts to Other Sectors

With reliable, affordable access to wi-fi, residents of Puerto Rico will have more ways to participate in every sector of society, including:

- community planning and capacity building: When all citizens can access the internet, they can participate in planning activities, share their needs and views on proposed actions, and feel more included.
- economic: Internet access is one of the most important drivers for economic recovery by providing information and opportunity.
- health and social services: Internet access can tangibly improve health.[8] Many hospitals and health plans (including those serving Medicaid patients) are expanding the use of internet-based communications tools to improve their members' health. In addition, new sensor technology can send health indicators to medical professionals and thereby improve compliance with doctors' recommendations.

---

[2] City of Vancouver, 2019.

[3] Visit Berlin, Public Wi-Fi Berlin, webpage, undated.

[4] Freifunk, Wikipedia, webpage, August 8, 2018.

[5] Julia Love, "Google Brings Free WiFi to Mexico, First Stop in Latin America," Reuters, March 13, 2018.

[6] Jagmeet Singh, "Google Station Free Public Wi-Fi to Go Beyond Railway Stations in India to Reach Cities," *Gadgets 360*, December 5, 2017.

[7] "EU to Give Free Wi-Fi Hotspots to Cities," *Euronews*, March 20, 2018.

[8] Sweeney, 2017.

## Potential Costs

Potential up-front costs: $0.8 million in estimated up-front costs[9]
Potential recurring costs: $8.8 million in estimated recurring costs (11 years)[10]
Potential total costs: $9.6 million in total estimated costs[11]

The estimate for up-front costs for municipal hotspots is based on comparable projects in the United States.[12] For each town center hotspot, 1 per municipality within each of the 78 municipalities, the cost was assumed to include a set-up cost of $10,000 in addition to an annual grant of $10,000 to each municipality to subsidize internet access. Thus the up-front and the annual cost per municipality are both equal to $10,000.[13] Therefore for 78 municipalities, the total up-front cost is estimated to be $0.8 million, and the recurring cost over 11 years is estimated to be $8.8 million.

## Potential Funding Mechanisms

Community Development Block Grant Disaster Recovery, U.S. Department of Commerce Economic Development Administration, Federal Communications Commission

This effort might utilize funds from a variety programs including:

- Economic Development Administration's Facilities and Public Works program
- Library Services and Technology Act Grants to States
- Housing and Urban Development Choice Neighborhood Implementation Grants
- Housing and Urban Development Broadband Infrastructure Grants
- National Telecommunications and Information Administration Broadband grants.

## Potential Implementers

Office of the Chief Innovation Officer, Telecommunications Bureau, government of Puerto Rico agencies, municipal governments

---

[9] The recovery plan represents the potential up-front cost at $1.6 million in estimated costs. That estimate is being corrected here. The set-up cost for each hotspot was double-counted previously, and is updated here, reducing the estimated cost.

[10] The recovery plan represents the potential recurring cost as $16 million in estimated costs. That recurring cost is being corrected here. The annual cost inadvertently represents twice the value estimated.

[11] The recovery plan represents the total potential cost as $18 million. The cost to acquire and maintain each hotspot has been corrected, resulting in the cost decreasing by half the estimate in the recovery plan.

[12] The comparable projects are Tribal Digital Village from the Southern California Tribal Chairmen's Association and Red Hook_WiFi (in New York).

[13] In the recovery plan, the cost per municipality was inadvertently overestimated to be $20,000 up front and $20,000 per year.

## Potential Pitfalls

Although Telecommunications Bureau officials said that they have managed government-sponsored hotspots since 1999, the Telecommunications Bureau could face resistance expanding the program into areas covered solely by the private telecommunications market, a situation encountered by some municipalities in the continental United States. However, a senior Telecommunications Bureau official explained that although the office pays for the initial installation of wi-fi equipment and internet service in the municipalities involved in the program, the municipalities pay the internet service provider directly after approximately 2 years. Thus, private telecommunications providers are paid for their services in rural or disadvantaged areas where they might not otherwise provide service. Also, in the future, the Telecommunications Bureau will need to make citizens aware of the lack of security associated with public wi-fi and establish content policies for use. In order to realize the full potential of affordable wi-fi access, individuals must also have access to digital devices and the digital skills to make use of the technology that connectivity makes possible.

## Likely Precursors

The Telecommunications Bureau would need to make citizens aware of the lack of security associated with public wi-fi and would need to establish content policies for use. Precursor COAs are *CIT 21 Government-Owned Fiber-Optic Conduits to Reduce Aerial Fiber-Optic Cable and Incentivize Expansion of Broadband Infrastructure*, and *CIT 25 Evaluate and Implement Alternative Methods to Deploy Broadband Internet Service Throughout Puerto Rico*.

## References

City of Minneapolis, "Wireless Minneapolis," website, June 6, 2019. As of June 22, 2019:
http://www.minneapolismn.gov/wireless/index.htm

City of Vancouver, "Vancouver's Digital Strategy," website, 2019. As of June 22, 2019:
https://vancouver.ca/your-government/digital-strategy.aspx

Euronews, "EU to Give Free Wi-Fi Hotspots to Cities," *Euronews*, March 20, 2018. As of November 6, 2018:
https://www.euronews.com/2018/03/20/eu-cash-funds-free-hi-speed-wifi-hotspots

Love, Julia, "Google Brings Free WiFi to Mexico, First Stop in Latin America," *Reuters*, March 13, 2018, 1:32 PM. As of November 6, 2018:
https://www.reuters.com/article/us-mexico-google/google-brings-free-wifi-to-mexico-first-stop-in-latin-america-idUSKCN1GP2VU

Singh, Jagmeet, "Google Station Free Public Wi-Fi to Go Beyond Railway Stations in India to Reach Cities," *Gadgets 360*, December 5, 2017, 19:28 IST. As of November 6, 2018:
https://gadgets.ndtv.com/telecom/news/google-station-railway-stations-india-smart-cities-1783972

Sweeney, Evan "AMIA sees internet access as a social determinant of health," *Fierce Healthcare*, May 24, 2017. As of April 25, 2019:
https://www.fiercehealthcare.com/mobile/amia-views-broadband-access-as-a-social
-determinant-health

Visit Berlin, Public Wi-Fi Berline, webpage, undated. As of November 6, 2018:
https://www.visitberlin.de/en/public-wi-fi-berlin

Wikipedia, Freifunk, webpage, August 8, 2018. As of November 6, 2018:
https://en.wikipedia.org/wiki/Freifunk

# CIT 20
## Continuity of Business at PRIDCO Sites

### Sectors Impacted

Communications and Information Technology, Economic, Municipalities

### Issue/Problem Being Solved

The resiliency of essential business activities delivered by the Puerto Rico Industrial Development Company (PRIDCO) in the aftermath of a disaster needs improvement.

### Description

This course of action (COA) would maintain key business activities at PRIDCO sites when primary communications methods are degraded or unavailable to provide continuity of services related to disaster recovery. This COA would establish multiple alternative business processes to leverage several platforms for telecommunications and information systems, including, but not limited to, the use of fiber, satellite, microwave systems, and cloud-based or hosted services and information systems. The 2017 storm season halted operations in several areas and had a cascading impact that reached well beyond Puerto Rico. Critical operational data were no longer accessible due to loss of communications and power at the sites. [1]

### Potential Benefits

PRIDCO facilitates economic development across a wide range of sectors in several key ways, including investment incentives, design of business proposals, assistance with the regulatory and permitting process, project management, and facility selection. Continuity of business at PRIDCO sites would directly benefit recovery options.

Business enterprises at PRIDCO sites in Puerto Rico are major contributors to the U.S. economy and span multiple domains, including, for example, biomedical devices and aerospace services. Enabling redundant communications networks and power systems at PRIDCO sites will help mitigate the risks arising from outages in upstream services, to which they are currently prone.

The improved availability of communications at PRIDCO sites can potentially be extended to local and rural communities during disasters. PRIDCO sites are distributed geographically to improve the economy as a whole, especially in areas that are not yet population centers. Through

---

[1] PRIDCO representatives, interview with authors, discussion regarding the impact of hurricanes on operational activities for businesses and on local communities, June 1, 2018.

improved, on-site communications network, PRIDCO may be able to further support these communities in the event of a disaster.[2]

## Potential Spillover Impacts to Other Sectors

Recovery efforts across sectors—including health and social services, housing, transportation, energy, and Communications and Information Technology—could be accelerated by leveraging PRIDCO services if they are available during a disaster or outage.

## Potential Costs

Potential up-front costs: $24 million in estimated up-front costs
Potential recurring costs: —
Potential total costs: $24 million in total estimated costs

Estimates to prepare microgrids at 5 PRIDCO sites range from $11 million to $37.5 million. The recovery plan used the midpoint of this range as the basis for the up-front costs. Estimates for a microwave-based site-to-site communications network range from $75,000 to $125,000, or $15,000 to $25,000 per site for equipment. The costs would vary substantially for selected sites and the requirements for telecommunication needs and backup power. According to a rough estimate drawn from Berkeley Labs case studies of microgrids,[3] a 5 Megawatt (MW) natural gas system has an estimated construction cost of $7.5 million (completed in 1 year).

## Potential Funding Mechanisms

U.S. Department of Commerce Economic Development Administration, Community Development Block Grant Disaster Recovery, government of Puerto Rico, private insurance

## Potential Implementers

Puerto Rico Industrial Development Company

## Potential Pitfalls

None

## Likely Precursors

This COA would need reliable power systems and may require infrastructural upgrades to accommodate modernized communication systems. Precursor COAs are *CIT 5 Implement Public Safety/Government Communications Backup Power*, *CIT 17 Puerto Rico Data Center*, and *CIT 23 Data Collection and Standardization for Disaster Preparedness and Emergency Response*.

---

[2] PRIDCO representatives, interview with authors, discussion regarding the impact of hurricanes on operational activities for businesses and on local communities, June 1, 2018.

[3] Microgrids at Berkley Lab, *Microgrids, DER Case Studies*, webpage, undated.

# References

Microgrids at Berkeley Lab, *Microgrids, DER Case Studies*, webpage, undated. As of November 5, 2018:
https://building-microgrid.lbl.gov/projects/der-cam/case-studies

PRIDCO representatives, interview with authors, discussion regarding the impact of hurricanes on operational activities for businesses and on local communities, June 1, 2018.

# CIT 21
# Government-Owned Fiber-Optic Conduits to Reduce Aerial Fiber-Optic Cable and Incentivize Expansion of Broadband Infrastructure

## Sectors Impacted

Communications and Information Technology, Education, Economic, Energy, Health and Social Services, Housing, Municipalities, Transportation

## Issue/Problem Being Solved

Increasing the resilience of telecommunications services while reducing costs to telecommunications providers for burying cable are priorities.[1] Facilitating the burial of aerial fiber-optic cable and the provision of broadband infrastructure throughout Puerto Rico need to be addressed. In the process, the need for multiple roadway disturbances to lay cable for separate firms should be minimized, the current permitting and rights of way (ROW) procedures, which are time-consuming and expensive, should be streamlined, and telecommunications providers should be offered the opportunity to pull fiber through government-owned conduit.

## Description

This course of action (COA) would provide trenching and conduit adequate for accommodating other utilities. This COA would also expedite permitting and ROW processes, which are time-consuming and expensive, and it would offer telecommunications providers the opportunity to pull fiber through government-owned conduit.

The design for the deployment of conduit for fiber-optic cable would be developed by the government of Puerto Rico in consultation with telecommunications providers. The government would trench and lay empty conduit according to the design. The government would own the conduit, but telecommunications providers would install and own their own fiber-optic cable.

In a May 11, 2018, meeting at the Puerto Rico Telecommunications Regulatory Board, Puerto Rico telecommunications providers endorsed the idea of using conduit that would be buried by the government of Puerto Rico, as long as they could have input about where the conduit would be located.[2] The government of Puerto Rico would be covering the most expensive part of burying fiber-optic cable—namely, the permitting and ROW approval process and the trenching costs (labor and equipment). For this reason, an incentive, rather than a

---

[1] This COA is important both to attain resilience (by reducing aerial fiber-optic cable) and to incentivize the expansion of broadband infrastructure, as explained in Chapter 5 of this report. Also, we note that both (buried) fiber conduit and aerial fiber-optic cable can be used to expand broadband infrastructure.

[2] The Puerto Rico Telecommunications Regulatory Board's name was changed to the Telecommunications Bureau of the Public Service Regulatory Board in August 2018, after the publication of the recovery plan.

governance change, might be the first approach to expedite the burial of fiber-optic cable in conduit.

Implementation of this COA will require considering market issues as well as potential "sticks" in addition to the "carrots" (incentives) discussed in this COA description. These issues are discussed in Appendix B of this report.

## Potential Benefits

If the government of Puerto Rico provides the trenching and conduit for fiber-optic cable, it would greatly reduce the cost in time and money for telecommunications providers to deploy buried instead of aerial fiber-optic cable. Buried fiber-optic cable is far more resilient to disasters. This COA would minimize the need for multiple roadway disturbances to lay cable for separate firms.

This COA would create incentives for telecommunications providers to decide to restore lost aerial fiber-optic cable with buried fiber. Going forward, it could alter the cost differential between pole attachments for aerial fiber and the cost of buried fiber in conduit. If the cost of burying fiber was reduced, it might also significantly increase the speed of deployment of fiber-optic cable throughout Puerto Rico to provide broadband internet connectivity. If conduit was available throughout Puerto Rico, the cost to pull fiber through the conduit, rather than the combined expense of permitting and trenching to bury fiber, might incentivize telecommunications providers to deploy fiber-optic cable to geographic areas that have adverse terrain or to disadvantaged areas. Commercial telecommunications providers evaluate the return on investment (ROI) for capital expenses such as extending their fiber-optic cable network. A conduit network that was available to all telecommunications providers throughout Puerto Rico could change the ROI of deploying fiber and thus speed deployment activity.

The Federal Communications Commission (FCC) recognized the importance of accelerating wireline broadband deployment by removing barriers to infrastructure investment in WC Docket No. 17-84. The FCC Fact Sheet for WC Docket No. 17-84 states:

> High-speed broadband is an increasingly important gateway to jobs, health care, education, information, and economic development. Access to high-speed broadband can create economic opportunity, enabling entrepreneurs to create businesses, immediately reach customers throughout the world, and revolutionize entire industries. But all too often, regulatory barriers increase the cost of deploying next-generation networks, resulting in higher prices, less competition, and worse service for consumers and small businesses, especially in rural America.[3]

A network of conduit provided by the government of Puerto Rico could circumvent the delays and expense imposed on telecommunications providers in Puerto Rico by the existing

---

[3] FCC, "Accelerating Wireline Broadband Deployment by Removing Barriers to Infrastructure Investment Report and Order, Declaratory Ruling, and Further Notice of Proposed Rulemaking - WC Docket No. 17-84," fact sheet, October 26, 2017.

permitting and ROW approval process. A network of conduit could be created in parallel to instituting regulatory reforms concerning permitting and ROW approval.

## Potential Spillover Impacts to Other Sectors

This COA would require coordination with the Puerto Rico Department of Transportation and with municipalities, if conduit for fiber is built along the highways. Municipalities would obtain access to broadband internet services and potentially also to health care services and education.

The President's Council of Economic Advisors Issue Brief of March 2016, stated that broadband access has a significant impact on economic growth, wages, medical care, and education, among other benefits.[4] Specifically, the brief stated:

> Addressing the digital divide is critical to ensuring that all Americans can take advantage of the many well-documented socio-economic benefits afforded by Internet connections. These benefits are most evident when consumers have access to the Internet at speeds fast enough to be considered broadband; these speeds are required to facilitate full interaction with advanced online platforms.

The brief cited the economic impact of broadband connectivity

> By 2006—before the widespread availability of streaming audio and video—broadband Internet accounted for an estimated \$28 billion in U.S. GDP. . . . Nearly half of this total was due to households upgrading from dial-up to broadband service. By 2009, broadband Internet accounted for an estimated \$32 billion per year in net consumer benefits. . . . These findings are broadly consistent with studies that cover other countries. Broadband expansion is also associated with local economic growth in some cases.

The brief added that "growth is particularly concentrated in industries that are more IT-intensive and in areas with lower populations. In addition, insofar as it allows a person to participate more fully in the economy, developing Internet skills may even positively affect a person's wages."

In terms of medical outcomes, the brief continued:

> Broadband has made medical care and medical information more convenient and more accessible. . . . Broadband-enabled virtual visits with trained medical professionals can improve patient outcomes at lower cost and with a lower risk of infection than comes with conventional care provided in person. . . . Telemedicine is particularly valuable for rural patients who may lack access to medical care, as telemedicine allows them to receive medical diagnoses and patient care from specialists who are located elsewhere. Broadband can also be used to more accurately track disease epidemics. Various studies have demonstrated how large datasets from search engines and social media can be exploited in this way.

---

[4] Council of Economic Advisors, 2016.

The brief also described the positive impact of broadband internet on education:

> Broadband also enables access to lower-cost online education. . . . A 10 percent increase in college students taking all their courses online is associated with a 1.4 percent decline in tuition. The importance of the role that the computer and broadband more specifically play in enabling students to do their homework is evidenced by the fact that nearly half of 14 to 18 year olds report that they use a library computer, commonly for homework. . . . This finding suggests that library computers can provide a crucial source of access for students who would not otherwise have the ability to get online.

The brief cited other benefits of broadband connectivity, including "supporting entrepreneurship and small businesses, promoting energy efficiency and energy savings, improving government performance, and enhancing public safety, among others. In addition, broadband has become a critical tool that job seekers use to search and apply for jobs."

## Potential Costs

Potential up-front costs: $1.3 billion in estimated up-front costs
Potential recurring costs: —
Potential total costs: $1.3 billion in total estimated costs

The up-front cost estimate is based on all federal aid for nonmunicipal roads. The cost of this COA would depend on whether federal grants are obtained to design the plan for the conduit and the trenching to bury the conduit throughout Puerto Rico.

If telecommunications providers have input in the design of the conduit network, it may take a course apart from interstates, freeways and arterial roads. For example, the conduit might extend to municipal centers in all 78 municipalities. An estimate for the cost of this design is $1.3 billion. If a comprehensive plan to deploy broadband for Puerto Rico is devised, as explained in *CIT 25 Evaluate and Implement Alternative Methods to Deploy Broadband Internet Service Throughout Puerto Rico*, existing resources for conduit may be included in the conduit network and thus reduce overall cost.

See Appendix A for a more detailed explanation of the cost estimate methodology used for this COA.

## Potential Funding Mechanisms

Community Development Block Grant Disaster Recovery, U.S. Department of Commerce Economic Development Administration, public-private partnership, Federal Communications Commission

Federal assistance might be available for the government of Puerto Rico to design, trench, and lay conduit for fiber to incentivize private telecommunications providers to bury fiber and thereby increase the resiliency of Puerto Rico telecommunications infrastructure. Additional funding may be available from the FCC, since Puerto Rico falls significantly below all other states in broadband deployment. The FCC stated in its "2018 Broadband Deployment Report":

> In the time since the last report, the Commission has acted aggressively "to accelerate deployment of [advanced telecommunications capability] by removing barriers to infrastructure investment and by promoting competition in the telecommunications market." . . . We are hard at work facilitating deployment— for instance, by reducing regulatory barriers to the deployment of wireline and wireless infrastructure, reforming the universal service program to make it more efficient and accessible to new entrants, modernizing the business data service rules to facilitate facilities-based competition, [and] freeing up additional spectrum for terrestrial and satellite services.[5]

### Potential Implementers

Telecommunications Bureau, Puerto Rico Department of Transportation and Public Works (Puerto Rico Highway and Transportation Authority), Federal Communications Commission, private telecommunications companies

### Potential Pitfalls

Maintenance of the conduit network by the government of Puerto Rico would be important to its resilience and to its use by telecommunications providers. If telecommunications providers pay a reasonable fee for using the conduit network, those funds could be escrowed for maintenance of the conduit network by the government of Puerto Rico.

Telecommunications providers probably would want to have a service level agreement with the government of Puerto Rico concerning the conduit network that would provide for immediate repairs to the conduit in case of damage or service interruption.

It could be a potential pitfall to have to do with the design of the conduit network. One is that telecommunications providers might not agree on the design if they are permitted to have input into the design process. Another is that it may be difficult to design a conduit network that will be perceived as competitively neutral by all telecommunications providers.

### Likely Precursors

This COA could be implemented along with *CIT 13 Streamline the Permitting and Rights of Way Processes for Towers and the Deployment of Fiber-Optic Cable*. The precursor COA is *CIT 25 Evaluate and Implement Alternative Methods to Deploy Broadband Internet Service Throughout Puerto Rico*.

---

[5] FCC, 2018a, section 706, para. 96; citation for quotation omitted.

# References

Council of Economic Advisors Issue Brief, "The Digital Divide and Economic Benefits of Broadband Access," March 2016. As of April 29, 2019:
https://obamawhitehouse.archives.gov/sites/default/files/page/files/20160308_broadband_cea_issue_brief.pdf

FCC Broadband Deployment Report, GN Docket No. 17-199, FCC 18-10, February 2, 2018. As of July 3, 2019:
https://www.fcc.gov/document/fcc-releases-2018-broadband-deployment-report

Federal Communications Commission, "Accelerating Wireline Broadband Deployment by Removing Barriers to Infrastructure Investment Report and Order, Declaratory Ruling, and Further Notice of Proposed Rulemaking - WC Docket No. 17-84," fact sheet, October 26, 2017.

# CIT 22
## Use Federal Programs to Spur Deployment of Broadband Internet Island-Wide

### Sectors Impacted

Communications and Information Technology, Education, Municipalities

### Issue/Problem Being Solved

Providing broadband internet services to students throughout Puerto Rico would allow them to participate in digital learning and thus help meet the objectives for digital transformation and a digital economy.

### Description

This course of action (COA) would develop a program to obtain broadband internet services for schools and libraries in all of Puerto Rico's 78 municipalities using the Federal Communications Commission (FCC)'s E-Rate program and other federal programs. This COA would build on the new pilot project established by Puerto Rico legislation for the use of E-Rate funding in approximately 20 schools in central Puerto Rico, as well as the Puerto Rico Bridge Initiative funded by National Telecommunications and Information Administration (NTIA).[1] The Puerto Rico Department of Education, assisted by the Telecommunications Bureau, has been designated by the new Puerto Rico legislation to administer the pilot project.

A task force could work with the FCC and Puerto Rico agencies (such as the Puerto Rico Department of Education, municipal boards of education, and the Telecommunications Bureau) to develop an island-wide program to obtain broadband internet services for schools and libraries in 78 municipalities using the FCC's E-Rate program. This program would build on the new pilot project established by Puerto Rico legislation for the use of E-Rate funding in approximately 20 schools in central Puerto Rico, as well as the Puerto Rico Bridge Initiative funded by NTIA. The Puerto Rico Department of Education (PRDE), assisted by the Telecommunications Bureau, will administer the pilot project, and could use experience with the pilot project to develop an island-wide program. If additional assistance was needed on a short-term basis, a contractor could provide it. For example, a task force that included employees of the Telecommunications Bureau and the PRDE could assist Puerto Rico schools and libraries with applications to receive funding for broadband internet services through the E-Rate program, as well as with the extensive follow-up compliance work required by the E-Rate program. This program would extend over at least a 2-year period to reach schools and libraries in 78 municipalities.

---

[1] The recent legislation is Resolucion Conjunta 40-2018.

The FCC describes the equipment and services the E-Rate program will cover, as well as the sliding scale of financial support that is available to schools and libraries:

> Eligible schools, school districts and libraries may apply individually or as part of a consortium. Funding may be requested under two categories of service: category one services to a school or library (telecommunications, telecommunications services and Internet access), and category two services that deliver Internet access within schools and libraries (internal connections, basic maintenance of internal connections, and managed internal broadband services). Discounts for support depend on the level of poverty and whether the school or library is located in an urban or rural area. The discounts range from 20 percent to 90 percent of the costs of eligible services. E-Rate program funding is based on demand up to an annual Commission-established cap of $3.9 billion.[2]

Potentially, Puerto Rico schools could apply for E-Rate funding as part of a consortium of all Puerto Rico schools. Although many schools and libraries in the central and rural parts of Puerto Rico may qualify for 90% funding under the program, additional non-FCC funding will be required for Puerto Rico schools and libraries to participate in the E-Rate program. The E-Rate program is structured so that schools and libraries contribute a portion of the cost for services. The amount is determined by a sliding scale of economic need. In Puerto Rico, some schools cannot afford to pay teachers for an entire school year, and so they have no funds available for the E-Rate program. Additional funding sources to supplement the E-Rate program may include Housing and Urban Development (HUD) Community Development Block Grants, the U.S. Department of Agriculture, contributions by the PRDE, and nongovernment sources, such as charitable foundations.

## Potential Benefits

This COA would allow Puerto Rico schools to participate in the E-Rate program by obtaining funding from additional federal sources to address such issues as upgrading the electricity grid at schools and providing personal computers and laptops for students, security for equipment, and air conditioning for equipment. The lack of some of these preconditions would prevent a school from participating in the E-Rate program. In addition, supplemental funding for many schools would be required, even if schools qualify for funding at the 90% level.[3]

In terms of potential benefits, the FCC states that the "E-Rate program makes telecommunications and information services more affordable for schools and libraries. With funding from the Universal Service Fund, E-Rate provides discounts for telecommunications,

---

[2] FCC, "E-Rate," 2018.

[3] FCC's website states, "eligible schools and libraries may receive discounts on telecommunications, telecommunications services and Internet access, as well as internal connections, managed internal broadband services and basic maintenance of internal connections. Discounts range from 20 to 90 percent, with higher poverty and rural schools and libraries. Recipients must pay some portion of the service costs."

Internet access and internal connections to eligible schools and libraries."[4] The FCC also notes that

> the ongoing proliferation of innovative digital learning technologies and the need to connect students, teachers and consumers to jobs, life-long learning and information have led to a steady rise in demand for bandwidth in schools and libraries. In recent years, the FCC refocused E-Rate from legacy telecommunications services to broadband, with a goal to significantly expand Wi-Fi access. These steps to modernize the program are helping E-Rate keep pace with the need for increased Internet access."[5]

The FCC addresses benefits that are available under the E-Rate program:

> Eligible schools and libraries may receive discounts on telecommunications, telecommunications services and Internet access, as well as internal connections, managed internal broadband services and basic maintenance of internal connections. Discounts range from 20 to 90 percent, with higher discounts for higher poverty and rural schools and libraries. Recipients must pay some portion of the service costs.[6]

The FCC's E-Rate program has been successful in bringing broadband internet services to schools and libraries in remote, rural, and underprivileged areas of the mainland U.S. since 1998. The FCC's "2016 Broadband Progress Report" showed that 98% of Puerto Rico residents in rural areas lacked broadband internet connections.[7]

The President's Council of Economic Advisors Issue Brief of March 2016 stated that broadband access has a significant impact on economic growth, wages, medical care, and education, among other benefits.[8] Specifically, the brief stated that

> addressing the digital divide is critical to ensuring that all Americans can take advantage of the many well-documented socio-economic benefits afforded by Internet connections. These benefits are most evident when consumers have access to the Internet at speeds fast enough to be considered broadband; these speeds are required to facilitate full interaction with advanced online platforms.

## Potential Spillover Impacts to Other Sectors

The availability of reliable, high-speed broadband internet to schools and libraries across Puerto Rico will have major impacts on other sectors such as education and municipalities.

---

[4] FCC, 2018d.

[5] FCC, 2018d.

[6] FCC, 2018d.

[7] FCC, 2016, Appendix D.

[8] Council of Economic Advisors, 2016.

Telecommunications and broadband internet services being made available to schools and libraries in a remote or disadvantaged area may incentivize telecommunications and broadband internet providers to offer services to residential and retail customers in those areas.[9]

## Potential Costs

Potential up-front costs: $1.25 million in estimated up-front costs (2 years)
Potential recurring costs: $37.8 million–$66.4 million in estimated recurring costs (11 years)
Potential total costs: $39.0 million–$67.6 million in total estimated costs

The estimate of $1.25 million for up-front costs is for a contractor (using a team of 5 staff) to assist with implementation over a 2-year period. The recurring cost is based upon a best estimate for the annual costs ranging from $3.6 million to $6.3 million per year. The FCC's E-Rate program would pay the telecommunications provider directly for supplying internet connectivity to a school or library. The FCC's website explains how the E-Rate program funding works:

> An eligible school or library identifies services it needs and submits a request for competitive bids to the Universal Service Administrative Company. USAC posts these requests on its website for vendors' consideration. After reviewing its offers, the school or library selects its preferred vendor(s) and applies to USAC for approval for the desired purchases. Next, USAC issues funding commitments to eligible applicants. When a vendor provides the selected services, either the vendor or the applicant submits requests to USAC for reimbursement of the approved discounts.[10]

The E-Rate program provides financial support to schools and libraries on a sliding scale dictated by economic need. The greatest amount that can be funded by the E-Rate program is 90% of the services requested by the school or library. To make up the difference, it is possible that Puerto Rico schools and libraries might qualify for another source of federal funding, such as HUD Community Development Block Grants. There would also be the cost of the team that would work with the FCC and Puerto Rico agencies (such as PRDE, municipal boards of education, and the Puerto Rico Telecommunications Regulatory Board) to develop an island-wide plan for Puerto Rico and assist with applications and approvals for Puerto Rico schools and libraries.

## Potential Funding Mechanisms

Federal Communications Commission, U.S. Department of Agriculture, U.S. Department of Housing and Urban Development, U.S. Department of Commerce Economic Development Administration, National Telecommunications and Information Administration

---

[9] Council of Economic Advisors, 2016.

[10] FCC, 2018d.

NTIA has already provided $25.7 million for the Puerto Rico Bridge Initiative, which was designed to build 11 towers to provide broadband connectivity to 1,500 schools. *CIT 25 Evaluate and Implement Alternative Methods to Deploy Broadband Internet Service throughout Puerto Rico*, which calls for the development of a comprehensive plan for broadband deployment throughout Puerto Rico, would include an analysis of existing broadband infrastructure, such as this network, in developing an overall plan for Puerto Rico. The FCC states that "eligible schools and libraries may receive discounts on telecommunications, telecommunications services and Internet access, as well as internal connections, managed internal broadband services and basic maintenance of internal connections. Discounts range from 20% to 90%, with higher discounts for higher poverty and rural schools and libraries. Recipients must pay some portion of the service costs."[11] Discussions with senior officials of the Universal Service Administrative Company (USAC) indicate that Puerto Rico schools and libraries might also be eligible to take advantage of the FCC's High Cost Program.[12]

In addition to the FCC's E-Rate and High Cost Programs, funding may be available from these funding sources: the U.S. Department of Agriculture (USDA)'s Rural Utility Service and Rural Broadband programs, HUD's Community Development Block Grants and Infrastructure Block Grants, Economic Development Administration, and additional funding from the National Telecommunications and Information Administration. In addition, funding may also be available from economic development facilities and public works and Library Services and Technology Act grants to states.

Regarding HUD funding, department states: "HUD continues its efforts to narrow the digital divide in low-income communities served by HUD by providing, where feasible and with HUD funding, broadband infrastructure to communities in need of such infrastructure." Although HUD plans to issue regulations that will formalize its steps for narrowing the digital divide, current Community Development Block Grants (CDBG) funds can be used for broadband installation infrastructure and service delivery.[13]

## Potential Implementers

Telecommunications Bureau, Federal Communications Commission, Puerto Rico Department of Education

## Potential Pitfalls

It is possible that a school that obtains funding from the E-Rate program could subsequently be closed by PRDE. Some schools in Puerto Rico are administered by the PRDE, and some

---

[11] FCC, 2018d.

[12] Senior officials at the Universal Service Administrative Company, teleconference with authors, about the funding available to Puerto Rico through E-Rate and related FCC programs, July 19, 2018.

[13] HUD, "CDBG Broadband Infrastructure FAQs," HUD Exchange, Resources, website, January 2016.

schools are administered by municipal boards of education. According to new legislation in Puerto Rico, the Department of Education will decide which schools in Puerto Rico should be closed and which schools in central Puerto Rico should participate in the pilot project with the E-Rate program. The Department of Education will need to select the schools in the target municipalities that will participate in the pilot program. The Department of Education would need to determine which schools should receive E-Rate funding in a broader program. It would be a pitfall to obtain funding from the E-Rate program for a school that was subsequently closed. Therefore, the PRDE should have oversight of this COA, assisted by the Puerto Rico Telecommunications Regulatory Board (PRTRB), as stated in the new legislation.

Although the FCC has taken steps to simplify the E-Rate application and compliance process, it remains time-consuming and complicated. Some states, such as Florida, have hired a full-time E-Rate coordinator for schools. Potentially, a small team at the Telecommunications Bureau could serve as a central source of support for schools and libraries that engage in the E-Rate program.

### Likely Precursors

The precursor COA is *CIT 25 Evaluate and Implement Alternative Methods to Deploy Broadband Internet Service Throughout Puerto Rico.*

### References

Council of Economic Advisors Issue Brief, "The Digital Divide and Economic Benefits of Broadband Access," March 2016. As of April 29, 2019: https://obamawhitehouse.archives.gov/sites/default/files/page/files/20160308_broadband_cea_issue_brief.pdf

Federal Communications Commission, "E-Rate: Universal Service Program for Schools and Libraries," webpage, February 9, 2018. As of June 18, 2018: https://www.fcc.gov/consumers/guides/universal-service-program-schools-and-libraries-e-rate

———, "2016 Broadband Progress Report," FCC 16-6, January 29, 2016. As of April 29, 2019: https://docs.fcc.gov/public/attachments/FCC-16-6A1.pdf

U.S. Department of Housing and Urban Development, "CDBG Broadband Infrastructure FAQs," *HUD Exchange, Resources*, website, January 2016. As of June 3, 2018: https://www.hudexchange.info/resource/4891/cdbg-broadband-infrastructure-faqs/

# CIT 23
## Data Collection and Standardization for Disaster Preparedness and Emergency Response

### Sectors Impacted

Communications and Information Technology, Community Planning and Capacity Building, Energy, Water, Economic, Health and Social Services, Municipalities

### Issue/Problem Being Solved

Puerto Rico lacks quality data to inform the public and the policymaking process. Updated data on day-to-day information are helpful under normal circumstances—and critical after a disaster. Some studies commissioned to review conditions of U.S. states exclude Puerto Rico from their scope while others subsume Puerto Rico into data for the U.S. when compared with other countries.[1] This ambiguous position makes it crucial that Puerto Rico itself prioritize the collection and publishing of data in order to allow independent analysis of its digital progress and position within the United States and the world.

In the wake of Hurricane Maria, the government of Puerto Rico recognized the need for publicly available up-to-date information about basic services in Puerto Rico. The Governor's office launched www.status.pr to update the media, public, and first responders about conditions across Puerto Rico.[2] Through personal outreach, crowdsourcing, and manual effort, status.pr kept track of how many ATMs, gas stations, supermarkets, and pharmacies were open; how many people were without water, power, or communications; and how many people were in shelters. Status.pr became an important source of public information for the recovery effort.

Media outlets frequently used status.pr to inform their stories and compare with other sources of information. The *Washington Post* created a Twitter bot to provide an hourly update on service delivery in Puerto Rico, pulling information directly from status.pr.[3] A member of the diaspora of Puerto Rico living in Connecticut created a scraper to check status.pr every hour for updates.[4] Federal Emergency Management Agency (FEMA) itself referred journalists to status.pr for up-to-date information on the restoration of basic needs in Puerto Rico.[5] Data quality remains

---

[1] A list of U.S. federal agencies that have studies that do not include Puerto Rico on par with states includes the Census Bureau, National Center for Education Statistics, Bureau of Labor Statistics, Bureau of Justice Statistics, National Center for Health Statistics, Substance Abuse and Mental Health Services Administration, Bureau of Economic Analysis. See FOMB, 2107; Congressional Task Force on Economic Growth in Puerto Rico, 2016.

[2] Jenny McGrath, "Puerto Rico Website Is Keeping Track of the Island's Re-Emerging Infrastructure," *Digital Trends*, October 4, 2017.

[3] Phillip Bump, "We've Created a Twitter Bot that Provides Hourly Updates on the Situation in Puerto Rico," *Washington Post*, October 11, 2017.

[4] jpadilla, tracking-status-pr, GitHub, 2017.

[5] Johnson, 2017.

an important issue given the importance of providing the public and media outlets with reliable information.

## Description

This course of action (COA) will support the ongoing data quality enhancement and expansion of the status.pr website, improve the dissemination of data, and include partnerships with private businesses to enable ongoing situational awareness.

The COA will be led by the Puerto Rico Innovation and Technology Service (PRITS) and will build upon PRITS's ongoing efforts to digitize government data, form data-sharing partnerships, and utilize "smart" devices such as internet of things (IoT) sensors to continue providing status updates to policymakers, the media, and the public and to provide increased transparency even if some entities chose not to participate. This COA will consider:

- partnerships with commercial entities such as banks, gas stations, pharmacies, and hospitals, to develop a mechanism for maintaining updated information on outages or service disruptions
- data-sharing partnerships with utilities and government service providers to enable application programming interface (API) delivery of status information to maintain a public-facing dashboard of service delivery and performance measures
- pilot programs to install or access sensors that could provide real-time updates on hours of operation, outages, or service disruptions
- working with the private sector to design ways for citizens to participate with status reports or updates on important issues, potentially drawing on existing solutions such as SeeClickFix or Waze for citizen reporting.[6]

In addition to the public-facing www.status.pr, PRITS will work with the medical community, especially with the 86 federally qualified health centers (FQHC) across Puerto Rico on potential solutions for mapping and maintaining contact with critical care patients in an emergency, while maintaining the security of data protected under the Health Insurance Portability and Accountability Act (HIPAA). In this regard, implementation of this COA should be closely coordinated with implementation of *CIT 29 Health Care Connectivity to Strengthen Resilience and Disaster Preparedness*. In particular, joint implementation of these two COAs should explore ways to address patients in need of insulin, oxygen, dialysis, or other critical medical issues that require priority attention during an emergency response.

## Potential Benefits

Creating a platform for publicly sharing data in a standardized, user-friendly format (possibly through the use of mobile applications) will provide valuable information for policymakers, the media, and emergency responders. The platform will provide access to standardized, machine-

---

[6] See ClickFix, homepage, 2017; Waze, homepage, 2019.

readable information that can be utilized for the development of new applications based on the collected data.

The COA builds on the success of status.pr as an information source for citizens, media, first responders, and policymakers.

## Potential Spillover Impacts to Other Sectors

This COA will provide better information and situational awareness that will aid the community planning and capacity building, economic, municipality, health and social services, energy, and water sectors.

## Potential Costs

Potential up-front costs: $0.8 million in estimated up-front costs[7]
Potential recurring costs: $2.7 million in estimated recurring costs[8]
Potential total costs: $3.5 million in total estimated costs[9]

Costs range depending on the level of technical build-out for the data-sharing platform, which can be managed either by an internal team (as described in *CIT 16 Government Digital Reform Planning and Capacity Building*) or contracted out. The potential cost basis for these estimates are included in Table C.2. The annual labor costs for a project manager and a data manager are approximately $249,000—thus, $2.7 million over 11 years in recurring costs. The up-front costs for the low-end estimate include $30,000 in equipment and $100,000 for partnerships and outreach. The mid-range and high-end estimates include additional one-time development costs for 4 and 10 programmers, respectively, to implement additional capabilities and deployment of the data-sharing platform. The up-front cost presented for the COA is the average of the low and high up-front costs provided in the table ($130,000 and $1.4 million, respectively), and rounded off.

**Table C.2. Basis for Range of Estimated Cost for Course of Action *Data Collection and Standardization for Disaster Preparedness***

| Low Cost | Intermediate Cost | High Cost |
|---|---|---|
| • Maintain status.pr as is with staff to provide updates | • Maintain status.pr as is with staff to provide updates | • Relaunch status.pr with real-time dashboard for citizens, policymakers, and media |

---

[7] The recovery plan represents the potential up-front cost as $100,000. That estimated up-front cost is corrected here to include equipment and labor. The potential cost is the average of the low and high up-front estimated costs ($130,000 and $1.4 million) and rounded off.

[8] The recovery plan presented no recurring cost for this COA. The potential recurring costs are corrected here to include the annual cost of 2 staff members.

[9] The recovery plan represents the potential to cost as $100,000. That estimated total cost is corrected here to account for equipment and labor inadvertently not included in the potential cost in the recovery plan.

| | | |
|---|---|---|
| • Gradually incorporate real-time API updates as government data comes online<br>• Integrate commercial APIs and data sources such as Waze | • Create self-reporting platform for banks, gas stations, pharmacies, etc. to update status<br>• Integrate commercial APIs and data sources such as Waze<br>• Engage developer community for crowdsourcing tool, public reporting | • Access IoT sensors for real-time information for banks, gas stations, pharmacies, etc.<br>• Integrate commercial APIs and data sources such as Waze<br>• Engage developer community for crowdsourcing tool, public reporting |
| Up-front cost:<br>Partnerships and Outreach $100,000<br>Equipment $30,000 | Up-front cost:<br>Partnerships and Outreach $100,000<br>Equipment $30,000 4 programmers | Up-front cost:<br>Partnerships and Outreach $100,000<br>Equipment $30,000<br>10 programmers |
| Annual staff costs:<br>Project Manager $124,600<br>Data Management $124,600 | Annual staff costs:<br>Project Manager $124,600<br>Data Management $124,600 | Annual staff costs:<br>Project Manager $124,600<br>Data Management $124,600 |
| Up-front costs: $130,000<br>Recurring costs: $2.7 million<br>Total costs: $2.8M | Up-front costs: $630,000<br>Recurring costs: $2.7 million<br>Total: $3.3 million | Up-front costs: $1.4 million<br>Recurring costs: $2.7 million<br>Total costs: $4.1 million |

## Potential Funding Mechanisms

Community Development Block Grant Disaster Recovery

## Potential Implementers

Office of the Chief Innovation Officer, Puerto Rico Emergency Management Agency, government of Puerto Rico agencies

## Potential Pitfalls

Open data, including crowdsourced data, always carry perceived political risk because they provide a view on actual service delivery, which can reveal unsuccessful efforts and thus provoke backlash or public critique.

## Likely Precursors

Precursor COAs are *CIT 16 Government Digital Reform Planning and Capacity Building* and *CIT 18 Data Storage and Data Exchange Standards for Critical Infrastructure.*

## References

Bump, Phillip, "We've Created a Twitter Bot that Provides Hourly Updates on the Situation in Puerto Rico," *Washington Post*, October 11, 2017. As of June 20, 2019: https://www.washingtonpost.com/news/politics/wp/2017/10/11/weve-created-a-twitter-bot -that-provides-hourly-updates-on-the-situation-in-puerto-rico/?noredirect=on&utm_term =.fd7d83c2265a

Congressional Task Force on Economic Growth in Puerto Rico, *Congressional Task Force on Economic Growth in Puerto Rico Report to the House and Senate*, 114th Congress, December 20, 2016. As of June 20, 2019:
https://www.finance.senate.gov/imo/media/doc/Bipartisan%20Congressional%20Task%20Force%20on%20Economic%20Growth%20in%20Puerto%20Rico%20Releases%20Final%20Report.pdf

Financial Oversight and Management Board for Puerto Rico, *Annual Report Fiscal Year 2017*, July 30, 2107. As of June 20, 2019:
https://www.google.com/url?sa=t&rct=j&q=&esrc=s&source=web&cd=3&cad=rja&uact=8&ved=2ahUKEwjkh4y6uPjiAhXRMd8KHUp9B5YQFjACegQIARAC&url=https%3A%2F%2Fcases.primeclerk.com%2Fpuertorico%2FHome-DownloadPDF%3Fid1%3DNzAxMjIy%26id2%3D0&usg=AOvVaw2W6GiVrtyeuqDxbprDca_A

Johnson, Jenna, "FEMA Removes—then Restores—Statistics about Drinking Water Access and Electricity in Puerto Rico from Website," *Washington Post*, October 6, 2017. As of June 20, 2019:
https://www.washingtonpost.com/news/post-politics/wp/2017/10/05/fema-removes-statistics-about-drinking-water-access-and-electricity-in-puerto-rico-from-website/?utm_term=.5a6f7186382d

jpadilla, tracking-status-pr, GitHub, 2017. As of June 20, 2019:
https://github.com/jpadilla/tracking-status-pr

McGrath, Jenny, "Puerto Rico Website is Keeping Track of the Island's Re-emerging Infrastructure," *Digital Trends*, October 4, 2017. As of June 20, 2019:
https://www.digitaltrends.com/home/puerto-rico-hurricane-maria-site/

SeeClickFix, homepage, 2017. As of June 20, 2019:
https://seeclickfix.com/

Waze, homepage, 2019. As of June 20, 2019:
https://www.waze.com/

# CIT 24
## Establish Puerto Rico Communications Steering Committee

### Sectors Impacted

Communications and Information Technology, Municipalities

### Issue/Problem Being Solved

Facilitate governance, coordination, and information-sharing among relevant communications stakeholders. In February 2018, U.S. Department of Homeland Security (DHS)'s Office of Emergency Communications (OEC) reported on the need for Puerto Rico to address communications governance issues.[1] Specifically, the report noted that "it is critical that there is coordination between emergency communications personnel, governance structures, and other decision-making offices, bodies, and individuals." The report also said that a previous governance body—the Puerto Rico Interoperability Committee for Emergency Communications— had been dormant for 2 years.

### Description

This course of action (COA) proposes the establishment of a new Puerto Rico Communications Steering Committee to develop clear governance in the Communications and Information Technology sector. This will allow key public safety and commercial communications stakeholders to better organize planning efforts, prioritize requirements, and coordinate among themselves. The Communications Steering Committee would be led by a rotating chairperson.

The Governor could establish this new communications steering committee through Executive Order. The purpose of the committee would be to better organize planning efforts and coordinate among key public safety and commercial communications stakeholders. The Committee would provide strategic guidance, policy, direction and standards associated with Puerto Rico communications networks.[2] The discrete responsibilities and tasks of the committee would be outlined in a charter signed by the Governor. Membership on the committee could include officials from commonwealth agencies, such as the Telecommunications Bureau; Puerto Rico Emergency Management Agency; Office of the Chief Information Officer; Office of the Chief Innovation Officer; Department of Public Safety including the Puerto Rico Police Department, emergency medical service (EMS), and the Fire Department; and municipal

---

[1] DHS OEC, Interoperable Communications Technical Assistance Program, "Puerto Rico Public Safety Communications Summary and Recommendations Report," February 2018.

[2] The communications steering committee would not be involved in tactical decisions such as reforms detailed in *CIT 13 Streamline the Permitting and Rights of Way Processes for Towers and the Deployment of Fiber Optic Cable* that deal with centralizing rights of way and permitting approval authority.

officials as relevant. The Governor may also include officials from key private-sector entities to ensure their input is represented. The committee chair would rotate among the committee members twice a year. The rotating chair would need authority to set direction and facilitate decisions.

## Potential Benefits

The Communications Steering Committee proposed by this COA could allow for proper planning, governance, and collaboration for effective and efficient maintenance and recovery of the communications infrastructure in the event of a future disaster. It could also mitigate the potential for interoperability and duplication of effort issues in the event of a future disaster.

According to the February 2018 Puerto Rico Public Safety Communications Report, collaboration is essential to improve and ensure future interoperable communications. Coordination is also important to achieve operable, interoperable, and reliable communications across jurisdictions and disciplines.[3]

## Potential Spillover Impacts to Other Sectors

Given the important role that municipalities play in disaster response, participation by municipalities in the Communications Steering Committee would be a critical factor to its success. Conversely, municipalities can benefit from improved disaster preparation and response by helping shape governance and other efforts led by the Communications Steering Committee. Participation would require that municipalities invest time and effort, which may create challenges because of competing priorities and demands.

## Potential Costs

Potential up-front costs: $1 million in estimated up-front costs
Potential recurring costs: $6 million in estimated recurring costs (11 years)
Potential total costs: $7 million in total estimated costs

The up-front costs capture the cost to form the committee, including office space and permanent staff.[4] This rough estimate is based on estimates of congressionally mandated commissions, but the level of expertise of the committee staff and other administrative costs will impact the costs of such committee. The estimate assumes that government officials will not be compensated for their participation as such activities are within their current responsibilities.

---

[3] DHS OEC, 2018.

[4] This rough order of magnitude cost is based on Congressional Budget Office estimates to establish congressionally mandated commissions. Such commissions may have similar startup and maintenance costs to the proposed committee. See also Congressional Research Service, "Congressional Commissions: Overview, Structure, and Legislative Considerations," R40076, November 17, 2017. However, additional work must be performed to determine whether these similarities are relevant to Puerto Rico.

## Potential Funding Mechanisms

Government of Puerto Rico

## Potential Implementers

Office of the Governor, Puerto Rico Emergency Management Agency, Telecommunications Bureau, Puerto Rico Department of Public Safety, Office of the Chief Information Officer, Office of the Chief Innovation Officer, Municipalities

## Potential Pitfalls

Potential pitfalls would be if the Communications Steering Committee lacked discrete responsibilities, measurable goals and objectives, authorization to make decisions, and effective leadership. The Committee would not be positioned for success if the rotating chairperson lacked authority to set direction and facilitate decisions. A prior emergency communications committee did not meet for 2 years.

## Likely Precursors

None

## References

Congressional Research Service, *Congressional Commissions: Overview, Structure, and Legislative Considerations*, R40076, updated November 17, 2017. As of April 29, 2019: https://www.everycrsreport.com/reports/R40076.html

U.S. Department of Homeland Security, Office of Emergency Communications/Interoperable Communications Technical Assistance Program, *Puerto Rico Public Safety Communications Summary and Recommendations Report*, February 2018.

# CIT 25
## Evaluate and Implement Alternative Methods to Deploy Broadband Internet Service throughout Puerto Rico

### Sectors Impacted

Communications and Information Technology, Economic, Energy, Health and Social Services, Housing, Public Buildings, Municipalities, Transportation

### Problem Being Solved

Speed deployment of broadband internet services across Puerto Rico to accomplish the objectives for a digital economy.

### Description

This course of action (COA) would create a comprehensive plan for deployment of broadband internet in Puerto Rico. A high-profile panel of nationally recognized subject matter experts (SMEs), industry leaders, senior government officials, and civil society representatives, including disability community advocates, would seek to obtain political and industry support for a broadband deployment plan within Puerto Rico. This blue-ribbon panel might be convened by the Governor and assisted by an advisory board and an outside contractor. For example, the Federal Communications Commission (FCC) could serve on an advisory board to the blue-ribbon panel. The blue-ribbon panel could work with carriers and regulators to gather additional information that would be important to analysis and decisionmaking, subject to reasonable protections. Forming a blue-ribbon panel would have the additional benefit of its being able to adjudicate among competing priorities and interests, especially among different government agencies, municipalities, and telecommunications providers, which may have different perspectives on the best path to recovery.

### Potential Benefits

This COA is a crucial first step to swiftly deploying broadband internet for the benefit of emergency medical services (EMS), education, health care, social services, the visitor economy, and others.

A comprehensive plan is required to deploy broadband internet effectively, using the existing fiber resources on Puerto Rico and leveraging available federal funding for broadband internet infrastructure. The benefit of a high-profile team that includes nationally recognized experts, senior industry members, civil society representatives, including disability community advocates, as well as government of Puerto Rico officials, would be that the plan might gain support from many stakeholders and actually be implemented. A team of experts from the private telecommunications sector, or from the Telecommunications Bureau, will not be sufficient to gain the broad political and industry support that would be needed to implement a comprehensive plan.

## Potential Spillover Impacts to Other Sectors

The availability of reliable, high-speed broadband internet across Puerto Rico would have major impacts on multiport and transportation services, energy, telecommunications, emergency services, local integrated services, entrepreneurship, housing, health and social services, education, human capital, the visitor economy, investments, and the ocean economy.

The President's Council of Economic Advisors Issue Brief of March 2016, stated that broadband access has a significant impact on economic growth, wages, medical care, and education, among other benefits.[1] Specifically, the brief stated that:

> Addressing the digital divide is critical to ensuring that all Americans can take advantage of the many well-documented socio-economic benefits afforded by Internet connections. These benefits are most evident when consumers have access to the Internet at speeds fast enough to be considered broadband; these speeds are required to facilitate full interaction with advanced online platforms.

The brief cited the economic impact of broadband connectivity:

> By 2006—before the widespread availability of streaming audio and video— broadband Internet accounted for an estimated $28 billion in U.S. GDP. . . . Nearly half of this total was due to households upgrading from dial-up to broadband service. By 2009, broadband Internet accounted for an estimated $32 billion per year in net consumer benefits. . . . These findings are broadly consistent with studies that cover other countries. Broadband expansion is also associated with local economic growth in some cases.

The brief added that "growth is particularly concentrated in industries that are more IT-intensive and in areas with lower populations. In addition, insofar as it allows a person to participate more fully in the economy, developing Internet skills may even positively affect a person's wages."

In terms of medical outcomes, the brief continued:

> Broadband has made medical care and medical information more convenient and more accessible. . . . Broadband-enabled virtual visits with trained medical professionals can improve patient outcomes at lower cost and with a lower risk of infection than comes with conventional care provided in person. . . . Telemedicine is particularly valuable for rural patients who may lack access to medical care, as telemedicine allows them to receive medical diagnoses and patient care from specialists who are located elsewhere. Broadband can also be used to more accurately track disease epidemics. Various studies have demonstrated how large datasets from search engines and social media can be exploited in this way.

The brief also described the positive impact of broadband internet on education:

---

[1] Council of Economic Advisors, 2016.

Broadband also enables access to lower-cost online education. A 10 percent increase in college students taking all their courses online is associated with a 1.4 percent decline in tuition. The importance of the role that the computer and broadband more specifically play in enabling students to do their homework is evidenced by the fact that nearly half of 14 to 18 year olds report that they use a library computer, commonly for homework. . . . This finding suggests that library computers can provide a crucial source of access for students who would not otherwise have the ability to get online.

The brief cited other benefits of broadband connectivity, including "supporting entrepreneurship and small businesses, promoting energy efficiency and energy savings, improving government performance, and enhancing public safety, among others. In addition, broadband has become a critical tool that job seekers use to search and apply for jobs."

## Potential Costs

Potential up-front costs: $900,000 in estimated up-front costs[2]
Potential recurring costs: —
Potential total costs: $900,000 in total estimated costs

The estimate for the up-front costs includes a Puerto Rico–wide comprehensive plan for broadband internet deployment. The cost of this COA would depend on the size of the team. A 5-member blue-ribbon panel working for 18 months is estimated to cost $935,000. This team would need to engage with senior members of the government of Puerto Rico, municipal officials, and private telecommunications providers to identify existing resources and gain support for broadband deployment to schools, libraries, and health clinics. The blue ribbon panel or the Telecommunications Bureau would request tower companies to provide information on the sites that broadcast Wireless Emergency Alerts (WEAs).

## Potential Funding Mechanisms

Community Development Block Grant Disaster Recovery, Federal Communications Commission

A blue-ribbon panel will need to identify and evaluate all funding sources to support deployment of broadband infrastructure throughout Puerto Rico. It is possible the Community Development Block Grants for Broadband Infrastructure sponsored by Housing and Urban Development (HUD) might fund, or partially fund, a blue-ribbon panel for a broadband infrastructure deployment plan for Puerto Rico. The FCC might also contribute funding.

---

[2] The recovery plan represents the cost estimate using 1 significant figure. The cost is rounded as compared with the estimated cost described.

## Potential Implementers

Telecommunications Bureau, FCC, private telecommunications companies

## Potential Pitfalls

Municipal mayors may create challenges to deploying broadband internet in their municipalities by delaying rights of way (ROW) and permitting approvals.

A comprehensive plan for broadband deployment could be derailed by disagreements among senior government of Puerto Rico officials or telecommunications providers, or municipalities that do not wish to lose revenue from ROW and permitting approvals. For this reason, a highly respected team of experts that has political and industry support is needed to devise the plan.

## Likely Precursors

No precursors. This COA would serve as a precursor for *CIT 19 Municipal Hotspots*, *CIT 22 Use Federal Programs to Spur Deployment of Broadband Internet Island-Wide*, *CIT 26 Wi-fi Hotspots in Public Housing and Digital Stewards Program*, and *CIT 29 Health Care Connectivity to Strengthen Resilience and Disaster Preparedness*.

## Reference

Council of Economic Advisors Issue Brief, "The Digital Divide and Economic Benefits of Broadband Access," March 2016. As of June 3, 2018:
https://obamawhitehouse.archives.gov/sites/default/files/page/files/20160308_broadband_cea_issue_brief.pdf

## CIT 26
## Wi-fi Hotspots in Public Housing and Digital Stewards Program

### Sectors Impacted

Communications and Information Technology, Community Planning and Capacity Building, Economic, Health and Social Services, Public Buildings

### Issue/Problem Being Solved

Many public housing residents lack options for internet access without expensive data plans. The availability of government-sponsored wi-fi in public housing will enable reliable delivery of internet services for these residents and allow for postdisaster priority connection points. Moreover, government-sponsored internet training will allow some of those residents to be responsible for maintaining the sites and for coordinating technology-based activities for the benefit of their communities.

### Description

This course of action (COA) will establish a Digital Stewards program in Puerto Rico. Digital Stewards are public housing residents who are trained in using the internet from a technical perspective, as well as in using it for education and entrepreneurial purposes in the community. Digital Stewards are taught about best practices for cybersecurity and privacy and are trained to install and service wi-fi hotspots in public housing (and other public projects). Finally, Digital Stewards serve as points of contact for the local community for questions and training on how to use the internet. Digital Stewards programs have been already implemented in Michigan and New York.[1]

The COA consists of creating wi-fi hotspots with routers and repeaters for public housing in Puerto Rico. Already, legislation authorizing funds for rebuilding damaged public housing requires that "any substantial rehabilitation . . . or new construction of a building with more than four rental units must include installation of broadband infrastructure."[2]

The program will work as follows. Public housing residents will be invited to apply to become Digital Stewards. The residents accepted as Digital Stewards will gain technical skills and knowledge during an 8-month training program. After successful completion of the training,

---

[1] Allied Media Projects, undated; Red Hook Initiative, 2019.

[2] HUD, 2018. The complete quotation is as follows: "*f. Broadband infrastructure in housing:* Any substantial rehabilitation, as defined by 24 CFR 5.100, or new construction of a building with more than four rental units must include installation of broadband infrastructure, except where the grantee documents that: (a) The location of the new construction or substantial rehabilitation makes installation of broadband infrastructure infeasible; (b) the cost of installing broadband infrastructure would result in a fundamental alteration in the nature of its program or activity or in an undue financial burden; or (c) the structure of the housing to be substantially rehabilitated makes installation of broadband infrastructure infeasible" (HUD, 2018).

they will coordinate the maintenance and promotion of wi-fi connectivity in their communities and coordinate technology-related projects and events. Program participants will learn valuable skills, build employment experience, and act as community liaisons for internet connectivity and maintenance of hotspots in their communities.

## Potential Benefits

Government-sponsored wi-fi access (and the ability to access computers, tablets, and smartphones) can help decrease the "digital divide" and expand opportunities for public housing residents. [3] It can also provide a priority point of connection and coordination after disasters. The Digital Stewards program creates a way to improve connectivity options in public housing and provides valuable digital skills and employment experience for people in those communities.

The Digital Stewards program began in Detroit.[4] It was also implemented in Red Hook, New York, where, in 2017 "92% of Digital Stewards agreed or strongly agreed that they learned skills that allow them to succeed in the workplace and to make a difference in their neighborhood" and "77% remained employed or were actively pursuing further education within six months of completing the program."[5]

Following Hurricane Sandy in 2012, the Red Hook Initiative—the nonprofit that leads the Red Hook Digital Stewards program—became a community hub of disaster response.[6] Such a hub provided a gathering place for residents to fill out Federal Emergency Management Agency (FEMA) forms and check in with family members. "At the peak of the crisis, over 1,200 people were coming through the Red Hood Initiate [RHI] to charge phones, get a hot meal, pick up supplies, receive medical or legal support, and offer to help. Many of these individuals had never been to RHI before the storm, but found a place where they felt cared for and where their needs were met" [7] RHI helped organize volunteers, posted updates on social media, and increased the reach of its mesh wi-fi to serve more than 1,000 people per day. The Red Hook community wi-fi program's contribution after Sandy was recognized at a White House–hosted FEMA roundtable on emergency response best practices.

The following are some of the lessons learned from the Hurricane Sandy experience:

---

[3] Gideon Lewis-Kraus, "Inside the Battle to Bring Broadband to New York's Public Housing," *Wired*, November 3, 2016.

[4] Allied Media Projects, undated.

[5] Red Hook Initiative, "Impact Report 2017," undated.

[6] Red Hook Initiative, 2013.

[7] Red Hook Initiative, 2013.

- It is important to have the relationships and the wireless nodes in place prior to a disaster to facilitate rapid network deployment, including
  - already established relationships with key community stakeholders
  - a heightened level of technological literacy in the community
  - pre-positioned wireless network equipment in the neighborhood
- The most challenging investment is in the initial organizing and design phase before any value is realized
- Community-designed applications add value to a local network, even at a small scale.

The success of the program can be measured through metrics such as the number of public wi-fi hotspots that are made available, the number of individuals who gain access to the internet (breaking this number into residents and tourists, for example), and the frequency of access.

## Potential Spillover Impacts to Other Sectors

With reliable, affordable access to wi-fi, public-housing residents would have more ways to participate in other sectors, including community planning and capacity building (wi-fi would enable residents to participate in planning activities and share needs and views on proposed actions); economic (wi-fi would give residents access to job opportunities); health and social services (wi-fi would allow residents to apply for social services and update their information); and public buildings (wi-fi would expand digital security systems, send updates to residents, and increase the ability to monitor building condition, maintenance needs, and energy use).

## Potential Costs

Potential up-front costs: $1 million in estimated up-front costs
Potential recurring costs: $39 million in estimated recurring costs (11 years)[8]
Potential total costs: $40 million in total estimated costs[9]

The estimate for up-front costs include hotspot equipment, computers, and mobile devices for Digital Stewards program.[10] These estimates assume 200 installations and 20 Digital Stewards. The recurring costs include the annual costs for internet access, Digital Stewards program management, periodic internet training in the local community, and a salary for each Digital Steward.[11]

---

[8] The recovery plan published a potential recurring cost of $20 million. The recovery plan cost did not include the labor of the Digital Stewards at a cost consistent with other COAs. The cost estimate is updated here to reflect an estimate for the anticipated labor for this action.

[9] The recovery plan published a potential total cost of $20 million. That estimate is updated here to reflect the updated recurring cost estimated.

[10] Estimates from comparable projects were used for the basis, such as Southern California Digital Village (Tribal Digital Village, 2017), and Red Hook Wifi (2019).

[11] The annual recurring costs include $2.5 million for the 20 Digital Stewards and $1 million for other costs, for a total of $3.5 million in annual recurring costs.

## Potential Funding Mechanisms

Community Development Block Grant Disaster Recovery, Puerto Rico Department of Housing

This effort could utilize funds from a variety of mechanisms, including:

- Community Development Block Grant Disaster Recovery Program (CDBG-DR) funds to extend broadband connectivity
- CDBG-DR funds for workforce development and skills training, especially for low income residents of public housing
- Housing Authority Operational Funds to finance monthly internet access cost

## Potential Implementers

Office of the Chief Innovation Officer, U.S. Department of Housing and Urban Development, Puerto Rico Department of Housing

## Potential Pitfalls

There are several risks associated with the program:

- It will be important to ensure awareness of online safety, security, and privacy and establish content policies for use. Such awareness needs to be instilled first on the Digital Stewards and then on the community users.
- Since the public housing wi-fi hot spots will be sponsored by the government, it is also important that existing commercial internet providers not be displaced.
- Equipment has the risk of being stolen or vandalized, and security measures need to be implemented.

The management of the Digital Stewards program should be responsible for addressing these risks and for implementing the program according to best practices and lessons learned from similar efforts undertaken elsewhere.

In order to realize the full potential of affordable wi-fi access, community users must also have access to digital devices and the digital skills to make use of the technology. To ensure its success the Digital Stewards program also has to sufficiently identify and address community needs depending on the conditions of the residences where the wi-fi systems will be based. This could include establishing a publicly accessible "lab" for residents to use, or refurbishing devices for use by residents. Identifying and addressing those needs constitute an additional challenge, but will also present opportunities for the Digital Stewards to display ingenuity and entrepreneurship.

## Likely Precursors

This COA could be implemented together with *CIT 19 Municipal Hotspots*. Precursor COAs are *CIT 16 Government Digital Reform Planning and Capacity Building, CIT 22 Use Federal*

*Programs to Spur Deployment of Broadband Internet Island-Wide, CIT 21 Government-Owned Fiber-Optic Conduits to Reduce Aerial Fiber-Optic Cable and Incentivize Expansion of Broadband Infrastructure,* and *CIT 25 Evaluate and Implement Alternative Methods to Deploy Broadband Internet Service Throughout Puerto Rico.*

## References

Allied Media Projects, Detroit Community Technology Project, Digital Stewards Training, webpage, undated. As of February 20, 2019:
https://www.alliedmedia.org/dctp/digitalstewards

Federal Emergency Management Agency, Ninth FEMA Think Tank conference call, transcript, September 19, 2013. As of June 15, 2019:
https://www.fema.gov/media-library-data/20130919-1717-27928-0769/transcripts20130919-27928-12i8txg.txt

Federal Register, Department of Housing and Urban Development, Docket No. FR–6066–N–01, Vol. 83, No. 28, February 9, 2018. As of June 15, 2019:
https://www.govinfo.gov/content/pkg/FR-2018-02-09/pdf/2018-02693.pdf

Lewis-Kraus, Gideon, "Inside the Battle to Bring Broadband to New York's Public Housing," *Wired*, November 3, 2016. As of June 15, 2019:
https://www.wired.com/2016/11/bringing-internet-to-new-york-public-housing/

Red Hook Initiative, "Digital Stewards," website, 2019. As of June 12, 2019:
http://rhicenter.org/programs-2/youth-development/digital-stewards/

———, "Impact Report 2017," website, undated. As of June 15, 2019:
http://rhicenter.org/wp-content/uploads/2017/10/RHI-ImpactReport-2017_WEB.pdf

———, "Red Hook Initiative: A Community Response to Hurricane Sandy," website, 2013. As of June 12, 2019:
https://rhicenter.org/wp-content/uploads/2013/10/RHI-Hurricane-Report-6_2013.pdf

Red Hook Wifi, "Mission: Resilience, Opportunity, Community and Social Justice," website, 2019. As of June 25, 2018:
https://redhookwifi.org/about/mission/

Tribal Digital Village, homepage, website, 2017. As of June 25, 2018:
https://sctdv.net/about-tdv/

# CIT 27
## Study Feasibility of Digital Identity

### Sectors Impacted

Communications and Information Technology, Economic, Health and Social Services

### Issue/Problem Being Solved

The creation of a secure digital identity enables digital transactions, reduces transaction costs, and decreases the potential for fraud and identity theft. The Puerto Rico Chief Innovation Officer has described digital identity as "the key" to successful digital services implementation and has identified this as a top priority.[1]

### Description

This course of action (COA) will undertake a study of existing approaches for—and will assess potential acceptance by the people of Puerto Rico and the business community of— creating a secure, strong digital identity, based on a resilient power and communications infrastructure, to facilitate government and private-sector transactions. The study will be designed to identify potential pitfalls, including concerns of privacy, security, and public perception.

The project can be led by a dedicated team within the Puerto Rico Innovation and Technology Service (PRITS), and the work can be conducted by a private contractor with relevant expertise. The scope will include surveying existing approaches for digital identity and undertaking field work within Puerto Rico to test cultural perceptions and public concerns and to identify obstacles to adoption.

Establishing digital identity for the people of Puerto Rico is motivated by reducing transaction costs and establishing secure means to validate identity for government services. Any transaction between two or more parties requires identity verification. "As the number of digital services, transactions and entities grows, it will be increasingly important to ensure the transactions take place in a secure and trusted network where each entity can be identified and authenticated."[2]

Two experts identified "a unique, uniform digital ID" as one of "three pillars of digital transformation."[3] Jurisdictions at the vanguard of digital transformation have prioritized digital identity through a variety of approaches ranging from (1) centralized and owned by the government (India), (2) decentralized and managed jointly by the government and the private

---

[1] Puerto Rico Chief Innovation Officer, interview with authors, May 31, 2018.

[2] World Economic Forum, 2018.

[3] Eggers and Hurst, 2017.

sector, mainly banks (Estonia), and (3) systems established by the government that rely on nongovernment entities to provide validation (UK).

Beyond these approaches, a fourth option is emerging (Canada, Switzerland) that is truly distributed, allows individuals to own data, and allows people to share information as they choose in a trusted, valid, effective system. In this new approach, individuals can collect credentials from multiple entities and share at a granular level the verification of certain data without exposing unrelated data.

This COA will leverage lessons learned from these projects:

- Michigan's MILogin—single sign-on for state agencies[4]
- BC Services—British Columbia's chip card used to identify and authenticate citizens for access to all digital government services[5]
- GOV.UK Verify—public-private partnership for identity verification in the United Kingdom[6]
- e-Estonia identity card[7]
- Zug-ID (Zug, Switzerland)—the first publicly verified blockchain identity.[8]

The government of Puerto Rico has identified identity verification as a key challenge for citizens and an "essential" element of the digital transformation of government.[9]

## Potential Benefits

A secure digital identity is a key component to digital transformation, facilitating financial transactions, contracts, and government services. A secure digital identity can increase accuracy and reduce the costs associated with validation and access to government services, especially in disaster recovery, when paper records can be inaccessible.[10]

The proposed study will identify existing approaches, implementation options, opportunities, challenges so that the government of Puerto Rico can make a decision based on the best fit for the specific conditions on the ground in Puerto Rico.

In Estonia, considered the digital leader among nations, 95% of all transactions in the country are performed digitally, thanks to its secure, trusted digital identity and digital signatures. The system enables verification of a wide range of information pertaining to the individual (with accessibility permissions tightly controlled). The ID can function as:

---

[4] Michigan Department of Health and Human Services, "What Is My Login?" website, undated.

[5] Government of British Columbia, "BC Services Card," website, undated.

[6] Government of the United Kingdom, "GOV.UK Verify," Introducing GOV.UK Verify, website, March 14, 2019.

[7] e-Estonia, "e-Identity," website, undated.

[8] Paul Kohlhaas, "Zug ID: Exploring the First Publicly Verified Blockchain Identity," *Medium*, December 6, 2017.

[9] Puerto Rico Chief Innovation Officer, interview with authors, May 31, 2018.

[10] This assumes that the telecommunications and information technology infrastructure that supports digital identity is still operational after the disaster.

- a driver's license
- a virtual ticket on public transportation
- a travel document
- a credential to enable electronic voting from anywhere in the world
- a health insurance card for picking up prescriptions or accessing health records
- a government identity for accessing services or paying taxes.[11]

Estonia estimates that this system "lifts annual GDP by 2% while saving a lot of paperwork and creating opportunities for business."[12]

In 2015, when Canada launched the effort that led to its identity pilot program, it compared the potential benefits of digital identity to the first intercontinental railroads:

> In the same way that our ancestors built a national railway that linked communities from across the country and created new markets, broad adoption of a modern, robust digital identification and authentication ecosystem will link us online and create new ways of interacting with each other and with others around the world.
>
> In the same way that building the railway stimulated new engineering and led to spin off and ancillary business opportunities, the move toward digital authentication will create greater demand for skilled resources and new technologies. In the same way that the railway increased employment opportunities and enabled Canadians to participate in society on a more equal footing, digital identification will create a more level playing field across Canada.[13]

## Potential Spillover Impacts to Other Sectors

A secure digital identity will allow for a secure login for government services, prevent duplication, and allow for an "ask/update once" so that information is accurate and updated across government services. A secure digital identity will also allow for more-efficient business processes and transactions, eliminating the need for in-person authentication.

## Potential Costs

Potential up-front costs: $2 million in estimated up-front costs
Potential recurring costs: —
Potential total costs: $2 million in total estimated costs

The up-front cost would be to engage contractors to provide overview of existing approaches, technologies, and programs. The contractors will also review costs, security issues, and feasibility estimates of various implementation options and perform an on-the-ground assessment

---

[11] Marie Sansom, "National Identity Card for Australians? Digital Government Lessons from Estonia," *Government News*, November 1, 2016.

[12] Sansom, 2016.

[13] Digital Identification and Authentication Council of Canada, "Building Canada's Digital Identity Future," May 2015.

of attitudes and potential for adoption within Puerto Rico. Depending on the contractors selected, 2 individual studies could be conducted, one focused on technical assessment and the other on public attitudes, concerns, and adoption potential on the ground in Puerto Rico. The cost for this COA was estimated based upon employing approximately 3 full-time contractors for 3 years, for a total estimate of approximately $2 million.[14]

These cost estimates are for assessing the feasibility of implementing digital identity in Puerto Rico, including lessons learned from other successful implementations. They do not include procuring this capability throughout the agencies of the government of Puerto Rico.

## Potential Funding Mechanisms

Government of Puerto Rico, public-private partnerships

## Potential Implementers

Office of the Chief Innovation Officer, government of Puerto Rico agencies

## Potential Pitfalls

At the completion of this COA, the government of Puerto Rico should be ready to undertake the larger effort of adopting digital identity throughout its agencies. For that larger effort Puerto Rico will need to implement cybersecurity best practices; and the costs to meet cybersecurity requirements—including systems, personnel, insurance overhead—can be significant. Because of persistent concerns related to cybersecurity and privacy, the feasibility study should address cybersecurity best practices and the potential costs to implement cybersecurity requirements. Incorporating these issues into the feasibility study will help the government of Puerto Rico proactively prepare for digital identity implementation.

This COA focuses on engaging in a rigorous study before deciding on a technical approach to be applied across the government of Puerto Rico in order to mitigate the potential pitfalls associated with this emerging technology.

Private sector services (e.g., banking) are grappling with similar issues and deploying their own solutions, which could lead to a future with multiple identity schemes. The proposed study will examine the different options and identify the best fit for Puerto Rico.

## Likely Precursors

Precursor COAs are *CIT 16 Government Digital Reform Planning and Capacity Building, CIT 21 Government-Owned Fiber-Optic Conduits to Reduce Aerial Fiber-Optic Cable and Incentivize Expansion of Broadband Infrastructure*, and *CIT 25 Evaluate and Implement Alternative Methods to Deploy Broadband Internet Service Throughout Puerto Rico.*

---

[14] Throughout the recovery plan, a contractor based in continental United States (CONUS) is estimated to have an annual cost of $227,300 (including travel to Puerto Rico).

## References

Digital Identification and Authentication Council of Canada, "Building Canada's Digital Identity Future," May 2015. As of June 25, 2018:
http://www.diacc.ca/wp-content/uploads/2015/05/DIACC-Building-Canadas-Digital-Future-May5-2015.pdf

e-Estonia, "ID-card," Solutions, e-Identity, website, undated. As of June 19, 2019:
https://e-estonia.com/solutions/e-identity/id-card/

Eggers, William D., and Steve Hurst, "Delivering the Digital State: What If State Government Services Worked Like Amazon?" *Deloitte Insights*, November 14, 2017. As of June 20, 2019:
https://www2.deloitte.com/insights/us/en/industry/public-sector/state-government-digital-transformation.html#endnote-5

Government of British Columbia, "BC Services Card," British Columbians and Our Governments, Government ID, website, undated. As of June 16, 2019:
https://www2.gov.bc.ca/gov/content/governments/government-id/bc-services-card

Government of the United Kingdom, "GOV.UK Verify," Introducing GOV.UK Verify, website, March 14, 2019. As of June 16, 2019:
https://www.gov.uk/government/publications/introducing-govuk-verify/introducing-govuk-verify

Kohlhaas, Paul, "Zug ID: Exploring the First Publicly Verified Blockchain Identity," *Medium*, December 6, 2017. As of June 16, 2019:
https://medium.com/uport/zug-id-exploring-the-first-publicly-verified-blockchain-identity-38bd0ee3702

Michigan Department of Health and Human Services, "What is MyLogin?" Doing Business with MDHHS, Health Care Providers, website, 2019. As of June 16, 2019:
https://www.michigan.gov/mdhhs/0,5885,7-339-71551_2945_72165---,00.html

Puerto Rico Chief Innovation Officer, interview with authors, May 31, 2018.

Sansom, Marie, "National identity card for Australians? Digital government lessons from Estonia," *Government News*, November 1, 2016. As of June 16, 2019:
https://www.governmentnews.com.au/25432/

World Economic Forum, "On the Threshold of a Digital Identity Revolution," January 2018. As of Jun 16, 2019:
http://www3.weforum.org/docs/White_Paper_Digital_Identity_Threshold_Digital_Identity_Revolution_report_2018.pdf

# CIT 28
## Innovation Economy/Human Capital Initiative

### Sectors Impacted

Communications and Information Technology, Community Planning and Capacity Building, Economic, Education

### Issue/Problem Being Solved

The issue being addressed is to provide people in Puerto Rico with the skills to work and participate in an increasingly digital society and to cultivate a culture of entrepreneurship. The term "innovation economy" describes a shift over the past two decades from capital-intensive industries like traditional manufacturing to investments in research and development, increased productivity, and better products and services. "Innovation industries" include:

- aerospace[1]
- information and communications technology product and component manufacturing
- other high-tech production and manufacturing
- medical devices[2]
- biotechnology and pharmaceuticals[3]
- internet and information services[4]
- software.

Some of the ways the government of Puerto Rico can help lay the foundation for a growing innovation economy include:

- ensuring a modern infrastructure, especially reliable power and broadband internet
- ensuring that citizens have the right skills for digital jobs

---

[1] "Puerto Rico has become a magnet for some of the world's leading aviation and aerospace companies. With a long history of manufacturing expertise and a strong pipeline of engineering talent, the island has attracted multimillion-dollar investments by these and other major firms during recent years" including Pratt & Whitney, Lockheed Martin, Honeywell Aerospace, Hamilton Sundstrand, Florida Turbine and ESSIG Research, according to PRIDCO, Industries, Aerospace, website, 2018.

[2] "Puerto Rico [is] home to over 70 medical devices manufacturing plants—surgical and medical instruments, ophthalmic goods, dental equipment and supplies, orthodontic goods, dentures and appliances, laboratory apparatus and furniture" (PRIDCO, Industries, Medical Devices, website, 2018).

[3] "With 49 FDA-approved pharmaceutical plants scattered across the island, Puerto Rico is home to top multinational pharmaceutical companies, including Astra Zeneca, Abbott-Abbvie, Bristol-Meyers Squibb, Merck, Pfizer, Eli Lilly and numerous others" (PRIDCO, Industries, Pharmaceutical, website, 2018).

[4] "Top players from the IT and communications segment have strong presence in Puerto Rico. Take for example Microsoft Corporation, which first established its presence on the island in 1990 and has continued to expand, making Puerto Rico the company's principal manufacturing center for all Office and Vista products intended for the U.S. market. Cisco, Oracle, Hewlett-Packard and numerous other software and hardware makers also have major operations in Puerto Rico" (PRIDCO Industries, Information Technology, website, 2018).

- encouraging entrepreneurship through training and investment and creating an atmosphere that fosters startups and new technologies
- creating a unified data strategy.

This course of action (COA), combined with several COAs required for the recovery and development the telecommunications infrastructure of Puerto Rico,[5] will address many of these issues, building on Puerto Rico's established industries such as medical devices, aerospace, and pharmaceuticals, to open up new opportunities in the digitally based innovation economy.

## Description

This COA will create a public-private initiative to provide digital skills training, entrepreneurship programs, and access to digital technologies for people throughout Puerto Rico by means of a network of innovation hubs and entrepreneur centers, training partnerships with schools, and outreach via mobile labs to rural and underserved areas. This initiative will support the message that Puerto Rico is "open for business" and intends to engage in a digitally based innovation economy. According to the Information Technology and Innovation Foundation, signs of an emerging innovation economy include:

- entrepreneurs taking risks to start new ventures
- companies funding breakthrough research
- regional clusters forming to foster innovation
- research institutions transferring knowledge to companies through patents
- policies fostering widespread adoption of new technologies.[6]

This COA proposes a public-private initiative to develop a digitally based innovation economy in Puerto Rico following successful models such as those implemented in Tennessee.

As an example, the initiative could be led by the Puerto Rico Innovation & Technology Service (PRITS), in cooperation with the Departamento de Desarrollo Económico y Comercio (DDEC), as a public-private partnership. The initiative itself could create a nonprofit entity, with a board that would include representatives from government, the private sector, the investment community, and academia,[7] as well as providing for the appointment of a CEO and staff to make the initiative a success.

This Puerto Rico Innovation Economy Initiative (the "Innovation Initiative") will examine successful models in other parts of the country and establish an action plan for Puerto Rico, to

---

[5] *CIT 21 Government-Owned Fiber Optic Conduits to Reduce Aerial Fiber-Optic Cable and Incentivize Expansion of Broadband Infrastructure, CIT 22 Use Federal Programs to Spur Deployment of Broadband Internet Island-Wide, CIT 25 Evaluate and Implement Alternative Methods to Deploy Broadband Internet Service Throughout Puerto Rico, CIT 2 Puerto Rico GIS Resource and Data Platform, CIT 17 Puerto Rico Data Center*, and *CIT 18 Data Store and Data Exchange Standards for Critical Infrastructure.*

[6] Information Technology and Innovation Foundation, undated.

[7] Board members for a similar program in Tennessee are appointed by the governor and state legislature. See Launch Tennessee, 2018.

include a network of innovation hubs, innovation partnership with schools for digital skills and coding courses, and mobile innovation labs.

A network of innovation hubs across Puerto Rico would provide residents with the opportunity to learn to use new digital technologies, access to training, tools, and co-work space, and foster the creation and scaling of new technology businesses. Following other successful models, they could also include makerspace and co-work space and regularly host events entrepreneurship trainings, mentorship, and competitions.

Puerto Rico already has several successful private co-work and makerspaces that could be candidates for joining the network of innovation hubs. The key difference between existing private hubs and those that choose to participate in the innovation network could be eligibility for matching resources.[8]

Innovation partnership with schools would sponsor a digital-skills-training platform that will allow students to sign on, learn skills, and earn badges at their own pace. Such an arrangement would involve:

- teams based at innovation hubs partnering with local schools to support teachers who serve as leads for the coding program
- curriculum that provides for a school-wide ranking of skills attained by the students, which can in turn be used to recognize and reward student coding champions with prizes such as donated devices, awards ceremony, and/or scholarships[9]
- ultimately extend availability of the digital learning platform to all residents of Puerto Rico to provide the digital skills required for jobs in technical fields.

Mobile units that bring training and innovation opportunities to rural areas, public housing developments, schools, festivals, and town centers would include:

- "biz buses" that offer training in areas such as business formation, accounting, and using the internet, social media, marketing, and creating a website
- "innovation labs" that allow participants to explore and use new technologies, including computer numeric control (CNC) router, laser etcher, 3D printers, robots, drones, science, technology, engineering, and mathematics (STEM) kits, and virtual reality sets.

This COA could leverage lessons from a program that was successfully launched in Tennessee as "TN Driving Innovation."[10]

---

[8] These matching resources could be tied to metrics that pertain to measuring training, outreach, and technology access for the public; new businesses created or incubated as a result of the hubs; and opportunities to partner with local schools for digital skills and coding training programs.

[9] The "Dev Catalyst" has been running successfully in Tennessee since 2013. Dev Catalyst, website, undated.

[10] Driving Innovation in Tennessee, website, undated.

## Potential Benefits

A broad, coordinated push to provide technology access and digital and coding skills training will help develop a digitally literate employment pool for recruiting or expanding tech-reliant industries, consistent with the idea of the "human cloud"—a skilled digital workforce that can work from Puerto Rico with companies around the world. This initiative will reinforce the message that Puerto Rico is "open for business" and intends to welcome new technologies and a digital workforce.

The Innovation Initiative will create a framework for opening the opportunities of the innovation economy to people in Puerto Rico through access to physical innovation hubs in their communities. These hubs will feature activities that demonstrate new technologies and training and provide mentorship to encourage the creation of new products and businesses.

The partnership with schools will facilitate learning digital skills, increase the number of skilled workers on Puerto Rico, and provide greater opportunity for residents to learn how to invent new technologies, start their own businesses, and contribute to a modern, digital Puerto Rico.

Mobile innovation labs will help address the growing disparity between those who are exposed to and understand emerging technologies and those without access.

## Potential Spillover Impacts to Other Sectors

Providing access and the skills for residents of Puerto Rico to use emerging technologies will create a positive impact in several sectors, including community planning and capacity building, education, and economic. Thus, for example, educational attainment for people who take online courses using the network of innovation hubs will increase; products created with new digital tools will provide economic benefits; and resources provided at innovation hubs will help launch new businesses.

## Potential Costs

Potential up-front costs: $1 million–$4 million in estimated up-front costs
Potential recurring costs: $30 million–$70 million in estimated recurring costs (11 years)
Potential total costs: $30 million–$70 million in total estimated costs[11]

The estimate for up-front costs scale with the number of hubs or mobile labs. The estimate for annual costs also scale depending on the number of hubs, labs, schools, and teachers participating.[12] Table C.3 shows that the potential up-front costs vary from $1.3 million to $4.2 million, depending on the assumptions. The estimated recurring costs range from

---

[11] The recovery plan potential costs for this course of action are represented with only 1 significant figure. The costs described are rounded in the recovery plan.

[12] Estimates based on a similar project in Jackson, Tennessee.

$2.9 million to $6 million per year (from $32 million to $66 million over 11 years). The recovery plan used the range from the low cost estimate to the high cost estimate as the potential cost.

The cost basis includes the following assumptions:

- The cost of the Innovation Initiative is estimated assuming an annual operating cost that includes a chief executive officer, a board, and staff.
- The cost of each innovation hub is estimated assuming $250,000 for setting it up and $350,000 in annual operating costs.
- The cost of the schools partnership is estimated assuming $500,000 for setup for the low-cost case and $1 million for the high-cost case. The annual cost is assumed to scale with the number of participating schools and teachers.
- Each mobile lab is estimated to cost between $300,000 and $450,000 for setting up and $200,00 in annual operating costs per bus.

**Table C.3. Range of Cost Estimates to Establish and Maintain Hubs for Course of Action**
***Innovation Economy/Human Capital***

| Low Cost | Intermediate Cost | High Cost |
|---|---|---|
| $1.5M (board, CEO, & staff)<br>$0.5M setup, $0.7M annual (2 hubs)<br>$0.5M setup, $.5M annual (20 schools)<br>$0.3M setup, $0.2M annual (1 bus) | $1.5M (board, CEO, & staff)<br>$1.25M setup, $1.75M annual (5 hubs)<br>$0.5M setup, $1M annual (50 schools)<br>$0.9M setup, $0.4M annual (2 lab & biz bus) | $2M (board, CEO, & staff)<br>$2.25M setup, $3.15M annual (9 hubs)<br>$1M setup, $1.5M annual (75 schools)<br>$0.9M setup, $0.4M annual (2 lab & biz bus) |
| **$1.3M Up-front**<br>**$2.9M Annual** | **$2.7M Up-front**<br>**$4.7M Annual** | **$4.2M Up-front**<br>**$6.0M Annual** |

## Potential Funding Mechanisms

U.S. Department of Commerce Economic Development Administration, National Science Foundation, U.S. Department of Education, nongovernment sources

Funding would need be provided through government sources, and would be a good fit for those that encourage STEM training. In addition, organizers could seek private sponsorship, especially for coding competitions (i.e., for "the [Corporation Name] coding championship"). Hubs should be encouraged to pursue entrepreneurial models, including selling memberships to the makerspace, rental of dedicated desks or office space in the co-work space, and special events.

Access to public funds can be tied to performance and reach, which would include:

- number of students participating
- number of hours spent coding
- percentage of participants who go on to college
- percentage of participants who go into STEM fields.

245

## Potential Implementers

Office of the Chief Innovation Officer, government of Puerto Rico agencies, universities, municipal governments

## Potential Pitfalls

There are various risks and concerns associated with this program:

- Physical equipment will need to be protected from theft and extreme weather.
- This COA will require sustained leadership and relationships with stakeholders in Puerto Rico and in the tech community worldwide.
- The initiative will require close coordination across school districts and classrooms. There may be cases in which some schools do not have teachers who feel comfortable enough with technology to be the lead for the project (this can be addressed through training and organizational support).
- Participation in the program requires connectivity and access to a desktop or laptop computer. The success of the program will depend heavily on other COAs that offer government-sponsored wi-fi and expanded access to computers, including the ability to check out computers and wi-fi hotspots from public libraries.

## Likely Precursors

Precursor COAs are *CIT 16 Government Digital Reform Planning and Capacity Building, CIT 21 Government-Owned Fiber-Optic Conduits to Reduce Aerial Fiber-Optic Cable and Incentivize Expansion of Broadband Infrastructure, CIT 22 Use Federal Programs to Spur Deployment of Broadband Internet Island-Wide*, and *CIT 25 Evaluate and Implement Alternative Methods to Deploy Broadband Internet Service Throughout Puerto Rico*.

## References

Dev Catalyst, website, 2015. As of June 11, 2019:
https://www.devcatalyst.com/

Driving Innovation in Tennessee, website, undated. As of June 11, 2019:
https://www.tndrivinginnovation.com/

Information Technology and Innovation Foundation: Innovation Economics: The Economic Doctrine for the 21st Century, undated. As of June 2, 2018:
https://itif.org/innovation-economics-economic-doctrine-21st-century

Interview with Puerto Rico Industrial Development Company staff member, undated.

Launch Tennessee, "Launch Tennessee Announces New Board Members," March 22, 2018. As of June 11, 2019:
https://launchtn.org/2018/03/launch-tennessee-announces-new-board-members/

Puerto Rico Industrial Development Company, Industries, Aerospace, website, 2018. As of June 18, 2019:
http://www.pridco.com/industries/Pages/Aerospace.aspx

———, Information Technology, website, 2018. As of June 18, 2019:
http://www.pridco.com/industries/Pages/Information-Technology.aspx

———, Medical Devices, website, 2018. As of June 18, 2019:
http://www.pridco.com/industries/Pages/Medical-Devices.aspx

———, Pharmaceutical, website, 2018. As of June 18, 2019:
http://www.pridco.com/industries/Pages/Pharmaceutical.aspx

# CIT 29
# Health Care Connectivity to Strengthen Resilience and Disaster Preparedness

## Sectors Impacted

Communications and Information Technology, Community Planning and Capacity Building, Health and Social Services, Municipalities

## Issue/Problem Being Solved

Data connectivity is critical to clinical care, access to patient data both inside and outside a clinical facility is a necessity, and additional bandwidth enhances the delivery of health and social services. In particular, loss of data connectivity during and after a disaster could mean losing track of the location of patients with critical drug and treatment needs and spending time and effort locating the vulnerable and elderly. Rural and mountainous regions account for 80% of the geography of Puerto Rico, and lack of physical access to these regions during and after a disaster accentuates the need for effective communications to better direct relief efforts.

This course of action (COA) addresses the need for data connectivity across the 86 community health clinics (known as Federally-Qualified Health Centers [FQHCs]) that span Puerto Rico and provide care to 10% of Puerto Rico's residents, with over 700,000 annual visits.[1] It also builds on existing initiatives to improve access to clinical data throughout Puerto Rico.

The need for health care is even more acute during a natural disaster or pandemic and can require accessing information from multiple sources, possibly including remote expertise via telemedicine or teleconsult. This situation is not unique to Puerto Rico. After Hurricane Sandy, the medical system of Manhattan was reeling from the increase in patient volume, due in part to the loss of hospitals in the region.[2]

This COA, combined with several COAs required for the recovery and development of the telecommunications infrastructure of Puerto Rico,[3] addresses the following needs:

- disaster resilience, arising from robust, resilient, secure communications, with access to clinical data that can be maintained during a catastrophic disaster, enable clinical support during and after the disaster, and provide a conduit for real-time status data collection

---

[1] National Association of Community Health Centers, "Puerto Rico Health Center Fact Sheet," January 2018.

[2] Nishant Kishore et al., "Mortality in Puerto Rico after Hurricane Maria," *New England Journal of Medicine*, Vol. 379, July 12, 2018, pp. 162–170.

[3] *CIT 21 Government-Owned Fiber Optic Conduits to Reduce Aerial Fiber-Optic Cable and Incentivize Expansion of Broadband Infrastructure, CIT 22 Use Federal Programs to Spur Deployment of Broadband Internet Island-Wide, CIT 25 Evaluate and Implement Alternative Methods to Deploy Broadband Internet Service Throughout Puerto Rico, CIT 2 Puerto Rico GIS Resource and Data Platform, CIT 17 Puerto Rico Data Center, CIT 18 Data Store and Data Exchange Standards for Critical Infrastructure,* and *CIT 5 Implement Public Safety/Government Communications Backup Power.*

- response, using communications and clinical data access prioritized to support medical response and data needs, including remote triage, patient monitoring, requests for medications and support, and status updates to decisionmakers at different levels
- recovery, providing 24/7 communications support and feedback, including support for remote locations and for those displaced or sheltered in place
- restoration, including increased access to health records and teleconsults, which will help facilitate sustained care by health professionals for disaster survivors suffering lasting physical, mental, and emotional effects.

In nondisaster times, additional communications bandwidth can be used to enhance services. Examples include perinatal care and many programs under Health Resources and Services Administration (HRSA) and Health and Human Services (HHS) Office of Minority Health (OMH).[4]

## Description

This COA has two complementary objectives. First, it aims at providing robust, resilient, multimodal "mesh" communications connectivity to the 86 community clinics across Puerto Rico, using satellite and low-power radio and line-of-site technologies, to complement connectivity that is available through the telecommunications infrastructure or to provide redundancy when such infrastructure is damaged. Each clinic will have a satellite uplink to the internet plus radio connectivity to other clinics for continuity of operations and coordination during emergencies and provide the ability to connect to existing hospitals.

Second, this COA will use the increased connectivity and information technology (IT) to ensure real-time access to clinical data—including mobile and telehealth—from many access points to improve clinical care delivery and to better adapt to disaster impacts. This COA will also take advantage of the increased connectivity and IT to support situational awareness, behavioral health, environmental monitoring, and social services, as well as other social services delivery when bandwidth permits and during hours when clinics are closed.

Efforts to implement this COA need to be coordinated with health authorities and with ongoing private-sector efforts by hospitals and insurance companies.

The COA will also address some gaps in health care communications availability by:

- building on other health care IT efforts including those sponsored under the American Recovery and Reinvestment Act (ARRA) and Health Information Technology for Economic Health (HITECH) Act to encourage the use of electronic health records and CHC Public Health and Emergency Preparedness Cooperative Agreement[5]

---

[4] Health Resources and Services Administration, Federal Office of Rural Health Policy, website, undated; U.S. Department of Health and Human Services, Office of Minority Health, website, undated.

[5] Health Care Information and Management Systems Society, HIPAA, ARRA/HITECH Act and Meaningful Use Compliance Resources, website, 2019; Center for Disease Control and Prevention, State and Local Readiness, webpage, April 1, 2019.

- accelerating the impact of other efforts such as the selection of Health Gorilla by the community health clinics to provide access to clinical data throughout Puerto Rico.[6]

This COA will deploy a "mesh" communications safety net in the FQHCs for resilient recovery. The FQHCs are clinics that provide care to the most vulnerable and underserved populations, who are often the hardest to reach in disasters. The robust mesh radio network will have nodes at each of the 86 community clinics and will have the ability to connect to existing hospital-based infrastructure. This network, if desired and funded, could be extended to other health care and social service locations across Puerto Rico.

This COA lays out a plan for each clinic to have a satellite uplink to the internet with low-powered radio connectivity to all other clinics. The design will enable a resilient network during a disaster since the clinic that maintains connectivity will share the network with others that might have lost connectivity. Line-of-site and wi-fi will be supported to enable deployments in multibuilding facilities or to connect to other critical health care infrastructure. During a disaster or loss of power or connection to the outside world (i.e., a massive failure of the telecommunications infrastructure of Puerto Rico), the mesh network will be capable of operating independently as a series of communications hubs for the local community, while waiting to reconnect with the outside world.

## Potential Benefits

This COA will leverage the network of FQHCs spanning Puerto Rico and improve care and emergency response capabilities. It will enable medical innovation and provide real-time clinical electronic health record access and telehealth. The mesh network will bolster access to local services and situational awareness during and after a disaster.

The mesh communications capability can be put in place within weeks and months, and will enable better preparedness for the next disaster. As a hurricane approaches, the satellite small dishes can be taken down quickly and left in storage until the hurricane has passed. They can be then quickly reinstalled so that the mesh radio network can facilitate connectivity even if part of the telecommunications infrastructure of Puerto Rico has failed. Potential examples of posthurricane services include access to medication information, prioritization and routing of dialysis patients, and enabling remote monitoring of individuals who cannot be brought to the clinic.

During recovery and on an on-going basis the mesh radio network can support other needs such as behavioral health teleconsultations with mainland specialists and diagnostic testing of individuals, as well as testing of water sources for Zika, dengue, and chikungunya.

---

[6] EMR Industry, "Puerto Rico Primary Care Association Network Selects Health Gorilla Clinical Network as Their Data Exchange Platform," April 16, 2018.

## Potential Spillover Impacts to Other Sectors

Mesh network capacity could enhance access to local government services (municipalities). This COA could also leverage clinic connectivity for resilience, education, and emergency response (community planning and capacity building).

This COA impacts all other COAs since health and social services is the primary resource for the health and wellness of the entire island, including the workforce involved in response, recovery and restoration efforts.

## Potential Costs

Potential upfront costs: $7.6 million–$16 million in estimated upfront costs [7]
Potential recurring costs: $140 million–$260 million in estimated recurring costs (11 years)
Potential total costs: $150 million–$280 million in total estimated costs[8]

The up-front cost estimate is based on implementation of a multimodal mesh network for 86 FQHCs in Puerto Rico, with 2 short-cycle assessment and evaluation phases. The recurring cost estimate includes annual costs for ongoing connectivity, usage, and maintenance and operations. The estimated costs allow for a full range of implementation options to be selected for information systems, network infrastructure, and workforce and sustainment needs, as well as secondary uses for broader health-related initiatives that can leverage a clinic-to-clinic mesh. Table C.4 summarizes the estimated costs for three scaling options.

Given a wide range of options for forming and maintaining a multimodal mesh network, the up-front costs are estimated to range from $7.6 million to $15.5 million, or approximately $88,000 to $180,000 per FQHC.[9] Similarly, per FQHC annual costs are approximated as ranging from $145,000 to $279,000; these include workforce and sustainment needs.[10] Costs will vary considerably depending on the detailed implementation strategies that are decided on. The government of Puerto Rico will also need to ensure protection of citizen privacy and strong cybersecurity during implementation of this COA.

---

[7] The recovery plan represents the potential up-front costs as $5.6 million–$12 million. That number inadvertently did not include the assessments costs intended to be captured in the estimated costs. This is being corrected here.

[8] The recovery plan represents the potential total costs as $140 million–$280 million. This is being corrected here to account for the increase to the potential up-front costs. This cost is represented using 2 significant figures.

[9] Notice that up-front costs include both *assessment and evaluation* and *up-front*/deploy, as shown in Table C.4.

[10] The figure of $12.5 million per year accounts for 1 engineer per clinic (at a cost of $124,600 each), plus additional staff of 14 engineers to serve the needs of all the FQHCs; $24 million per year accounts for 2 engineers per clinic (at the same annual cost per engineer), plus additional staff of 20 engineers to serve the needs of all the FQRCs.

**Table C.4. Cost Range Basis for Course of Action *Health Care Connectivity to Strengthen Resilience***

| Level | Low Cost | Intermediate Cost | High Cost |
|---|---|---|---|
| Elements | • Assessment and evaluation<br>• Basic connectivity in all situations<br>• Satellite and radio only<br>• Limited utility for other purposes | All low-cost elements plus<br>• Stand-up/down satellite<br>• Line of site and wi-fi<br>• Interop with Health Gorilla<br>• Additional bandwidth<br>• Solar power for communications | All intermediate cost elements plus<br>• Enhanced mesh resilience<br>• Support for mobile/portable clinical sites<br>• Additional interop with government health and mainland |
| Up-front: assessment & evaluation | $2,000,000 | $2,500,000 | $3,500,000 |
| Up-front: deployment costs | $5,600,000 | $9,500,000 | $12,000,000 |
| Recurring: annual operating costs[a] | $12,500,000 | $19,500,000 | $24,000,000 |

[a] As noted in the text, these include more than just data connection. Additional costs are for satellite and other data service; IT, communications, and health equipment maintenance, replacement and upgrades; data connection support, updates, and enhancements; staff and facility to monitor, continuously test, and troubleshoot; staff and facility to support enhanced health care delivery, health and social services community education and engagement, clinic and technical staff, exercises and continuous training; staff to train, educate, and support nondisaster services to optimize use of paid-for bandwidth, cost-sharing, cost-reallocation; and additional services.

## Potential Funding Mechanisms

Community Development Block Grant Disaster Recovery, Office of the Assistant Secretary for Preparedness and Response, Federal Communications Commission, U.S. Department of Housing and Urban Development, U.S. Department of Veterans Affairs, U.S. Department of Defense, government of Puerto Rico

## Potential Implementers

Office of the Chief Innovation Officer, Puerto Rico Emergency Management Agency, Telecommunications Bureau, Puerto Rico Department of Health

## Potential Pitfalls

This COA will require coordination with external medical expertise, including remote, continental U.S.–based health care and social and behavioral services, and it must leverage the resilience of clinics' connectivity to mitigate risk.

## Likely Precursors

The primary precursor COA is *CIT 21 Government-Owned Fiber-Optic Conduits to Reduce Aerial Fiber-Optic Cable and Incentivize Expansion of Broadband Infrastructure.*

## References

Center for Disease Control and Prevention, State and Local Readiness, webpage, April 1, 2019. As of July 3, 2019:
https://www.cdc.gov/phpr/readiness/phep.htm

EMR Industry, "Puerto Rico Primary Care Association Network Selects Health Gorilla Clinical Network as Their Data Exchange Platform," April 16, 2018. As of July 3, 2019:
http://www.emrindustry.com/puerto-rico-primary-care-association-network-selects-health-gorilla-clinical-network-as-their-data-exchange-platform/

Framingham Heart Study, homepage, 2019. As of April 29, 2019:
http://www.framinghamheartstudy.org/

Health Care Information and Management Systems Society, HIPAA, ARRA/HITECH Act and Meaningful Use Compliance Resources, website, 2019. As of July 3, 2019:
http://www.himss.org/library/healthcare-privacy-security/risk-assessment/compliance

Health Resources and Services Administration, Federal Office of Rural Health Policy, website, undated. As of July 3, 2019:
https://www.hrsa.gov/rural-health/index.html

Kishore, Nishant M.P.H., et al., "Mortality in Puerto Rico after Hurricane Maria," The New England Journal of Medicine, Vol. 379, pp. 162–170, July 12, 2018. As of July 3, 2019:
https://www.nejm.org/doi/full/10.1056/NEJMsa1803972

National Association of Community Health Centers, "Puerto Rico Health Center Fact Sheet," January 2018. As of July 3, 2019:
http://www.nachc.org/wp-content/uploads/2018/01/PR_18.pdf

U.S. Department of Health and Human Services, Office of Minority Health, website, undated. As of June 25, 2018:
https://minorityhealth.hhs.gov/

# CIT 30
## Resiliency Innovation Network Leading to Development of a Resiliency Industry

### Sectors Impacted

Communications and Information Technology, Community Planning and Capacity Building, Economic

### Issue/Problem Being Solved

During 2017 Hurricane Maria alone caused over $100 billion in damage to Puerto Rico.[1] Similar hurricane seasons are expected in the future.[2] Puerto Rico is vulnerable to natural disasters and needs innovative capacity development approaches to enhance resiliency and overcome human capital and investment constraints. Puerto Rico needs to adopt tools and create businesses that can enhance Puerto Rico's resilience. This should be part of Communications and Information Technology courses of action (COAs) since "almost all societies (including Puerto Rico) have become heavily dependent on the internet, which has become the world's most important piece of infrastructure, and also the infrastructure on which all the other infrastructure relies."[3] As human and technical talent grow, Puerto Rico can leverage research and development (R&D) to create new resiliency products and services and to encourage new ventures for their commercialization. This could transform Puerto Rico into a significant producer, consumer, and exporter of these products and services.

### Description

This COA will create a Resiliency Innovation Network (RIN) across Puerto Rico to build on existing Puerto Rico Science, Technology, and Research Trust (PRSTRT) and university facilities to teach, test, and refine existing resiliency products and services and to develop new ones to enhance capability and stimulate new commercial ventures. RIN would accomplish a variety of objectives, including:

- establishing research priorities based on the comparative advantage of Puerto Rico researchers in areas of resiliency innovation, such as telecommunications, energy, water, and so on, and working closely with the Rockefeller Foundation's 100 Resilient Cities to expand public-private partnerships and private industry offerings for increasing resiliency in communities and cities[4]

---

[1] Masters, 2017.

[2] Chris Morris, "After a Record Hurricane Season in 2017, Forecasters Warn U.S. to Brace for More This Year," *Fortune*, May 14, 2018.

[3] Carl Bildt, "The Pandora's Box of the Digital Age," Project Syndicate, November 16, 2017.

[4] 100 Resilient Cities, website, undated.

- rolling out the Healthy Generations Project community resiliency model,[5] which trains community peers in such areas as understanding effects of trauma on individual and social well-being; building resiliency individually and community-wide; establishing benchmarks for growth and success; and developing communication strategies that lead to shared healing
- setting up makerspaces, hackerspaces, and business incubators at all 25 municipalities that host Puerto Rico universities outside of the San Juan metro area
- establishing two resiliency innovation labs, one at PRSTRT's San Juan headquarters and another at PRSTRT's Guanajibo Research and Innovation Park in Mayaguez
- leveraging PRSTRT's existing resources, such as entrepreneurial programs, corporate ties to encourage established companies to collaborate and invest in resiliency innovations, government ties for tax incentives and business credits; and the Technology Transfer Office to provide IP protection and negotiate licensing agreements
- establishing methods to institutionalize progress.

Ultimately, the aim of RIN would be to develop a resiliency industry in Puerto Rico that includes "resiliency maturity models,"[6] ties to the insurance and reinsurance industry,[7] and ways to engage volunteers.[8]

Puerto Rico researchers already have field experimentation sites throughout Puerto Rico, which can be used to test promising prototypes in its various microclimates and topographies. Where knowledge gaps exist, PRSTRT could use its Ciencia Puerto Rico (CienciaPR) network to identify subject matter experts (SMEs) from around the world in academia, private industry, and government to engage with local efforts. The diaspora of Puerto Rico could be a particularly valuable resource.

After testing and development, PRSTRT could also:

- work with academic researchers to foster entrepreneurship and help incubate resiliency innovations into sustainable businesses
- train academic researchers how to obtain tax credits, funding, and incentives
- collaborate with the Puerto Rico Trade and Export Company and the Puerto Rico Manufacturers Association to identify export and training opportunities for local resiliency entrepreneurs.

Finally, PRSTRT could work with the Resilient Puerto Rico Advisory Commission and Fomento's Puerto Rico Tourism Company to convene an annual resiliency innovation conference that would bring representatives from private industry, governments, nonprofits, and universities together in Puerto Rico to collaborate and share findings. The annual conference could present an opportunity to bring venture capitalists and the principals of private equity firms

---

[5] Healthy Generations Project, "About HGP," webpage, undated.

[6] Derived from Carnegie Mellon's cybercapability maturity models. See, for example, Caralli et al., 2010.

[7] As in CAMICO's model for CPA-related insurance and reinsurance. CAMICO, website, undated.

[8] As in the California Health Medical Reserve Corps (CH-MRC), website, undated.

to Puerto Rico, during which time they could help judge resiliency innovation startup competitions. The Puerto Rico Trade and Export Company could bring potential customers from all over the world to meet with the local innovators.

## Potential Benefits

This COA will help generate local companies and jobs to increase Puerto Rico's long-term resiliency to natural disasters through an innovation initiative and lower the impacts and costs of such events.

One expected long-term outcome is the establishment of a resiliency innovation cluster in Puerto Rico. A cluster is a geographically proximate group of interconnected companies and associated institutions in a particular field, linked by commonalities and complementarities.[9] "Clusters encompass an array of linked industries and other entities important to competition."[10]

Resiliency Center of Education and Innovation (RCOEI) could be a way to institutionalize progress. A major goal is to build upon the RIN to develop a resiliency industry that includes "resiliency maturity models,"[11] ties to the insurance and reinsurance industry,[12] and ways to engage volunteers.[13] Homegrown expertise in such areas could provide a significant competitive advantage, with potentially long term benefits.

An annual resiliency innovation conference in Puerto Rico could bring diverse representatives together to collaborate and share findings which should facilitate cross-sectoral "pollination." Resiliency innovation facilities at each university could help these facilities provide emergency services, local integration, and social support during and after natural disasters. Finally, new ventures would form, the local workforce would be educated and trained, and opportunities would emerge for nonlocal investors (including the diaspora) to invest locally.

Clusters could encourage established companies to set up new operations or expand in Puerto Rico.

Research suggests that governments can and should play a role in the establishment of clusters.[14]

---

[9] Michael E. Porter, "Location, Competition, and Economic Development: Local Clusters in a Global Economy," *Economic Development Quarterly*, Vol. 14, No. 1, February 1, 2000, pp. 15–34.

[10] Porter, 2000.

[11] Derived from CMMI Institute's capability maturity models for cybersecurity; CMMI Institute, undated.

[12] An example of a mutual insurance model is CAMICO's CPA-related insurance and reinsurance. CAMICO, website, undated.

[13] The California Health Medical Reserve Corp is one such example, as noted previously.

[14] As Porter argues, "A role for government cluster development . . . should not be confused with the notion of industrial policy. . . . Industrial policy aims to distort competition in favor of a particular location, cluster theory focuses on removing obstacles, relaxing constraints, and eliminating inefficiencies to productivity and productivity growth. The emphasis in cluster theory is not on market share but rather on dynamic improvement" (Porter, 2000).

## Potential Spillover Impacts to Other Sectors

The RIN will have an impact on different sectors. Its primary contribution will be in the Communications and Information Technology sector, where it will provide an ecosystem for developing and testing new resilient telecommunications-related technologies, including those that apply to emergency services. It will contribute to the community planning and capacity building sector by teaching innovation and entrepreneurship and enhancing Puerto Rico's educational institutions. The RIN will also help build skills in "resiliency" technology that can be shared with other disaster-prone areas beyond Puerto Rico. These could be solidified in an RCOEI. In the economic sector, RIN will teach planning and business development skills that could attract investment. The distributed, integrated RIN should also make more rural and municipal areas attractive to investors.

The RIN could be also useful to the energy sector, by providing an ecosystem for developing and testing new resilient energy-related technologies; housing and public buildings, by providing an ecosystem for developing and testing new resilient and sustainable housing-related technologies and building methods (Ponce's School of Architecture also can help); health and social services, by teaching skills that can support health care tools and services, as well as service delivery; and municipalities, by supporting local integrated services in both cities and small communities.

## Potential Costs

Potential up-front costs: $2.2 million in estimated up-front costs
Potential recurring costs: $26 million in estimated recurring costs (11 years)
Potential total costs: $29 million in total estimated costs

Potential costs are $2.2 million for up-front and $2.4 million annually for the RIN to expand testing, teaching, and applications, while leveraging the existing infrastructure of PRSTRT and Puerto Rico universities, plus starting the Healthy Generations Project community resiliency model and establishing RCOEI. Rough order-of-magnitude cost estimates are show in Table C.5.

## Potential Funding Mechanisms

U.S. Department of Commerce Economic Development Administration; National Science Foundation; Puerto Rico Science, Technology, and Research Trust; Puerto Rico Industrial Development Company; private sector

**Table C.5. Details of Cost Estimates for Course of Action *Resiliency Innovation Network***

| Category | Up-Front | Justification | Annual | Justification |
|---|---|---|---|---|
| Set up 2 PRSTRT labs in San Juan and Mayagüez | 2 × $100,00 = $200,000 | Average cost of a PRSTRT lab, per PRSTRT staff | 2 × $100,000 = $200,000 | Maintenance of equipment and staff salaries |
| Set up innovation spaces & business incubators at other municipalities (25) with universities | 25 × $50,000 = $1.25 million | Average cost of a typical Texas Instruments lab, per PRSTRT staff | 25 × $50,000 = $1.25 million | Maintenance of equipment and staff salaries |
| Increase PRSTRT staff to increase bandwidth | 3 × $100,000 = $300,000 | Increase staff by 3 to increase tech transfer, entrepreneurship, & external coordination | 3 × $100,000 = $300,000 | Retain new staff |
| Annual conference | $50,000 | Average cost of this size conference per PR government event planner | $50,000 | Average cost of this size conference per PR government event planner |
| Healthy Generations "Train the Trainer" 3-day community workshops | $15,000 | Cost is based on past workshops (e.g., U.S. at-risk communities) and related expenses | 6 × $15,000 = $90,000 | Ignite resiliency from the ground up, led by local champions to ensure success of entire program |
| Develop RCOEI | 1 × $150,000, 2 × $75,000 = $300,000 | 1 professor and 2 support staff initially | 1 × $150,000 + 3 × $100,000 = $450,000 | Increase support staff on a sustaining basis |
| Miscellaneous | 5% of subtotal | To cover unanticipated expenses | 5% of subtotal | To cover unanticipated expenses |
| TOTAL | $2.2 million | | $2.4 million | |

NOTE: First column = What is paid; second and third column = nonrecurring costs; and fourth column = recurring costs.

Two major economic factors are driving the development of the resiliency industry: natural disaster–related damage with an estimated cost of $306 billion in 2017 in the United States, and cybersecurity-related damage estimated to cost $600 billion in 2017.[15]

The Federal Emergency Management Agency (FEMA) and the insurance and reinsurance industries have highlighted the need to make major new moves to prepare for and reduce the impact of disasters. Given the major cost impact of physical and cyber disasters (over $900 billion in 2017) and the fact that the resilience industry can mitigate risks and lower the cost of recovery and rebuilding, there is an important opportunity for commercial investment in

---

[15] U.S. National Oceanic and Atmospheric Administration, Office for Coastal Management, Weather Disaster and Costs, website, May 9, 2019; James Lewis, *Economic Impact of Cybercrime—No Slowing Down*, McAffee, February 2018.

resiliency devices, technology, and insurance models, which can contribute to preparation, response, recovery, and above all coordination and communication.

## Potential Implementers

Central Office of Recovery, Reconstruction, and Resiliency; Office of the Chief Innovation Officer, universities; Puerto Rico Science, Technology, and Research Trust; Resilient Puerto Rico Advisory Commission; Puerto Rico Department of Economic Development and Commerce; Puerto Rico Industrial Development Company

## Potential Pitfalls

Investment could be limited by Puerto Rico's austere fiscal situation, the participant pool could be limited by the ongoing "brain drain," and new businesses could be dissuaded by barriers. These are described in the Financial Oversight and Management Board's (FOMB's) "New Fiscal Plan for Puerto Rico: Restoring Growth and Prosperity."[16]

Potential pitfalls for RIN include the crowding out of private investment by public debt; the shrinking number of banks in Puerto Rico; and austerity measures forcing local ventures developing resiliency innovation to move.

## Likely Precursors

RIN can benefit from other COAs, such as *CIT 19 Municipal Hotspots* and *CIT 26 Wi-fi Hotspots in Public Housing and Digital Stewards Program*, but does not depend on them. RIN can use digital workforce skills and rural mesh networks (*CIT 28 Innovation Economy/Human Capital Initiative* and *CIT 29 Health Care Connectivity to Strengthen Resilience and Disaster Preparedness*). It also can draw on past and present research projects at PRSTRT and Universidad de Puerto Rico (UPR) and on existing resiliency labs.

## References

100 Resilient Cities, "What is Urban Resilience?" website, 2019. As of July 1, 2019:
http://100resilientcities.org/resources/

Bildt, Carl, "The Pandora's Box of the Digital Age," Project Syndicate, November 16, 2017. As of July 1, 2019:
https://www.project-syndicate.org/commentary/offensive-cyber-weapons-arms-race-by-carl-bildt-2017-11?barrier=accesspaylog

California Health Medical Reserve Corps, homepage, undated. As of July 3, 2019:
http://www.ch-mrc.org/

---

[16] *New Fiscal Plan for Puerto Rico: Restoring Growth and Prosperity, Certified by the Financial Oversight and Management Board for Puerto Rico*, June 29, 2018.

CAMICO, homepage, 2018. As of July 3, 2019:
http://www.camico.com/

CMMI Institute, "What is Cyberresilience?" Cybermaturity, overview, website, 2019. As of
July 3, 2019:
https://cmmiinstitute.com/products/cybermaturity

Drinker, Biddle and Leath, *New Fiscal Plan for Puerto Rico: Restoring Growth and Prosperity*,
certified by the Financial Oversight and Management Board for Puerto Rico, June 29, 2018.
As of August 29, 2018:
https://www.drinkerbiddle.com/-/media/files/services/bondholders/government-development
-bank-for-puerto-rico/new-fiscal-plan-for-puerto-rico-june-2018.pdf?la=en

Lewis, James, *Economic Impact of Cybercrime—No Slowing Down*, McAffee, February 2018.
As of July 3, 2019:
https://www.mcafee.com/enterprise/en-us/solutions/lp/economics-cybercrime.html

Masters, Jeff, "Hurricane Maria Damage Estimate of $102 Billion Surpassed Only by Katrina,"
Category 6, News & Blogs, Weather Underground, November 22, 2017. As of June 20,
2019:
https://www.wunderground.com/cat6/hurricane-maria-damages-102-billion-surpassed-only
-katrina

Morris, Chris, "After a Record Hurricane Season in 2017, Forecasters Warn U.S. to Brace for
More This Year," *Fortune*, May 14, 2018. As of June 20, 2019:
http://fortune.com/2018/05/14/2018-hurricane-season-forecast-possible-tropical-storm-in-gulf/

Porter, Michael E., "Location, Competition, and Economic Development: Local Clusters in a
Global Economy," *Economic Development Quarterly*, Vol. 14, Issue 1, pp. 15–34,
February 1, 2000. As of July 1, 2019:
http://journals.sagepub.com/doi/abs/10.1177/089124240001400105

Puerto Rico Science, Technology and Research Trust, homepage, undated. As of June 20, 2019:
https://prsciencetrust.org

U.S. National Oceanic and Atmospheric Administration, Office for Coastal Management,
Weather Disaster and Costs, website, May 9, 2019. As of July 3, 2019:
https://coast.noaa.gov/states/fast-facts/weather-disasters.html

# CIT 31
## Resilience/e-Construction Learning Lab

### Sector Impacted

Communications and Information Technology, Housing, Municipalities, Transportation

### Issue/Problem Being Solved

Leverage state-of-the-art resilient e-construction approaches to accelerate socioeconomic development. A streamlined, paperless construction-administration delivery process will facilitate legacy and new construction documentation and digital management in a secure environment. It will improve disaster response by conducting damage assessment in the field and sending assessments back to an operations center, and it will help with reports, scheduling, and predictions of what kinds of supplies and equipment will be needed.

Some of the needs that have to be addressed include implementing a paperless construction-administration delivery process; reducing construction time and cost; increasing efficiency through automation or digital communication; improving quality and service; increasing transparency; using a digital platform for facilitating construction-related workflows for business processes (such as documentation, collection, review, routing, electronic signatures for approvals, digital plans and specifications, materials tracking tools, construction inspection reports, damage, quality assurance, costs, timelines, and performance reporting tools).

### Description

This course of action (COA) will start by establishing a Resilience/e-Construction Learning Lab for a one-year pilot project to digitize assessment, permitting, and reporting processes in one Puerto Rico municipality. At the end of this pilot project, findings, including a cost-benefit analysis, will be presented to inform the feasibility for an e-permitting and e-construction ecosystem throughout Puerto Rico. The Resilient e-Construction Learning Lab will be staffed with a multidisciplinary teams of 7 subject matter experts (SMEs): an information technology (IT) lead (program manager); a government process SME; a finance analyst; an architect (Master Planner); an architect support staff member; a civil engineer; and an analyst (focused on successes/best practices out of Puerto Rico). The team to run the pilot program, under joint oversight by the Puerto Rico Department of Housing working with the Department of Transportation, academic institutions (with architecture and business programs), and a municipality.

The COA will establish one physical Resilient e-Construction Learning Lab that will engage in a 1-year investigation and assessment, including a practical on-the-ground project. The intention is to work with the government, faculty, and students and to involve local communities to digitize assessment, permitting, and reporting processes in one municipality in

Puerto Rico, and present the findings and a cost-benefit analysis to inform the feasibility of an island-wide e-permitting and e-construction ecosystem.

The pilot project could include:

- engaging in a 3-month assessment to gather input from government, industry, academia, and the public to identify strengths, weaknesses, threats, and opportunities
- identifying priority problem sets and providing recommendations aligned to achieving e-resilient construction, to include legacy issues (e.g., lack of documentation, rigorous identification system for title and property rights, rights of ways [ROW], housing and building ownership in the case of multiple shared owners, existing infrastructure repair, renovation, reinforcement and disaster-readiness) and new projects
- establishing reporting metrics (time, costs, efficiency, safety, and so on) and monitoring progress
- developing a paperless e-construction administration process by facilitating all construction documentation and digital management in a secure environment
- streamlining paperless permits for roads and buildings, while increasing transparency
- conducting market research, evaluating promising technical solutions, determining best solutions and costs for adoption, and providing rational justification of need
- delivering quarterly progress reports and an annual final report
- developing a 5-year resilient e-construction master plan.

Findings from the first-year pilot will inform continuation work for the remaining 10 years of implementation of this COA.

## Potential Benefits/

A streamlined paperless construction-administration delivery process facilitates all legacy and new construction documentation and digital management in a secure environment. It also saves money by decreasing paper use, printing, and document storage costs, and it saves time by decreasing communication delays and transmittal time, all while increasing transparency and tax collection.

The improvement to communication and the transparency of the process has the potential for eliminating questions, claims, and disputes as to when (or if) a document was submitted. All stakeholders can see the name of the document approver along with the exact timing of each recorded step.

In Michigan, the Department of Transportation estimates that it saves about $12 million in added efficiencies and saves 6 million pieces of paper annually by using electronic document storage. At the same time, it has reduced its average contract modification processing time from 30 days to 3 days.[1]

---

[1] Sharone Fisher, "E-Construction Paves the Way to Highway Success," Laserfiche, August 26, 2016.

## Potential Spillover Impacts to Other Sectors

This COA would accelerate the development of affordable and resilient homes and structures (housing sector). It would facilitate the development of roads and bridges, ensuring public safety and the continuity of essential government functions (transportation sector). And it would streamline paperless construction administration and increase transparency and tax collection (municipality sector).

The construction industry is responsible for substantial spillover effects in the broader economy, generating $86 of additional economic activity for every $100 of construction sector activity, according to figures from the U.S. Bureau of Economic Analysis for 2012.[2]

## Potential Costs

Potential upfront costs: $1.5 million–$6.0 million in estimated upfront costs[3]
Potential recurring costs: $9.6 million–$38 million in estimated recurring (11 years)[4]
Potential total costs: $11 million–$44 million in total estimated costs[5]

Once a successful pilot is completed, costs for expansion, barriers, and facilitators can be further refined. The costs of software tools and services can vary significantly depending on desired capabilities, which are to be defined and evaluated during the pilot project. The costs scale with the number of teams, as shown in Table C.6.

**Table C.6. Details of Cost Estimates for Course of Action *Resilience/e-Construction Learning Lab***

| Low Cost | Intermediate Cost | High Cost |
|---|---|---|
| Up-front costs: | Up-front costs: | Up-front costs: |
| e-Construction Software Tools, plus support services = $1.5 million | e-Construction Software Tools, plus support services = $3 million | e-Construction Software Tools, plus support services = $6 million |
| Recurring costs: Labor for Housing + Transportation + Municipality + Academia | Recurring costs: Labor for Housing + Transportation + Municipality + Academia | Recurring costs: Labor for Housing + Transportation + Municipality + Academia |
| 1 team of 7 staff (7 staff total) $870,000 per year | 2 teams of 7 staff (14 staff total) $1,740,000 per year | 4 teams of 7 staff (28 staff total) $3,490,000 per year |
| Summary: Up-front costs: $1.5 million Recurring costs: $9.6 million | Summary: Up-front costs: $3 million Recurring costs: $19.2 million | Summary: Up-front costs: $6 million Recurring costs: $38.4 million |

---

[2] Richard Florida, "The Housing Crisis Is a Building Crisis," CityLab, February 28, 2017.

[3] The recovery plan represents potential up-front costs as $500,000–$10 million. The up-front cost lower bound is updated here to include software tools.

[4] The recovery plan represents potential recurring costs as $20 million–$60 million. That estimate is updated here to reflect labor costs consistent with the other COAs.

[5] The recovery plan represents potential total costs as $20 million–$70 million. That estimate is updated here to reflect labor costs consistent with the other COAs.

| Low Cost | Intermediate Cost | High Cost |
|---|---|---|
| Total costs: $11 million | Total costs: $22.2 million | Total costs: $44.4 million |

## Potential Funding Mechanisms

Community Development Block Grant Disaster Recovery, U.S. Department of Commerce Economic Development Administration, U.S. Department of Transportation

A potential program that may provide funding includes National Infrastructure Investment at the Department of Transportation.

## Potential Implementers

Central Office of Recovery, Reconstruction, and Resiliency, Office of the Chief Innovation Officer, Puerto Rico Department of Housing

## Potential Pitfalls

This process requires coordination to organize a multidisciplinary team. The risks are minimized by conducting an initial pilot project to present findings and cost-benefit analysis before expansion. The pilot project must take steps to ensure protection of citizen privacy and strong cybersecurity.

An examination of previous construction projects included recommendations to avoid troublesome outcomes, including implementing electronic records and harnessing the latest technological developments.[6]

## Likely Precursors

Precursor COAs are *CIT 30 Resiliency Innovation Network Leading to Development of a Resiliency Industry* and *CIT 16 Government Digital Reform Planning and Capacity Building*. It would be important to empower the Puerto Rico Department of Housing with the resilient e-construction mission before establishing collaboration with the U.S. Department of Transportation, academia, and a municipality for the staffing of the e-Construction Learning Lab.

## References

Fisher, Sharon, "E-Construction Paves the Way to Highway Success," *Laserfiche*, August 26, 2016. As of April 29, 2019:
https://www.laserfiche.com/ecmblog/e-construction-paves-way-highway-success-2/

---

[6] Stuart Wilks, "The Century's Most Troublesome Construction Projects," *Global Construction Review*, October 5, 2015.

Florida, Richard, "The Housing Crisis Is a Building Crisis," *CityLab*, February 28, 2017. As of April 29, 2019:
https://www.citylab.com/equity/2017/02/solving-americas-construction-crisis/517968/

Jha, Abhas, "But what about Singapore?" Lessons from the best public housing program in the world," Sustainable Cities, *World Bank*, January 1, 2018. As of April 29,2019:
http://blogs.worldbank.org/sustainablecities/what-about-singapore-lessons-best-public
-housing-program-world

U.S. Department of Health and Human Services, *grants.gov*, Search Grants, using Infrastructure keyword, undated. As of April 29, 2019:
https://www.grants.gov/web/grants/search-grants.html?keywords=infrastructure

Wilks, Stuart, "The century's most troublesome construction projects," *Global Construction Review*, October 5, 2015. As of April 29, 2019:
http://www.globalconstructionreview.com/perspectives/centurys-most-troublesome
-construction-pr8oje8ct8s/

Yu, Leping (Tommy), "Using Technology to Enhance Government Transparency and Counter Corruption," Chicago Policy Review, June 20, 2017.

# CIT 32
## Digital Citizen Services

### Sectors Impacted

Communications and Information Technology, Community Planning and Capacity Building, Economic, Health and Social Services

### Issue/Problem Being Solved

Citizens increasingly expect their experience with government services to be as easy, quick, and simple to understand as the user interfaces for consumer services. This requires building on successful efforts for the provision of digital services at the federal and state levels as well as learning what has and has not worked when updating public-facing digital services.

The drive for improving the experience for citizens interacting with government is for the most part coming from citizens themselves. From booking an airline ticket, to applying for a credit card, or paying bills, individuals are increasingly used to a seamless, online experience that involves minimal wait and minimal complications.

While some government services have been gradually brought online in Puerto Rico, the experience from one government agency to the next is inconsistent. Some of the most important interactions, such as renewing a driver's license, still require an in-person paper application and can take hours. The same is true for many processes required for setting up a business or applying for government services. This has been exacerbated by budget-cutting moves that shut down district offices of the government of Puerto Rico and thus require people around Puerto Rico to travel to the capital of San Juan in order to process documents or submit applications. These inefficiencies not only are frustrating for citizens and visitors, but also drive up the cost of government and contribute to greater data inconsistencies.

### Description

This course of action (COA) seeks to expand the scope of the Puerto Rico Innovation and Technology Service (PRITS) to include a focus on citizen-centered digital services and prioritize a "one-stop-shopping" experience for accessing government services and information electronically in an easy-to-use fashion. The COA will be implemented by "expanded" PRITS teams (product teams) that will interact closely with Puerto Rico agencies. These teams will work across agencies to complete projects, with an emphasis on examining citizens' needs and experiences from start to finish of a given procedure or interaction and ensuring privacy and security. These teams will ensure that agile approaches are implemented at the start of each project and that projects allow for iterative product releases where appropriate (some policy efforts may require alternative project methodologies). The teams will help the agencies structure budgets and contract deliverables to be outcomes-focused and use data to drive decisions that improve the experiences of citizens engaging in government services.

The COA will implement best practices for ensuring digital inclusion and accessibility, such as the ability to access government services from mobile devices. One of the best-known government technical blunders was the catastrophic 2013 launch of healthcare.gov by the U.S. federal government, which crashed immediately and delayed signup for hundreds of thousands of people shopping for health insurance. The U.S. government spent $1.7 billion on the website (originally budgeted at $93.7 million) and subsequently recruited top technical talent to overhaul the way the government planned, budgeted, and built technology to embrace modern techniques, such as agile development, open-source tools, and data-driven decisions.[1]

The actions used to address some of the issues that were faced by healthcare.gov have become a model for state and local governments alike—a roadmap for applying agile principles to government technology with a focus on the citizen's experience and simple, clear language to improve understanding and interactions with government.[2] These digital services principles include:

- Understand what people need
- Address the whole experience, from start to finish
- Make it simple and intuitive
- Build the service using agile and iterative practices
- Structure budgets and contracts to support delivery
- Assign one leader and hold that person accountable
- Bring in experienced teams
- Choose a modern technology stack
- Deploy in a flexible hosting environment
- Automate testing and deployments
- Manage security and privacy through reusable processes
- Use data to drive decisions.

This COA embraces these principles for the digitization of government services. In practice this will require (1) learning from experts who have been a part of creating digital services in other jurisdictions; (2) communicating the new approach within government *and* valuing the expertise and experience of subject matter experts (SMEs) within agencies; (3) adhering to the digital services principles indicated above, and evaluating effectiveness, cultural acceptance, and success.

---

[1] U.S. Department of Health and Human Services, Office of Inspector General, "An Overview of 60 Contracts That Contributed to the Development and Operation of the Federal Marketplace," OEI-03-14-00231, August 24, 2014.

[2] U.S. Digital Service, undated.

## Potential Benefits

This COA could increase public trust, efficiency, transparency, and accountability; increase use and adoption of digital services tailored to citizens' needs and experiences; streamline internal government processes; and reduce the number of resources required for rote government services, thus allowing for more attention to human interaction and other challenges to improve the public experience overall.

By embracing the principles of citizen-centric digital services and modernization of legacy systems, governments can reduce the cost to maintain outdated systems and processes and gain long-term cost savings, increased public transparency, and better data for policymaking. Digital services also enable standardization, which can fuel automation, real-time analytics, data sharing, and process optimization.

An increasingly digitized government sector would also provide more opportunities for highly skilled graduates to remain in Puerto Rico after graduation, as well as opportunities for the members of the diaspora who want to apply their skills at home.[3] The people of Puerto Rico have been involved in the evolution of digital services in the United States for many years, as catalogued former CIO Giancarlo Gonzalez.[4]

## Potential Spillover Impacts to Other Sectors

Digitized government services would affect all sectors through their (1) better data for planning and evaluation, (2) transparency and accountability, (3) more efficient citizen interaction for services, and (4) possible cost savings, streamlining of reporting, and reducing of time to decision. The sectors potentially affected include community planning and capacity building, economic, and health and social services.

## Potential Costs

Potential up-front costs: $240,000 in estimated up-front costs[5]
Potential recurring costs: $25 million in estimated recurring costs (11 years)[6]
Potential total costs: $25 million in total estimated costs[7]

---

[3] Glade, 2017, observes that "the 2012–2013 Global Competitiveness Report from the World Economic Forum ranked Puerto Rico third in the availability of scientists and engineers. According to Lucy Crespo . . . 'Puerto Rico graduates 22,000 STEM students and 60 to 70 percent leave the island.'" See also Alvarez, 2017.

[4] Gonzalez, 2009–present.

[5] The recovery plan published a potential up-front cost of $400,000. That cost overestimated the cost of equipment for set-up cost. The estimated cost is corrected here, lowering the estimated cost.

[6] The recovery plan published a potential recurring cost of $33 million. The recurring costs have been corrected to have labor costs consistent with other COAs, reducing the estimated recurring costs. This update lowers the estimated recurring cost to $25 million.

[7] The recovery plan published a potential total cost of $33 million. The total potential cost is updated here to reflect labor costs consistent with other COAs, reducing the total estimated cost. This updated cost lowers the estimated recurring cost to $25 million, using 2 significant figures.

The potential total cost scales with the number of teams employed. The recovery plan estimated the cost for 2 teams, which is shown as the intermediate cost option. The up-front cost is $120,000 per team for equipment. The recurring costs are estimated for 2 teams and amount to approximately $1.1 million annually per team. The estimated cost scaled by the number of project teams is shown in Table C.7. Table C.8 details the estimated costs for a project team.

**Table C.7. Scaling of Cost with Number of Project Teams for Course of Action *Digital Citizen Services***

| Low cost | Intermediate cost | High cost |
|---|---|---|
| 1 team | 2 teams | 3 teams |
| $120,000 up-front $1.1 million annual | $240,000 up-front $2.2 million annual | $360,000 up-front $3.3 million annual |

**Table C.8. Project Team Composition and Cost for Course of Action *Digital Citizen Services***

| Project Team | Number per Team | Annual Cost per Staff Member | Annual Total Labor Cost | Staff Equipment (Laptop, Mobile, Wi-Fi, Data, Software) |
|---|---|---|---|---|
| Project manager | 1 | $124,600 | $124,600 | $10,000 |
| Team acquisition staff | 1 | $124,600 | $124,600 | $10,000 |
| Team technical staff | 4 | $124,600 | $498,400 | $40,000 |
| Team documentation staff | 4 | $62,300 | $249,200 | $40,000 |
| Team tracking and reporting | 2 | $62,300 | $124,600 | $20,000 |
| Total labor | | | $1,121,400 | |
| Total equipment | | | | $120,000 |

## Potential Funding Mechanisms

Community Development Block Grant Disaster Recovery, U.S. Department of Commerce Economic Development Administration, government of Puerto Rico

## Potential Implementers

Office of the Chief Innovation Officer, government of Puerto Rico agencies

## Potential Pitfalls

The COA may result in internal friction, if the new team members are not integrated well with the existing team members; if that is the case, they will also miss a chance to learn from each other. The COA may result in "change fatigue" or perceived concerns that only superficial procedures are being improved with the digital services, not more complex issues that also need to be addressed. The COA may result in actual workforce reduction, resistance to such reduction, or perceived loss of control from existing agencies.

Digital government efforts must take steps to ensure protection of citizen privacy and strong cybersecurity, in addition to getting off legacy systems and moving away from legacy processes, and all of this must be done within a fixed budget.

## Likely Precursors

Precursor COAs are *CIT 16 Government Digital Reform Planning and Capacity Building* and *CIT 27 Study Feasibility of Digital Identity*.

## References

Alvarez, Lisette, "As Others Pack, Some Millennials Commit to Puerto Rico," *New York Times*, August 5, 2017. As of April 29, 2019: https://www.nytimes.com/2017/08/05/us/puerto-ricans-millennials-start-ups-small -business.html

Dear Fiscal Board, homepage, undated. As of April 29, 2019: http://www.dearfiscalboard.com/blog-2/

Glade, Jim, "Puerto Rico turns to tech and entrepreneurialism to revitalize the economy," *TechCrunch*, January 15, 2017. As of April 29, 2019: https://techcrunch.com/2017/01/15/puerto-rico-turns-to-tech-and-entrepreneurialism-to -revitalize-the-economy/

U.S. Department of Health and Human Services, Office of Inspector General, "An Overview of 60 Contracts That Contributed to the Development and Operation of the Federal Marketplace," OEI-03-14-00231, August 24, 2014.

U.S. Digital Service, Digital Services Playbook, website, undated. As of April 29, 2019: https://playbook.cio.gov/r

# CIT 33
# Government Digital Process Reform

## Sector Impacted

Communications and Information Technology, Community Planning and Capacity Building, Economic

## Issue/Problem Being Solved

Puerto Rico is consolidating and reorganizing its government departments and agencies. This presents an opportunity to bridge traditional silos and introduce service cultures (reflecting the distinct mission of each agency) of coordination by design to include data-driven, outcomes-based, whole-systems governing to continuously improve service, be more effective in the money spent, improve service delivery, better serve the public, and make better policy. Such a venture is inward-facing and should work hand-in-glove with the digital services efforts of the Puerto Rico Innovation and Technology Service (PRITS).

Introducing service cultures of coordination by design, including data-driven, outcomes-based whole-systems governing, could help improve service and service delivery, spend resources more effectively, better serve the public, and make better policy. Capturing the potential of digital technology for better government requires a new, "whole government" approach and integration of systems. Effective use of data can lead to improved policy and delivery tailored services for citizens This data-driven and customer-oriented culture for providing services to people could improve Puerto Rico's ability to serve public needs and simultaneously directly inform policymaking. The culture would be continuously advanced through data-driven improvements that are informed by drawing input (e.g., quantitative data) from all of the governmental services being provided.

## Description

This course of action (COA) would (1) adopt a systems approach to government technology, with an emphasis on human-centered digital process design and data standardization to drive policy decisions; (2) establish people-centered digital design and data science teams within the government of Puerto Rico to tackle cross-cutting policy and operational issues, coordinating different projects with agencies (especially during the agency-consolidation process to ensure clear accountability); and (3) open up government services internally and externally through the use of application program interfaces (APIs), where appropriate, and for feedback to drive continuous improvement.

The work will consist of leveraging PRITS teams for a people-centered approach to improving back-end processes, digitizing processes (after a design evaluation), designing APIs and dashboards, implementing approaching security and privacy controls, updating the inventory of data assets held by each agency and the risk-relative-to-security controls assigned to each. Work will also include developing effective engagement and feedback approaches with citizens, local and federal agencies, national networks, and private sector players.

There is a risk of expertise and capabilities leaving Puerto Rico, especially among college graduates in Puerto Rico. To address this, this COA could incorporate partnerships with academic institutions and leverage internship programs offered by the Governor's office. Such partnerships could offer (1) incentives for recent graduates to stay in Puerto Rico in terms of job opportunities and placement immediately after graduation and (2) potential college scholarships in return for promising to work in Puerto Rico a certain number of years after graduation.

The COAs for *CIT 32 Digital Citizen Services* and *CIT 33 Government Digital Process Reform* address two different aspects of digital transformation. *Digital Citizen Services* focuses on improvements to public sector, public-facing government services that the residents of Puerto Rico are seeking. *Governmental Digital Process Reform* focuses on the internal governmental processes and digital information architecture dedicated to processing data for government-provided services. Both COAs will use digital transformation to dramatically improve the efficiency, experience, and long-term value of the governmental services provided to people, the activities required to deliver them to the public, and the performance of agencies with respect to governmental missions.

## Potential Benefits

Agency consolidation and reorganization are opportunities to establish a "whole-of-government," people-centered, digital design and data-driven approach that would improve technical aspects of service and service delivery, be more cost-effective, better serve the public, and make better policy.

Modern businesses use data analysis and machine learning to glean insights and improve their processes based on how their services are used. A recent McKinsey Global study found that if governments around the world seized upon best practices for improving productivity, they could save $3.5 trillion a year by 2021 while providing improved outcomes for citizens. The study notes an enormous efficiency gap between the top- and bottom-performing countries, which can be closed through smarter financial and procurement management and by leveraging digital technologies and advanced data analytics.[1]

In December 2017, the Governor of Puerto Rico signed Act 122, which "seeks the reduction of 118 agencies to 35 more efficient ones."[2] In April 2018 the Fiscal Board noted that with 116,500 employees across 114 Executive Branch government agencies, Puerto Rico's government is "outsized compared to the actual service needs of the people of Puerto Rico" and

---

[1] McKinsey Center for Government, *Government Productivity: Unlocking the $3.5 Trillion Opportunity*, April 2017, pp.17–18.

[2] La Fortaleza, 2017.

an outlier compared with U.S. states.[3] The Financial Oversight and Management Board agreed with the Governor about the need for agency consolidation and right-sizing "to deliver services in as efficient a manner as possible."[4] In any jurisdiction, fundamental agency restructuring would require a corresponding digital reform. In Puerto Rico, given the needs of the recovery and future disaster preparedness, strategic digital reform is essential.

Incorporating a digital and data-driven approach to agency reorganization allows for an outcomes-focused reframing of government priorities, rather than simply attempting to reinstitute the processes of the past in a new agency structure. While the transformation will require significant culture and change management, as well as systems redesign, it is an opportunity for real reform.

### Potential Spillover Impacts to Other Sectors

All sectors would benefit from a coordinated, effective, integrated, feedback-driven, and more holistic data-driven governing approach. Sectors that will benefit most directly from this COA are community planning and capacity building and economic.

### Potential Costs

Potential up-front costs: $300,000 in estimated up-front costs
Potential recurring costs: $70 million in estimated recurring costs (11 years)
Potential total costs: $70 million in total estimated costs[5]

Up-front cost involves preparation work by 2 subject matter experts (SMEs) at cost of $124,600 each for one year. The recurring cost estimate for each digital design and data team is $2 million annually. This cost estimate assumes that there will be 3 PRITS teams.

### Potential Funding Mechanisms

Community Development Block Grant Disaster Recovery, U.S. Department of Commerce Economic Development Administration, government of Puerto Rico

### Potential Implementers

Office of the Chief Innovation Officer, government of Puerto Rico agencies

---

[3] FOMB, 2018, pp. 65–66.

[4] FOMB, 2018, pp. 68–69.

[5] The recovery plan represents the estimated costs with 1 significant figure in this course of action. The costs described are rounded in the recovery plan.

## Potential Pitfalls

As with *CIT 32 Digital Citizen Services*, this COA could potentially lead to internal friction, "change fatigue," workforce reduction, shift or reduction in influence and authority by existing leadership, and perceived loss of control on the part of existing agencies. It will be crucial to ensure access and safety for the most vulnerable and underserved, as well as protection of citizen privacy and strong cybersecurity.

It should be noted that a design-centered approach will be needed to reexamine government processes and policy-related practices. Digitalization alone may not achieve the full benefits of this venture.

## Likely Precursors

Precursor COAs are *CIT 16 Government Digital Reform Planning and Capacity Building*, *CIT 27 Study Feasibility of Digital Identity*, and *CIT 32 Digital Citizen Services*.

## References

Financial Oversight and Management Board for Puerto Rico (Junta de Supervisión y Administración Financiera de Puerto Rico), "New Fiscal Plans for Puerto Rico," April 5, 2018.

La Fortaleza, Oficina del Gobernador, "Governor of Puerto Rico makes New Government bill into law," San Juan, Puerto Rico, December 18, 2017. As of November 11, 2018: https://www.fortaleza.pr.gov/content/governor-puerto-rico-makes-new-government-bill-law

McKinsey Center for Government, *Government Productivity: Unlocking the $3.5 Trillion Opportunity*, April 2017. As of April 29, 2019: https://www.mckinsey.com/~/media/McKinsey/Industries/Public%20Sector/Our%20Insights/ The%20opportunity%20in%20government%20productivity/Executive%20summary%20 Government%20productivity%20unlocking%20the%2035%20trillion%20opportunity.ashx

# References

100 Resilient Cities, "San Juan's Resilience Challenge," undated. As of July 25, 2019:
http://www.100resilientcities.org/cities/san-juan/

———, "What Is Urban Resilience?" website, 2019. As of July 1, 2019:
http://100resilientcities.org/resources/

"2016 Colorado Revised Statutes Title 24—Government—State Principal Departments Article 33.5—Public Safety Part 7—Emergency Management § 24-33.5-705.5. Auxiliary Emergency Communications Unit—Powers and Duties of Unit and Office of Emergency Management Regarding Auxiliary Communications—Definitions," Justia US Law, undated. As of July 25, 2019:
https://law.justia.com/codes/colorado/2016/title-24/principal-departments/article-33.5/part-7/section-24-33.5-705.5/

Abaffy, Luke, "Feds Fight to Firm Up Puerto Rico Dam," ENR Southeast, February 26, 2018. As of July 24, 2019:
https://www.enr.com/articles/44047-feds-fight-to-firm-up-puerto-rico-dam

"About PREPA Networks," PREPA Networks. As of December 10, 2019:
http://www.prepanetworks.net/?page_id=129

Abre Puerto Rico, "About Us," website, 2019. As of September 3, 2019:
https://www.abrepr.org/en/about-abre

Act 213 of 1996, Puerto Rico Telecommunications Act, September 12, 1996, as amended, revised May 10, 2012.

Allied Media Projects, Detroit Community Technology Project, Digital Stewards Training, webpage, undated. As of February 20, 2019:
https://www.alliedmedia.org/dctp/digitalstewards

Alvarez, Lisette, "As Others Pack, Some Millennials Commit to Puerto Rico," *New York Times*, August 5, 2017. As of April 29, 2019:
https://www.nytimes.com/2017/08/05/us/puerto-ricans-millennials-start-ups-small-business.html

AT&T, Comments of AT&T, *In the Matter of Response Efforts Undertaken During the 2017 Hurricane Season*, PS Docket No. 17-344, February 21, 2018.

Becker, Rachel, "Trying to Communicate After the Hurricane: 'It's As If Puerto Rico Doesn't Exist,'" *The Verge*, September 29, 2017. As of June 19, 2019: https://www.theverge.com/2017/9/29/16372048/puerto-rico-hurricane-maria-2017-electricity -water-food-communications-phone-Internet-recovery

Belson, David, "Internet Impacts of Hurricanes Harvey, Irma and Maria," *Oracle + Dyn* blog, September 25, 2017. As of June 19, 2018: https://dyn.com/blog/internet-impacts-of-hurricanes-harvey-irma-and-maria/

Bildt, Carl, "The Pandora's Box of the Digital Age," Project Syndicate, November 16, 2017. As of July 1, 2019: https://www.project-syndicate.org/commentary/offensive-cyber-weapons-arms-race-by-carl -bildt-2017-11?barrier=accesspaylog

Blessing, Gabriel Parra, "At PR Telephone, the Past Is Prologue," *Caribbean Business*, October 27, 2005.

Broadband Now, Internet Providers in San Juan, Puerto Rico, 2014–2018, undated. As of July 24, 2019: https://broadbandnow.com/Puerto-Rico/San-Juan

Broniatowski, D. A., M. J. Paul, and M. Dredze, "National and Local Influenza Surveillance through Twitter: An Analysis of the 2012–2013 Influenza Epidemic," *PLoS ONE*, Vol. 8, No. 12, December 9, 2013, e83672.

Brown, Jared T., "The Hurricane Sandy Rebuilding Strategy: In Brief," Washington, D.C.: Congressional Research Service, February 10, 2014.

Bump, Phillip, "We've Created a Twitter Bot That Provides Hourly Updates on the Situation in Puerto Rico," *Washington Post*, October 11, 2017.

California Health Medical Reserve Corps, website, undated. As of July 3, 2019: http://www.ch-mrc.org/

"Calling All TITANs: Nissan and the American Red Cross Mobilize Purpose-Driven Campaign with the Ultimate Service TITAN," *Nissan News*, October 4, 2018.

CAMICO, website, undated. As of July 3, 2019: http://www.camico.com/

Caralli, Richard, Julia Allen, Pamela Curtis, Davide White, and Lisa Young, *CERT Resilience Management Model, Version 1.0, Technical Report CMU/SEI-2010-TR-012*, Pittsburgh, Pennsylvania: Software Engineering Institute, Carnegie Mellon University, 2010. As of September 12, 2019: https://resources.sei.cmu.edu/library/asset-view.cfm?assetid=9479

Carpenter, Darryl, "Overcoming the Challenges of Digital Transformation—Lessons Learned from the NZ Government," University of Melbourne Power of Collaboration series, original February 6, 2018, updated September 19, 2018. As of April 29, 2019:
https://melbourne-cshe.unimelb.edu.au/lh-martin-institute/insights/overcoming-the-challenges -of-digital-transformation-lessons-learned-from-the-nz-government

Center for Disease Control and Prevention, State and Local Readiness, webpage, April 1, 2019. As of July 3, 2019:
https://www.cdc.gov/phpr/readiness/phep.htm

Central Office for Recovery, Reconstruction and Resiliency, *Transformation and Innovation in the Wake of Devastation: An Economic and Disaster Recovery Plan for Puerto Rico*, August 8, 2018. As of April 8, 2019:
http://www.p3.pr.gov/assets/pr-transformation-innovation-plan-congressional-submission -080818.pdf

City of Minneapolis, "Wireless Minneapolis," website, June 6, 2019. As of June 22, 2019:
http://www.minneapolismn.gov/wireless/index.htm

City of Vancouver, "Vancouver's Digital Strategy," website, 2019. As of June 22, 2019:
https://vancouver.ca/your-government/digital-strategy.aspx

ClickFix, homepage, 2017. As of June 20, 2019:
https://seeclickfix.com/

CMMI Institute, "What Is Cyber Resilience? A Step Beyond Compliance," undated. As of July 26, 2019:
https://cmmiinstitute.com/products/cybermaturity

CODAN Communications, website, undated. As of October 25, 2018:
https://www.codanradio.com/about-codan-radio/why-p25/

Colorado ARES, "New Colorado Law Creates Auxiliary Emergency Communications Unit within the State's Division of Homeland Security and Emergency Management," Colorado Auxcom, news release, June 6, 2016. As of April 29. 2019:
http://www.coloradoares.org/wordpress/links/colorado-auxcomm/

Colorado General Assembly, "HB16-1040 Auxiliary Emergency Communications," 2016 Regular Session. As of April 29, 2019:
https://leg.colorado.gov/bills/hb16-1040

Comments of Public Knowledge, *In the Matter of Public Safety and Homeland Security Bureau Seeks Comment on Response Efforts Undertaken During the 2017 Hurricane Season*, PS Docket No. 17-344, January 22, 2018.

Committee on Energy and Commerce, CTIA, "Wireless Network Resilience Cooperative Framework for Disasters and Emergencies, U.S.A. & Puerto Rico," press release, April 27, 2016.

Congressional Research Service, "Congressional Commissions: Overview, Structure, and Legislative Considerations," R40076, November 17, 2017.

Congressional Task Force on Economic Growth in Puerto Rico, *Congressional Task Force on Economic Growth in Puerto Rico Report to the House and Senate*, 114th Congress, December 20, 2016. As of September 7, 2019:
https://www.finance.senate.gov/imo/media/doc/Bipartisan%20Congressional%20Task%20Force%20on%20Economic%20Growth%20in%20Puerto%20Rico%20Releases%20Final%20Report.pdf

Connect Puerto Rico, "Galería de Mapas Puerto Rico," Washington, D.C.: Connected Nation, 2014. As of July 25, 2019:
http://www.connectpr.org/mapping/state

Council of Economic Advisors, "The Digital Divide and Economic Benefits of Broadband Access," Issue Brief, March 2016.

CTIA, Comments of CTIA, In the Matter of Response Efforts Undertaken During the 2017 Hurricane Season, PS Docket No. 17-344, January 22, 2018.

CTIA, Comments of CTIA Before the Federal Communications Commission in the Matter of Response Efforts Undertaken During the 2017 Hurricane Season, PS Docket No. 17-344, January 22, 2018.

Dev Catalyst, website, 2015. As of September 7, 2019:
https://www.devcatalyst.com/

DHS—*See* U.S. Department of Homeland Security.

DHS OEC—*See* U.S. Department of Homeland Security, Office of Emergency Communications.

Digital Identification and Authentication Council of Canada, "Building Canada's Digital Identity Future," May 2015. As of June 25, 2018:
http://www.diacc.ca/wp-content/uploads/2015/05/DIACC-Building-Canadas-Digital-Future-May5-2015.pdf

Driving Innovation in Tennessee, "Tennessee Is Driving Innovation and We're Bringing It to You," undated. As of June 11, 2019:
https://www.tndrivinginnovation.com/

DTOP—*See* Puerto Rico Department of Transportation and Public Works.

e-Estonia, "e-Identity," website, undated. As of June 19, 2019:
https://e-estonia.com/solutions/e-identity/id-card/

———, "We Have Built a Digital Society and So Can You," undated. As of July 25, 2019:
https://e-estonia.com/

Eggers, William D., and Joel Bellman, *The Journey to Government's Digital Transformation*, Deloitte University Press, 2015. As of April 29, 2019:
https://www2.deloitte.com/content/dam/insights/us/articles/digital-transformation-in -government/DUP_1081_Journey-to-govt-digital-future_MASTER.pdf

Eggers, William D., and Steve Hurst, "Delivering the Digital State: What If State Government Services Worked Like Amazon?" *Deloitte Insights*, November 14, 2017. As of June 20, 2019:
https://www2.deloitte.com/insights/us/en/industry/public-sector/state-government-digital -transformation.html#endnote-5

EMR Industry, "Puerto Rico Primary Care Association Network Selects Health Gorilla Clinical Network as Their Data Exchange Platform," April 16, 2018. As of July 3, 2019:
http://www.emrindustry.com/puerto-rico-primary-care-association-network-selects-health -gorilla-clinical-network-as-their-data-exchange-platform/

"ESF-2, Communications Annex," June 2016. As of July 24, 2019:
https://www.fema.gov/media-library-data/1473679033823-d7c256b645e9a67cbf09d3c 08217962f/ESF_2_Communications_FINAL.pdf

Estudios Tecnicos Inc., "About Us," website, 2019. As of September 3, 2019:
https://www.estudiostecnicos.com/etidisasterservice/

"EU to Give Free WiFi Hotspots to Cities," *Euronews*, March 20, 2018. As of September 7 2019:
https://www.euronews.com/2018/03/20/eu-cash-funds-free-hi-speed-wifi-hotspots

Expedient, "The Complete Data Center Build vs Buy Calculator," website, undated. As of April 29, 2019:
https://www.expedient.com/data-center-build-vs-buy-calculator/

EY, "EY's Spotlight on Telecommunications Accounting," No. 2, 2015. As of July 24, 2019:
https://www.ey.com/Publication/vwLUAssets/ey-spotlight-on-telecommunications -accounting/$FILE/ey-spotlight-on-telecommunications-accounting-issue2.pdf

FCC—*see* Federal Communications Commission.

Federal Communications Commission, "911 and E911 Services," undated. As of July 24, 2019:
https://www.fcc.gov/general/9-1-1-and-e9-1-1-services

———, "2016 Broadband Progress Report," FCC 16-6, January 29, 2016. As of April 29, 2019:
https://docs.fcc.gov/public/attachments/FCC-16-6A1.pdf

———, "2018 Broadband Deployment Report," GN Docket No. 17-199, FCC 18-10, February 2, 2018a. As of September 7, 2019:
https://www.fcc.gov/reports-research/reports/broadband-progress-reports/2018-broadband-deployment-report

———, "A Next Generation 911 Cost Study," September 2011. As of October 29, 2018:
https://www.911.gov/pdf/FCC_Next_Generation_911_Cost_Study_2011.pdf

———, "Chairman Pai Unveils $954 Million Plan to Restore and Expand Networks in Puerto Rico and U.S. Virgin Islands," March 6, 2018b.

———, "Communications Status Report for Areas Impacted by Hurricane Maria," March 21, 2018c.

———, "E-Rate: Universal Service Program for Schools and Libraries," February 9, 2018d. As of June 18, 2018:
https://www.fcc.gov/consumers/guides/universal-service-program-schools-and-libraries-e-rate

———, *FCC Solicits Nominations For New Disaster Response and Recovery Working Group of the Broadband Advisory Committee*, Public Notice, GN Docket No. 17-83, DA 18-837, August 9, 2018e. As of July 24, 2019:
https://docs.fcc.gov/public/attachments/DA-18-837A1.pdf

———, *In the Matter of Improving Resilience of Mobile Wireless Communications Networks*, PS Docket No. 13-239, and *In the Matter of Reliability and Continuity of Communications Networks, Including Broadband Technologies,* PS Docket No. 11-60, September 27, 2013. As of July 24, 2019:
https://docs.fcc.gov/public/attachments/FCC-16-173A1.pdf

———, Order and Notice of Proposed Rulemaking, FCC 18-57, May 29, 2018. As of July 26, 2019:
https://docs.fcc.gov/public/attachments/FCC-18-57A1.pdf

———, Report and Order, Declaratory Ruling, and Further Notice of Proposed Rulemaking in WC Docket No. 17-84, *In the Matter of Accelerating Wireline Broadband Deployment by Removing Barriers to Infrastructure Investment*, FCC-CIRC1711-04, November 29, 2017.

———, "Trends in Telephone Service," September 2010.

Federal Emergency Management Agency, Disaster Page Definitions, undated-a. As of July 24, 2019:
https://www.fema.gov/disaster-page-definitions

———, Emergency Management Performance Grant Program, website, June 7, 2018. As of December 4, 2018:
https://www.fema.gov/emergency-management-performance-grant-program

———, Emergency Operations Center Assessment Checklist, undated-b. As of July 26, 2019: https://www.fema.gov/media-library-data/20130726-1524-20490-0618/eocchecklist.pdf

———, Executive Order OE-2001-26. "Order Ejecutiva de la Gobernadora del Estado Libre Asociado de Puerto Rico para establecer la coordinacion de funciones ejecutivas en manejo de emergencias o desastres y derogar los boletines administrativos OE-1993-23 y 4974-E," June 25 2001.

———, Hazard Mitigation Grant Program, website, last updated September 19, 2018. As of December 3, 2018:

https://www.fema.gov/hazard-mitigation-grant-program

———, Hurricane Maria Task Force, *DR-4339-PR Consolidated Communications Restoration Plan*, October 30, 2017.

———, Integrated Alert and Warning System, website, undated-c. As of July 26, 2019: https://www.fema.gov/integrated-public-alert-warning-system

———, "Integrated Alert and Warning System (IPAWS)," Executive Order 13407, undated-d. As of July 24, 2019:

https://www.fema.gov/pdf/emergency/ipaws/ipaws_handouts_brochure_format_june %202010.pdf

———, National Incident Management System, October 2017. As of July 25, 2019: https://www.fema.gov/national-incident-management-system

———, NIMS Mobile Communications Command Centers, 2005. As of October, 25, 2018: https://www.fema.gov/media-library-data/20130726-1826-25045-1951/fema_508_2_typed _resource_definitions_incident_management_resources_2005.pdf

———, Ninth FEMA Think Tank Conference Call, transcript, September 19, 2013. As of July 26, 2019:

https://www.fema.gov/media-library-data/20130919-1717-27928-0769/transcripts20130919 -27928-12i8txg.txt

———, *Public Assistance Applicant Handbook FEMA P-323*, March 2010. As of July 24, 2019: https://www.fema.gov/pdf/government/grant/pa/fema323_app_handbk.pdf

Federal Exchange Commission, *In the Matter of Inquiry concerning Deployment of Advanced Telecommunications Capability to All Americans in a Reasonable and Timely Fashion*, GN Docket No. 17-199, February 2, 2018.

Federal Highway Administration, Office of Highway Policy Information, "Highway Performance Monitoring System (HPMS)," 2018. As of July 26, 2019: https://www.fhwa.dot.gov/policyinformation/hpms.cfm

FEMA—*See* Federal Emergency Management Agency.

FEMA Communications/IT Solutions-based Team, "Puerto Rico Communications/IT Solutions-based Team Report," June 30, 2018.

Financial Oversight and Management Board for Puerto Rico, *Annual Report Fiscal Year 2017*, July 30, 2017. As of June 20, 2019:
https://www.google.com/url?sa=t&rct=j&q=&esrc=s&source=web&cd=3&cad=rja&uact=8&ved=2ahUKEwjkh4y6uPjiAhXRMd8KHUp9B5YQFjACegQIARAC&url=https%3A%2F%2Fcases.primeclerk.com%2Fpuertorico%2FHome-DownloadPDF%3Fid1%3DNzAxMjIy%26id2%3D0&usg=AOvVaw2W6GiVrtyeuqDxbprDca_A

———, "New Fiscal Plans for Puerto Rico," April 5, 2018.

Finkelstein, S. M., S. M. Speedie, and S. Potthoff, "Home Telehealth Improves Clinical Outcomes at Lower Cost for Home Health Care," *Telemedicine journal of e-health*, Vol. 12, No. 2, April 2006, pp. 128–136.

Fiscal Agency and Financial Advisory Authority (Autoridad de Asesoría Financiera y Agencia Fiscal), *New Fiscal Plan for Puerto Rico*, April 5, 2018. As of August 23, 2018:
http://www.aafaf.pr.gov/assets/newfiscalplanforpuerto-rico-2018-04-05.pdf

Fisher, Sharon, "E-Construction Paves the Way to Highway Success," Laserfiche, August 26, 2016. As of April 29, 2019:
https://www.laserfiche.com/ecmblog/e-construction-paves-way-highway-success-2/

Florida, Richard, "The Housing Crisis Is a Building Crisis," CityLab, February 28, 2017. As of April 29, 2019:
https://www.citylab.com/equity/2017/02/solving-americas-construction-crisis/517968/

FOMB—*See* Financial Oversight and Management Board for Puerto Rico.

Framingham Heart Study, homepage, 2019. As of April 29, 2019:
http://www.framinghamheartstudy.org/

Freifunk, Wikipedia, August 8, 2018. As of November 6, 2018:
https://en.wikipedia.org/wiki/Freifunk

GAO—*See* U.S. Government Accountability Office.

Ginsberg, Jeremy, Matthew H. Mohebbi, Rajan S. Patel, Lynnette Brammer, Mark S. Smolinski, and Larry Brilliant, "Detecting Influenza Epidemics Using Search E Query Data," *Nature*, Vol. 457, February 19, 2009, pp. 1012–1014.

Glade, Jim, "Puerto Rico Turns to Tech and Entrepreneurialism to Revitalize the Economy," *TechCrunch*, January 15, 2017. As of April 29, 2019:
https://techcrunch.com/2017/01/15/puerto-rico-turns-to-tech-and-entrepreneurialism-to-revitalize-the-economy/

Gonzalez, Giancarlo, *Dear Fiscal Board* blog, 2009–present. As of July 25, 2019:
http://www.dearfiscalboard.com/blog-2/

———, "PR—Former CIO Team Is Recruited by USDS," *Dear Fiscal Board* blog, December 7, 2015. As of July 25, 2019:
http://www.dearfiscalboard.com/entire-puerto-rico-cio-team-recruited-by-usds/

Goss, Ernest P., and Joseph M. Phillips, "How Information Technology Affects Wages: Evidence Using Internet Usage as a Proxy for IT Skills," *Journal of Labor Research*, Vol. 23, No. 2, 2002, pp. 463–474.

Government of British Columbia, "BC Services Card," website, undated. As of June 16, 2019:
https://www2.gov.bc.ca/gov/content/governments/government-id/bc-services-card

Government of Puerto Rico, Directory of Municipalities of Puerto Rico, *pr.gov—Official Portal of the Government of Puerto Rico*, 2019.

———, Executive Order OE-2019-012, Responsibilities of the Chief Innovation and Information Officer and Chief Technology Officer, March 13, 2019. As of July 24, 2019:
https://prits.pr/assests/OE-2019-012.pdf

———, *Puerto Rico Disaster Recovery Action Plan: For the Use of CDBG-DR Funds in Response to 2017 Hurricanes Irma and Maria*, July 29, 2018. As of August 30, 2018:
https://www.cdbg-dr.pr.gov/wp-content/uploads/2018/07/HUD-Approved-Action-Plan_EN.pdf

———, "U.S. Federal Studies That Do Not Include Puerto Rico on Par with States," undated. As of September 30, 2019:
https://juntasupervision.pr.gov/wp-content/uploads/wpfd/50/597eb4ede89ad.pdf

Government of the United Kingdom, "GOV.UK Verify: Introducing GOV.UK Verify," March 14, 2019. As of June 16, 2019:
https://www.gov.uk/government/publications/introducing-govuk-verify/introducing-govuk-verify

———, "Taking GOV.UK Verify to the Next Stage," Government Digital Service, October 11, 2018. As of July 25, 2019:
https://gds.blog.gov.uk/2018/10/11/taking-gov-uk-verify-to-the-next-stage

"Governor Announces Appointments to Public Service Regulatory Board," *Caribbean Business*, August 20, 2018. As of July 24, 2019:
https://caribbeanbusiness.com/governor-announces-appointments-to-public-service-regulatory-board/

Governor of Puerto Rico, *Build Back Better Puerto Rico: Request for Federal Assistance for Disaster Recovery*, San Juan: Government of Puerto Rico, November 13, 2017. As of September 7, 2019:
https://www.governor.ny.gov/sites/governor.ny.gov/files/atoms/files/Build_Back_Better _PR.pdf

"Governor of Puerto Rico Makes New Government Bill into Law," La Fortaleza, Oficina del Gobernador, San Juan, December 18, 2017. As of July 26, 2019:
https://www.fortaleza.pr.gov/content/governor-puerto-rico-makes-new-government-bill-law

"Governor Signs Law to Establish Puerto Rico Public Service Regulatory Board Reorganization," *Caribbean Business*, August 13, 2018. As of July 24, 2019:
https://caribbeanbusiness.com/gov-signs-law-to-establish-puerto-rico-public-service -regulatory-board-reorganization/

Greenstein, Shane, and Ryan McDevitt. "The Broadband Bonus: Estimating Broadband Internet's Economic Value," *Telecommunications Policy*, Vol. 35, No. 7, August 2011, pp. 617–632.

Guimarães, Nathália, "Furacão Maria deixa conexão de internet lenta no Brasil," *LeiaJa*, September 9, 2017. As of June 19, 2018:
http://www.leiaja.com/tecnologia/2017/09/22/furacao-maria-deixa-conexao-de-internet -lenta-no-brasil/

Hardesty, Linda, "AT&T Taps Synergies Between Its 5G and FirstNet Build-Outs," SDX Central, December 7, 2018. As of July 25, 2019:
https://www.sdxcentral.com/articles/news/att-taps-synergies-between-its-5g-and-firstnet -build-outs/2018/12/

Health Care Information and Management Systems Society, HIPAA, ARRA/HITECH Act and Meaningful Use Compliance Resources, website, 2019. As of July 3, 2019:
http://www.himss.org/library/healthcare-privacy-security/risk-assessment/compliance

Health Resources and Services Administration, Federal Office of Rural Health Policy, website, undated. As of July 3, 2019:
https://www.hrsa.gov/rural-health/index.html

Healthy Generations Project, "About HGP," webpage, undated. As of August 18, 2020:
https://www.healthygenerationsproject.org/our-model

House Committee on Energy and Commerce, "CTIA & Pallone Announce 'Wireless Network Resilience Cooperative Framework' for Disasters and Emergencies," press release, April 27, 2016. As of July 24, 2018:
https://democrats-energycommerce.house.gov/newsroom/press-releases/ctia-pallone-announce -wireless-network-resiliency-cooperative-framework-for-disasters-and-emergencies

HUD—*See* U.S. Department of Housing and Urban Development.

Hurricane Sandy Rebuilding Task Force, Hurricane Sandy Rebuilding Strategy, August 2013. As of July 26, 2019:
https://www.hud.gov/sites/documents/HSREBUILDINGSTRATEGY.PDF

"Information Technology," Oxford Reference, undated. As of July 24, 2019:
http://www.oxfordreference.com/view/10.1093/oi/authority.20110803100003879

Information Technology and Innovation Foundation, "Innovation Economics: The Economic Doctrine for the 21st Century," undated. As of June 2, 2018:
https://itif.org/innovation-economics-economic-doctrine-21st-century

Jaresko, Natalie, Executive Director Financial Oversight and Management Board for Puerto Rico, "Examining Challenges in Puerto Rico's Recovery and the Role of the Financial Oversight and Management Board," written testimony before the House Committee on Natural Resources, November 7, 2017:
https://naturalresources.house.gov/imo/media/doc/testimony_jaresko.pdf

Jha, Abhas, "But What About Singapore? Lessons from the Best Public Housing Program in the World," World Bank, *Sustainable Cities*, blog, January 1, 2018. As of April 29, 2019:
http://blogs.worldbank.org/sustainablecities/what-about-singapore-lessons-best-public-housing-program-world

Johnson, Jenna, "FEMA Removes—Then Restores—Statistics about Drinking Water Access and Electricity in Puerto Rico from Website," *Washington Post*, October 6, 2017.

jpadilla, tracking-status-pr, GitHub, 2017. As of June 20, 2019:
https://github.com/jpadilla/tracking-status-pr

Juvare, WebEOCx, "Emergency Management Technology—Powered by Juvare Exchange," website, 2019. As of September 4, 2019:
https://www.juvare.com/webeoc/

Kishore, Nishant M. P. H., et al., "Mortality in Puerto Rico after Hurricane Maria," *New England Journal of Medicine*, Vol. 379, July 12, 2018, pp. 162–170.

Kohlhaas, Paul, "Zug ID: Exploring the First Publicly Verified Blockchain Identity," *Medium*, December 6, 2017. As of June 16, 2019:
https://medium.com/uport/zug-id-exploring-the-first-publicly-verified-blockchain-identity-38bd0ee3702

Kolko, Jed, "Broadband and Local Growth," *Journal of Urban Economics*, Vol. 71, No. 1, January 2012, pp. 100–113. As of July 26, 2019:
https://www.sciencedirect.com/science/article/pii/S0094119011000490?via%3Dihub

Launch Tennessee, "Launch Tennessee Announces New Board Members," March 22, 2018. As of June 11, 2019:
https://launchtn.org/2018/03/launch-tennessee-announces-new-board-members/

Lehman Brothers Collection, Harvard Business School, Baker Library Historical Collection, Cambridge, Mass.

Lewis, James, *Economic Impact of Cybercrime—No Slowing Down*, McAffee, February 2018. As of July 3, 2019:
https://www.mcafee.com/enterprise/en-us/solutions/lp/economics-cybercrime.html

Lewis-Kraus, Gideon, "Inside the Battle to Bring Broadband to New York's Public Housing," *Wired*, November 3, 2016.

Liberty Cablevision of Puerto Rico, website, undated. As of July 24, 2019:
https://www.libertypr.com/

Liberty Cablevision of Puerto Rico LLC, Comments of Liberty Cablevision of Puerto Rico LLC, *In the Matter of Accelerating Wireline Broadband Deployment by Removing Barriers to Infrastructure Investment*, WC Docket No. 17-84, June 15, 2017.

Love, Julia, "Google Brings Free WiFi to Mexico, First Stop in Latin America," Reuters, March 13, 2018.

Madory, Doug, "Puerto Rico's Slow Internet Recovery," Oracle Internet Intelligence, December 7, 2017. As of June 19, 2018:
https://blogs.oracle.com/Internetintelligence/puerto-ricos-slow-Internet-recovery

Magdelana Petrova, "Phone Service Can Mean Life or Death After a Disaster and AT&T and Verizon Are Using Drones That Could Help," CNBC, August 24, 2018.

Masters, Jeff, "Hurricane Maria Damage Estimate of $102 Billion Surpassed Only by Katrina," *Weather Underground*, November 22, 2017. As of June 20, 2019:
https://www.wunderground.com/cat6/hurricane-maria-damages-102-billion-surpassed-only-katrina

McGrath, Jenny, "Puerto Rico Website Is Keeping Track of the Island's Re-Emerging Infrastructure" *Digital Trends*, October 4, 2017. As of June 20, 2019:
https://www.digitaltrends.com/home/puerto-rico-hurricane-maria-site/

McKinsey Center for Government, *Government Productivity: Unlocking the $3.5 Trillion Opportunity*, April 2017. As of April 29, 2019:
https://www.mckinsey.com/~/media/McKinsey/Industries/Public%20Sector/Our%20Insights/The%20opportunity%20in%20government%20productivity/Executive%20summary%20Government%20productivity%20unlocking%20the%2035%20trillion%20opportunity.ashx

Michigan Department of Health and Human Services, "What Is MyLogin?" website, undated. As of July 26, 2019:
https://www.michigan.gov/mdhhs/0,5885,7-339-71551_2945_72165---,00.html

Microgrids at Berkeley Lab, *Microgrids, DER Case Studies*, webpage, undated. As of November 5, 2018:
https://building-microgrid.lbl.gov/projects/der-cam/case-studies

Miranda, Ismael, "Sabes que es un KP-4," October 13, 2013. As of September 7, 2019:
https://www.prarl.org/?p=7366

"Mobile Fact Sheet," Washington, D.C.: Pew Research Center, February 5, 2018. As of July 24, 2019:
http://www.pewinternet.org/fact-sheet/mobile/

Morgan, Jeffrey, "Digital Transformation in the Public Sector," *CIO Magazine*, January 11, 2018. As of April 29, 2019:
https://www.cio.com/article/3247305/digital-transformation-in-the-public-sector.html

Morris, Chris, "After a Record Hurricane Season in 2017, Forecasters Warn U.S. to Brace for More This Year," *Fortune*, May 14, 2018. As of June 20, 2019:
http://fortune.com/2018/05/14/2018-hurricane-season-forecast-possible-tropical-storm-in-gulf/

National Association for Amateur Radio, *2017 Annual Report*, October 4, 2017. As of July 25, 2019:
http://www.arrl.org/files/file/About%20ARRL/Annual%20Reports/ARRL%202017%20Annual%20Report.pdf

———, "Comments of ARRL, The National Association for Amateur Radio," *In the Matter of Response Efforts Undertaken During the 2017 Hurricane Season*, PS Docket No. 17-344, January 22, 2018. As of July 25, 2019:
https://ecfsapi.fcc.gov/file/10122279117760/2018%20January%20FINAL%20Comments%20PS%20Docket%2017-344.pdf

National Association of Community Health Centers, "Puerto Rico Health Center Fact Sheet," January 2018.

National Disaster Recovery Strategy, 6 U.S.C. Section 771, 2006. As of July 25, 2019:
https://www.govinfo.gov/app/details/USCODE-2010-title6/USCODE-2010-title6-chap2-subchapII-partC-sec771

National Emergency Number Association, Interconnection & Security Committee, NG9-1-1 Architecture Subcommittee, Emergency Services IP Network Design Working Group, *Emergency Services IP Network Design (ESIND) Information Document*, April 5, 2018. As of July 26, 2019:

https://cdn.ymaws.com/www.nena.org/resource/resmgr/standards/NENA-INF-016.2-2018
_ESIND_20.pdf

National Telecommunications and Information Administration, Broadband USA, undated. As of
July 24, 2019:
https://www2.ntia.doc.gov/puerto-rico

*New Fiscal Plan for Puerto Rico: Restoring Growth and Prosperity, Certified by the Financial
Oversight and Management Board for Puerto Rico*, June 29, 2018. As of July 26, 2019:
https://www.drinkerbiddle.com/-/media/files/services/bondholders/government-development
-bank-for-puerto-rico/new-fiscal-plan-for-puerto-rico-june-2018.pdf?la=en

NTIA—*See* U.S. Department of Commerce, National Telecommunications and Information
Administration.

Parra Blessing, Gabriel, "At PR Telephone, the Past Is Prologue," *Caribbean Business*,
October 27, 2005.

Peterson, John A., "Auxiliary Emergency Communications: Recognition of Its Support to
Public Safety," U.S. Department of Homeland Security, SAFECOM, July 11, 2016. As of
April 29, 2019:
https://www.dhs.gov/safecom/blog/2016/07/11/auxiliary-emergency-communications

PlugPower, *Comparing Backup Power Options for Communications*, undated. As of October 29,
2018:
https://www.plugpower.com/wp-content/uploads/2015/07/FCvGen_Stat_F1_101416.pdf

Porter, Michael E., "Location, Competition, and Economic Development: Local Clusters in a
Global Economy," *Economic Development Quarterly*, Vol. 14, No. 1, February 1, 2000,
pp. 15–34.

PPC, *The Complete Guide to Fiber to the Premises Deployment: Options for the Network
Operator*, 2017. As of July 1, 2019:
https://cdn2.hubspot.net/hubfs/2057289/eBooks/The_Complete_Guide_to_Fiber_to_the_Pre
mises_Deployment.pdf?__hssc=153509031.2.1550006317508&__hstc=153509031.ffe969
d0e05e645e40278fc050a986b3.1549665014026.1549665014026.1549665014026.1&__hsfp
=2356915907&hsCtaTracking=31a7166d-0259-4217-8e0b-56447ba29146%7C6027f796
-2fee-4e8d-bbca-e81963959964

PRIDCO—*See* Puerto Rico Industrial Development Company.

Public Law No. 115-123, Bipartisan Budget Act of 2018, February 9, 2018.

Public Safety and Homeland Security Bureau, Federal Communications Commission, *2017
Atlantic Hurricane Season Impact on Communications Report and Recommendations*, Public

Safety Docket No. 17-344, August 2018a. As of July 24, 2019:
https://docs.fcc.gov/public/attachments/DOC-353805A1.pdf

———, *Public Safety and Homeland Security Bureau Seeks Comments on Improving Wireless Network Resiliency to Promote Coordination Through Backhaul Providers*, PS Docket No. 11-60, December 10, 2018b. As of July 24, 2019:
https://docs.fcc.gov/public/attachments/DA-18-1238A1.pdf

———, *Public Safety and Homeland Security Bureau Seeks Comment on Response Efforts Undertaken During 2017 Hurricane Season*, PS Docket No. 17-344, Public Notice, 32 FCC Rcd 10245 (2017) (Hurricane Public Notice), December 10, 2018c. As of July 24, 2019:
https://docs.fcc.gov/public/attachments/DA-18-1238A1.pdf

Puerto Rico Department of Transportation and Public Works, "Puerto Rico 2040 Islandwide Long Range Transportation Plan," San Juan, 2013. As of April 27, 2018:
https://www.yumpu.com/en/document/view/30782909/puerto-rico-2040-islandwide-long -range-transportation-plan-dtop

Puerto Rico Executive Order OE-2001-26, "Executive Order of the Governor of the Commonwealth of Puerto Rico to Establish the Coordination of Executive Functions in Emergency or Disaster Management and to Repeal the Administrative Bulletins OE-1993-23 and 4974-E [Order Ejecutiva de la Gobernadora del Estado Libre Asociado de Puerto Rico para establecer la coordinacion de funciones ejecutivas en manejo de emergencias o desastres y derogar los boletines administrativos OE-1993-23 y 4974-E]," June 25, 2001.

Puerto Rico Industrial Development Company, Company Overview, undated. As of July 25, 2019:
http://www.pridco.com/who-we-are/Pages/Company-Overview.aspx

———, Industries, Aerospace, website, 2018. As of June 18, 2019:
http://www.pridco.com/industries/Pages/Aerospace.aspx

———, Industris, Information Technology, website, 2018. As of June 18, 2019:
http://www.pridco.com/industries/Pages/Information-Technology.aspx

———, Industries, Medical Devices, website, 2018. As of June 18, 2019:
http://www.pridco.com/industries/Pages/Medical-Devices.aspx

Puerto Rico Industrial Development Company, Industries, Pharmaceutical, website, 2018. As of June 18, 2019:
http://www.pridco.com/industries/Pages/Pharmaceutical.aspx

———, homepage, undated. As of July 24, 2019: https://prits.pr/#

Puerto Rico Science, Technology and Research Trust, website, undated. As of June 20, 2019:
https://prsciencetrust.org

Puerto Rico Telecommunications Regulatory Board, Comments of the Puerto Rico Telecommunications Regulatory Board, *In the Matter of Response Efforts Undertaken During the 2017 Hurricane Season*, PS Docket No. 17-344, January 22, 2018.

RAND Corporation, "Supporting Puerto Rico's Disaster Recovery Planning," webpage, undated. As of September 1, 2019:
https://www.rand.org/hsoac/puerto-rico-recovery.html

Red Hook Initiative, "Digital Stewards," website, 2019. As of June 12, 2019:
http://rhicenter.org/programs-2/youth-development/digital-stewards/

———, "Impact Report 2017," undated. As of September 7, 2019:
http://rhicenter.org/wp-content/uploads/2017/10/RHI-ImpactReport-2017_WEB.pdf

———, "Red Hook Initiative: A Community Response to Hurricane Sandy," website, 2013. As of September 7, 2019:
https://rhicenter.org/wp-content/uploads/2013/10/RHI-Hurricane-Report-6_2013.pdf

Red Hook Wifi, "Mission: Resilience, Opportunity, Community and Social Justice," website, 2019. As of June 25, 2018:
https://redhookwifi.org/about/mission/

ReImagina Puerto Rico, website, September 21, 2018. As of July 25, 2019:
http://www.resilientpuertorico.org/informes/

Resolucion Conjunta [Joint Resolution] 40-2018, R.C. de la C. 256, Puerto Rico Legislative Action, March 20, 2018.

Rose, Veronica, "Cost of Operating Different Types of Fire Departments," Connecticut General Assembly, Office of Legislative Research, 2014. As of April 29, 2019:
https://www.cga.ct.gov/2014/rpt/2014-R-0147.htm

Rua Jovet, Javier J., Chairman, Puerto Rico Telecommunications Regulatory Board, Letter to the Hon. Orrin Hatch, Chairman, U.S. Senate Committee on Finance, September 2, 2016.

Samaniego, David G., Division Director, Federal Emergency Management Agency, Communications/IT Sector, FEMA-4339-DR-PR, Memorandum for Mike Byrne, Federal Coordinating Officer/Disaster Recovery Manager, FEMA-4336/4339-DR-PR, April 8, 2018.

Sansom, Marie, "National Identity Card for Australians? Digital Government Lessons from Estonia," *Government News*, November 1, 2016. As of June 16, 2019:
https://www.governmentnews.com.au/25432/

SBA Communications Corporation, *About Us*, website, 2019.

Singh, Jagmeet, "Google Station Free Public Wi-Fi to Go Beyond Railway Stations in India to Reach Cities," *Gadgets 360*, December 5, 2017. As of November 6, 2018:

https://gadgets.ndtv.com/telecom/news/google-station-railway-stations-india-smart-cities
-1783972

Southern California Tribal Chairmen's Association, Tribal Digital Village, website, 2017. As of
June 25, 2018:
https://sctdv.net/about-tdv/

"Sprint, Open Mobile Complete Joint Venture to Serve PR/USVI," News Is My Business,
December 5, 2017. As of July 24, 2019:
http://newsismybusiness.com/sprint-complete-venture/

Stafford, Elaine, "The Suboptic Guide: Chapter 1 Planning, Contracting, Constructing, Owning
and Operating A Submarine Cable Network," Suboptic, 2013. As of April 29, 2019:
http://suboptic.org/wp-content/uploads/2014/10/The-Guide-1.pdf

State of New Jersey, Office of Homeland Security and Preparedness, website, undated. As of
July 25, 2018:
https://www.njhomelandsecurity.gov/jerseynet/

status.pr, website, undated. As of September 7, 2019:
http://www.status.pr/?dash=toHome#reporttitle

Submarine Cable Map, website, undated. As of November 16, 2018:
https://www.submarinecablemap.com/#/landing-point/san-juan-puerto-rico-united-states

Sweeney, Evan, "AMIA Sees Internet Access as a Social Determinant of Health," Fierce
Healthcare, May 24, 2017. As of April 25, 2019:
https://www.fiercehealthcare.com/mobile/amia-views-broadband-access-as-a-social
-determinant-health

"Telecommunications," Legal Information Institute, Cornell Law School, undated. As of July 24,
2019:
https://www.law.cornell.edu/uscode/text/47/153#50

Telecommunications Act of 1996, Pub. LA. No. 104-104, 110 Stat. 56 (1996), January 3, 1996.
As of July 24, 2019:
https://www.fcc.gov/general/telecommunications-act-1996

Telecommunications Industry Association, About Data Centers, ANSI/TIA-942 Quality
Standard for Data Centers, TIA-942.org, website, undated. As of September 12, 2019:
http://www.tia-942.org/content/162/289/About_Data_Centers

"Telecommunications Services," Legal Information Institute, Cornell Law School, undated. As
of July 24, 2019:
https://www.law.cornell.edu/uscode/text/47/153#53

T-Mobile, Comments of T-Mobile USA, INC., *In the Matter of Response Efforts Undertaken During the 2017 Hurricane Season*, PS Docket No. 17-344, January 22, 2018.

Torres, Joseph, Senior Director of Strategy and Engagement, Free Press, "Ex Parte Notice in WC Docket Nos. 18-143, 10-90, 14-58 and 17-287; and PS Docket No. 17-344," letter to Marlene H. Dortch, Secretary, FCC, July 20, 2018. As of July 25, 2019:
https://ecfsapi.fcc.gov/file/10720156823712/FP%20Rosenworcel%20Ex%20Parte%20re
%20Puerto%20Rico%20-%20July%2020%202018.pdf

Torres Lopez, Sandra E., Letter from the Puerto Rico Telecommunications Regulatory Board to Honorable Ajit Pai, Chairman of the Federal Communications Commission, January 22, 2018.

Unique Identifier Authority of India, webpage, undated. As of August 13, 2019:
https://uidai.gov.in/

U.S. Chamber of Commerce, Technology Section, "Data Centers: Jobs and Opportunities in Communities Nationwide," undated. As of July 25, 2019:
https://www.uschamber.com/sites/default/files/ctec_datacenterrpt_lowres.pdf

U.S. Code, Title 47, Section 153, Definitions, U.S. Government Printing Office, 2011. As of August 13, 2019:
https://www.govinfo.gov/content/pkg/USCODE-2011-title47/html/USCODE-2011-title47
-chap5-subchapI-sec153.htm

U.S. Department of Commerce, National Telecommunications and Information Administration, BroadbandUSA Grants Awarded, New Jersey, website, undated-a. As of April 29, 2019:
https://www2.ntia.doc.gov/new-jersey

———, "Next Generation 911," undated-b. As of July 25, 2019:
https://www.ntia.doc.gov/category/next-generation-911

U.S. Department of Health and Human Services, Search Grants, website, undated. As of April 29, 2019:
https://www.grants.gov/web/grants/search-grants.html?keywords=infrastructure

———, Office of Inspector General, "An Overview of 60 Contracts That Contributed to the Development and Operation of the Federal Marketplace," OEI-03-14-00231, August 24, 2014.

———, Office of Minority Health, website, undated. As of June 25, 2018:
https://minorityhealth.hhs.gov/

U.S. Department of Homeland Security, Office of Emergency Communications, Interoperable Communications Technical Assistance Program, *Puerto Rico Public Safety Communications Summary and Recommendations Report*, February 2018.

———, Office of Inspector General, "A Performance Review of FEMA's Disaster Management Activities in Response to Hurricane Katrina," OIG-06-32, March 2006. As of September 7, 2019:
https://www.oig.dhs.gov/assets/Mgmt/OIG_06-32_Mar06.pdf

———, Office of Inspector General, "Lessons Learned from Prior Reports on Disaster-Related Procurement and Contracting," OIG-18-29, December 5, 2017a. As of November 16, 2018:
https://www.oig.dhs.gov/sites/default/files/assets/2017-12/OIG-18-29-Dec17.pdf

———, Office of Inspector General, "Management Alert—FEMA Faces Significant Challenges Ensuring Recipients Properly Manage Disaster Funds," OIG-18-33, December 17, 2017b. As of November 16, 2018:
https://www.oig.dhs.gov/sites/default/files/assets/2017-12/OIG-18-33-Dec17.pdf

———, "DHS Puerto Rico and Virgin Islands (PR/VI) Reconstitution Effort Key Highlights and Talking Points," April 4, 2018.

———, "SAFECOM Guidance Frequently Asked Questions: Understanding P25 Standards and Compliance," undated. As of August 8, 2019:
https://www.dhs.gov/sites/default/files/publications/SAFECOM%20Guidance%20FAQ_P25%20Compliance_05-08-2017_508.pdf

U.S. Department of Housing and Urban Development, "Allocations, Common Application, Waivers, and Alternative Requirements for 2017 Disaster Community Development Block Grant Disaster Recovery Grantees," Docket FR-6066-N-01, *Federal Register*, Vol. 83, No. 8, February 9, 2018. As of July 25, 2019:
https://www.gpo.gov/fdsys/pkg/FR-2018-02-09/pdf/2018-02693.pdf

———, "CDBG Broadband Infrastructure FAQs," HUD Exchange, Resources, website, January 2016. As of June 3, 2018:
https://www.hudexchange.info/resource/4891/cdbg-broadband-infrastructure-faqs/

———, Office of the Inspector General, Audit Report: 2013-FW-0001. As of July 26, 2019:
https://www.hudoig.gov/sites/default/files/2013-FW-0001.pdf

U.S. Department of Transportation, Federal Highway Administration, "Rural Interstate Corridor Communications Study: Report to States," February 2009. As of July 26, 2019:
https://ops.fhwa.dot.gov/publications/fhwahop09021/fhwahop09021.pdf

———, Office of the Assistant Secretary for Research and Technology, Intelligent Transportation Systems Joint Program Office, Costs Database, webpage, undated. As of June 20, 2018:
https://www.itscosts.its.dot.gov/its/benecost.nsf/SubsystemCosts?ReadForm&Subsystem=Emergency+Response+Center+(ER)

U.S. Digital Service, Digital Services Playbook, website, undated. As of April 29, 2019:
https://playbook.cio.gov/r

U.S. Government Accountability Office, *Hurricane Sandy: An Investment Strategy Could Help the Federal Government Enhance National Resilience for Future Disasters*, GAO-15-515, July 30, 2015. As of November 16, 2018:
https://www.gao.gov/products/GAO-15-515

———, "Hurricanes Katrina and Rita Disaster Relief: Continued Findings of Fraud, Waste, and Abuse," GAO-07-300, March 2007. As of July 26, 2019:
https://www.gao.gov/assets/260/257650.pdf

———, "Telecommunications: FCC Should Improve Monitoring of Industry Efforts to Strengthen Wireless Network Resiliency," Report to the Ranking Member, Committee on Energy and Commerce, House of Representatives, GAO-18-198, Washington, D.C., December 2017a.

———, "Telehealth Use in Medicare and Medicaid," Statement of A. Nicole Clowers, Managing Director, Health Care, GAO 17-760T, Washington, D.C., July 20, 2017b. As of November 12, 2018:
https://www.gao.gov/products/GAO-17-760T

U.S. House of Representatives, Subcommittee on Oversight and Investigations of the Committee on Energy and Commerce, "Waste, Fraud, and Abuse Concerns with the E-Rate Program," 109th Congress, November 2005, accessed on November 15, 2018. As of July 26, 2019:
https://www.govinfo.gov/content/pkg/CPRT-109HPRT24466/html/CPRT-109HPRT24466.htm

U.S. National Oceanic and Atmospheric Administration, Office for Coastal Management, Weather Disaster and Costs, website, May 9, 2019. As of July 3, 2019:
https://coast.noaa.gov/states/fast-facts/weather-disasters.html

United Nations Economic and Social Council, "Strategic Framework on Geospatial Information and Services for Disasters," June 20, 2018. As of April 14, 2019:
https://undocs.org/E/2018/L.15

United States Digital Service, website, undated. As of July 25, 2019:
https://www.usds.gov/

Verizon, Comments of Verizon, *In the Matter of Response Efforts Undertaken During the 2017 Hurricane Season*, PS Docket No. 17-344, January 22, 2018.

Victory, Nancy, Chair, Independent Panel Reviewing the Impact of Katrina on Communications Networks, "Report and Recommendations to the Federal Communications of the Independent Panel Reviewing the Impact of Katrina on Communications Networks," letter to Kevin J. Martin, Chair, FCC, June 12, 2006. As of July 25, 2019:
https://transition.fcc.gov/pshs/docs/advisory/hkip/karrp.pdf

Visit Berlin, Public Wi-Fi Berlin, webpage, undated. As of November 6, 2018:
https://www.visitberlin.de/en/public-wi-fi-berlin

Waze, homepage, 2019. As of June 20, 2019:
https://www.waze.com/

Westerman, George, Didier Bonnet, and Andrew McAfee, "The Nine Elements of Digital Transformation," *MIT Sloan Management Review*, January 7, 2014. As of April 29, 2019:
https://sloanreview.mit.edu/article/the-nine-elements-of-digital-transformation/

Wilks, Stuart, "The Century's Most Troublesome Construction Projects," *Global Construction Review*, October 5, 2015. As of April 29, 2019:
http://www.globalconstructionreview.com/perspectives/centurys-most-troublesome
-construction-pr8oje8ct8s/

Wood, Colin, "What is 311?" Digital Communities, *Government Technology*, August 4, 2016.

World Economic Forum, "On the Threshold of a Digital Identity Revolution," January 2018. As of September 7, 2019:
http://www3.weforum.org/docs/White_Paper_Digital_Identity_Threshold_Digital_Identity
_Revolution_report_2018.pdf

Yu, Leping (Tommy), Using Technology to Enhance Government Transparency and Counter Corruption, Chicago Policy Review, June 20, 2017.

## Interpersonal Communications

Emergency Communications Coordinator for Region 2, U.S. Department of Homeland Security, telephone interview with the authors, March 2, 2018.

FEMA Comms/IT Sector, weekly Puerto Rico Comms/IT leadership teleconference, October 9, 2018.

Government of Puerto Rico telecommunications official, discussion with the authors, April 10, 2018.

Official of the Puerto Rico Department of Public Safety, telephone communication with the authors, October 19, 2018.

PRIDCO representatives, interview with authors, discussion regarding the impact of hurricanes on operational activities for businesses and on local communities, June 1, 2018.

Puerto Rico Chief Innovation Officer, interview with authors, May 11, 2018.

———, interview with authors, May 31, 2018.

Puerto Rico Public Buildings Authority representative, interview with the HSOAC Public Buildings team, March 29, 2018.

Representative of a Puerto Rico telecommunications carrier, interview with the authors, April 10, 2018.

Representative of a major Puerto Rico telecommunications carrier, interview with the authors, May 15, 2018.

Senior official of the Puerto Rico Department of Public Safety, telephone communication with authors, October 25, 2018.

———, telephone communication with authors, November 12, 2018.